Applied
Reservoir
Engineering

Volume 2

Applied Reservoir Engineering

Volume 2

Charles R. Smith

G. W. Tracy

R. Lance Farrar

OGCI Publications

Oil & Gas Consultants International, Inc.

Tulsa

CONTENTS, Volume 2

Nomenclature

Index

CONTENTS, Volume 1

3   FLUID PROPERTIES   3-1 — 3-124

6   GAS CONDENSATE RESERVOIRS   6-1 — 6-24

7   FLUID FLOW IN RESERVOIRS   7-1 — 7-28

8   OIL RESERVOIR DRIVE MECHANISMS

9   SOLUTION GAS DRIVE RESERVOIRS

12 ANALYSIS OF WATER DRIVE RESERVOIRS

INTRODUCTION

A substantial number of hydrocarbon reservoirs (oil and/or gas) have an associated aquifer which is an important source of reservoir energy. This energy provides a drive mechanism for the expulsion of fluids when the wells completed in the reservoir are placed on production. The existence of a significant number of water drive reservoirs in the world is believed to stem from the marine origin of most reservoirs. The most popular current theories consider the oil to have migrated to the reservoir some time after the reservoir was formed (originally entirely water filled). Therefore, pore volume not occupied by hydrocarbons will be occupied by water.

In a reservoir that is completely sealed (whether by shale-outs, permeability barriers, faults, or some other mechanism), if the aquifer volume is less than 10 times the hydrocarbon reservoir volume, then the water drive mechanism will be small. However, if the aquifer size is significantly larger (than 10x) or if the reservoir is not completely sealed and has a water recharge source available, then the water drive mechanism may be the principal source of reservoir energy.

When a well is produced, pressure is lowered in the reservoir. This creates a pressure differential across the oil/water (or gas/water) contact. Following the basic laws of fluid flow in porous media, the aquifer reacts by providing water encroachment across the original hydrocarbon/water contact. Some of the mechanisms by which this occurs are:

(1) The expansion of the water in the aquifer;

(2) Contraction of the aquifer pore volume caused by rock expansion attributable to the compressible nature of the rock;

(3) The transmission through the aquifer of the expansion of other hydrocarbon accumulations linked with a common aquifer;

(4) Artesian flow where the reservoir formation outcrops and is recharged by surface waters.

Within the hydrocarbon portion of the reservoir, there is also expansion of rock, hydrocarbon, and connate water. These effects also contribute to the displacement of hydrocarbons toward the producing wells. These expansion effects within the hydrocarbon portion of the reservoir system are usually called volumetric, expansion, or solution gas drive. Expansion drive is directly related to the time variance of the average pressure within the hydrocarbon portion of the reservoir.

Water drive is a function of the time varying behavior of the pressure at the boundary or hydrocarbon/water contact. Often the distinction between boundary pressure and average reservoir pressure is lost during the calculation procedures. Of course, the average reservoir pressure is less than the pressure between the reservoir and the aquifer.

Where discovery pressure exceeds the bubblepoint pressure in an oil reservoir, no gas cap is present. It is quite possible in a strong water drive oil reservoir (water influx rates high or production rates low) for the reservoir pressure to be maintained above the bubblepoint. For such a reservoir, the calculations are easier than if the reservoir pressure drops below the bubblepoint and a free gas phase develops. In this case a combination drive results.

ANALYSIS OF WATER DRIVE RESERVOIRS

The following are recovery efficiency ranges for some common types of oil and gas reservoirs. Although recovery can be outside the listed bounds, most reservoirs will fall within these limits.

Oil Reservoirs (°API > 20)	
Drive Mechanism	**Recovery Efficiency Range, % OOIP**
Solution Gas Drive	10 to 25
Water drive	35 to 65

Gas Reservoirs	
Drive Mechanism	**Recovery Efficiency Range, % OGIP**
Volumetric or Internal-Expansion Drive	70 to 90
Water drive	35 to 65

Considering the oil solution gas drive reservoir, the reservoir energy is comprised of the expansion of the fluids and the rock. This drive mechanism is relatively inefficient because of (1) the relatively high oil viscosity (compared to gas viscosity), and (2) gas/oil relative permeability effects. A low ultimate recovery results.

Considering a gas volumetric reservoir (where the pore volume available to the gas is relatively constant), the reservoir energy here mainly stems from the expansion of the gas itself. In this type of a reservoir, there are no relative permeability effects to dissipate the reservoir energy; and the gas viscosity is quite low. Therefore, the reservoir rock does not impede the gas flow much, and the ultimate recovery is high.

A water drive oil reservoir and a water drive gas reservoir will have similar ultimate recovery efficiencies. There are slightly higher capillary forces in the water-drive gas reservoir attributable to the higher interfacial tension that exists between gas and water than between oil and water. Thus, the residual gas saturation in the water-invaded zone may be higher on the average than the corresponding residual oil saturation in a water drive oil reservoir.

If the same core is used in the laboratory to generate oil/water and gas/water relative permeability relationships, the curves will typically resemble those shown in Figure 12-1. Notice that the two sets of curves have similar shapes. The relative permeability relationship is predominantly a function of the wetting phase characteristics, and water is usually the wetting phase. So, the shapes of the water/oil and the water/gas relative permeability curves normally are quite similar because these relationships are so strongly a function of wettability. The "wetting phase"—"nonwetting phase" aspect of the relationship has more influence than exactly what is the nonwetting phase.

Considering the water/oil relative permeability curves in Figure 12-1, notice that these curves are water-wet, imbibition curves. "Imbibition" indicates that the wetting-phase saturation is increasing with time. A "drainage" process is the reverse situation: the wetting phase is decreasing. Normally, we have imbibition taking place in a water drive reservoir, because water is usually the wetting phase.

Relative permeability curves provide insight into the wettability of the formation. For instance, consider the water-wet, imbibition, oil/water relative permeability curves shown in Figure 12-1. The oil relative permeability at the irreducible water saturation is approximately equal to one, but notice that the relative permeability to water at the residual oil saturation is considerably less than one.

$$k_{rw} (@ S_w = 1 - S_{or}) = k_{rw} (S_{or})$$

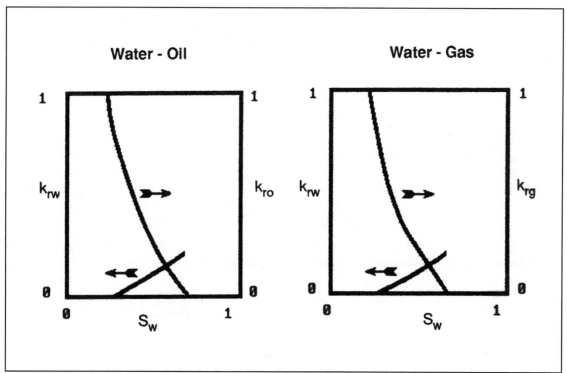

Fig. 12-1. *Relative permeability characteristics.*

The lower that this value is, then the more water-wet is the formation. For a strongly water-wet formation $k_{rw}(S_{or})$ will be 0.2 or less. As an average value for water-wet formations:

$$k_{rw}(S_{or}) \approx 0.3$$

If a "water-wet" formation has a $k_{rw}(S_{or}) \geq 0.5$, then the formation exhibits intermediate wettability or may be partly oil-wet. Whatever the wettability of the formation, an indication of the maximum water drive recovery possible (on a saturation basis) may be seen in the distance between the saturation end-points of the relative permeability relationship: $1 - S_{wi} - S_{or}$. This is the maximum value that the oil saturation can be decreased by water drive.

RECOGNITION OF WATER DRIVE

Where substantial quantities of water are being produced, the presence of mobile water in the reservoir is obvious, and water drive is the likely drive mechanism. However, many of the better oil reservoirs produce water-free for prolonged periods and have strong water drives. Some of the indicators used to determine the presence of water drive are:

    (1) The hydrocarbon (oil or gas) is underlain by water,

    (2) Sufficient permeability must exist to support water movement (usually at least 50 md),

    (3) With time or cumulative hydrocarbon production, water production is seen to be increasing,

    (4) Material balance is the best indicator.

The material balance equation for an oil reservoir without a gas cap is:

$$N = \frac{N_p \left[B_o + B_g (R_p - R_s) \right] + W_p B_w - W_e}{B_g (R_{si} - R_s) - (B_{oi} - B_o)} \tag{12-1}$$

Assuming that production data, static pressures, and PVT properties (vs pressure) are known, then there are only two unknowns: N and W_e. W_e represents the cumulative reservoir barrels of water that have entered the reservoir across the original oil/water contact. Of course, for the case of a water drive reservoir, W_e is not really just one unknown. Instead, it is a strictly increasing, positive function. So, if we know something about water influx, then the original oil in place, N can be calculated.

For a water drive reservoir, if the original oil in place is known, then it is possible to rearrange the material balance equation and solve for water encroachment:

$$\textit{Water influx, } W_e = \textit{Withdrawals - Expansion} \tag{12-2}$$

      where:

Withdrawals = cumulative withdrawals from the reservoir, Res. Bbls, and

   Expansion = the change in volume of the original hydrocarbon in place caused by the pressure changing from the original value, p_i, to the current static pressure, $\bar{p}$.

For an oil reservoir (with or without an initial gas cap):

$$Withdrawals = N_p \, [\, B_o + B_g \, (\, R_p - R_s \,) \,] + W_p \, B_w \qquad (12\text{-}3)$$

For an oil reservoir <u>without</u> an initial gas cap:

$$Expansion = N \, [\, B_g \, (\, R_{si} - R_s \,) - (\, B_{oi} - B_o \,) \,] \qquad (12\text{-}4)$$

Oil reservoir with an initial gas cap:

$$Expansion = N \left[B_g \, (\, R_{si} - R_s \,) - (\, B_{oi} - B_o \,) + m \frac{B_{oi}}{B_{gi}} \, (\, B_g - B_{gi} \,) \right] \qquad (12\text{-}5)$$

where:

$$m = \frac{G \, B_{gi}}{N \, B_{oi}}$$

G = initial gas cap size, (Mscf or scf),

B_{gi} = gas formation volume factor at p_i, (Bbl/Mscf or Bbl/scf, such that G x B_{gi} = Bbl)

For an undersaturated oil reservoir (where the entire reservoir history has been above the bubblepoint):

$$Expansion = N \, c_{oe} \, B_{oi} \, (\, p_i - \bar{p} \,) \qquad (12\text{-}6)$$

where:

c_{oe} = the effective oil compressibility (based on the oil volume, but takes into account the effects of the expanding rock and water), psi^{-1}

$$c_{oe} = c_o + \frac{c_w \, S_w}{1 - S_w} + \frac{c_f}{1 - S_w} \qquad (12\text{-}7)$$

where:

c_f = the pore volume compressibility, psi^{-1}, or

$$c_f = \left[\frac{1}{pore \; volume} \right] \frac{\Delta \, (\, pore \; volume \,)}{\Delta \, (\, pressure \,)} \; psi^{-1}$$

Note that the effective oil compressibility defined (used in the material balance equation for an undersaturated oil reservoir) is not the same as the effective or total compressibility used in various well testing equations.

For a gas reservoir:

$$Withdrawals = G_p B_g + W_p B_w \qquad (12\text{-}8)$$

$$Expansion = G(B_g - B_{gi}) \qquad (12\text{-}9)$$

To investigate the existence of water encroachment, return to the material balance equation arranged to solve for original hydrocarbon in place (N for an oil reservoir; G for a gas reservoir). To illustrate, consider the material balance equation for an oil reservoir without an initial gas cap:

$$N = \frac{N_p\,[B_o + B_g\,(R_p - R_s)] + W_p B_w - W_e}{B_g\,(R_{si} - R_s) - (B_{oi} - B_o)} \tag{12-10}$$

The initial problem is to determine if a water drive exists. The function, W_e, in equation 12-10 is unknown. To begin, W_e is set equal to zero, and the resulting N that is calculated is referred to as the apparent oil in place, N_a:

$$N_a = \frac{N_p\,[B_o + B_g\,(R_p - R_s)] + W_p B_w}{B_g\,(R_{si} - R_s) - (B_{oi} - B_o)} \tag{12-11}$$

If at least four or five years of production history and the corresponding static pressures are available, then N_a can be calculated using data corresponding to the end of each period of production history. More credence should be given to the later values of N_a, since more data is available; and small differences (or errors) in PVT property values will not influence the calculations as much. (The denominator of the material balance equation is quite small at early times; hence, errors are magnified).

If these calculations are performed, then we should have an N_a for each period of production history: N_{a1}, N_{a2}, N_{a3}, etc. Then, a graph is prepared for the apparent oil in place versus time or cumulative production.

If there is no water influx occurring, then the calculated N_a's will be fairly constant as shown in Figure 12-2. If water encroachment is occurring, then the N_a's will be increasing with time or cumulative production.

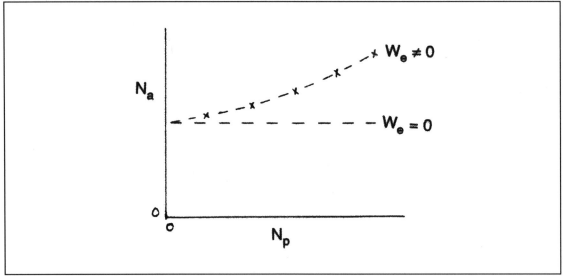

Fig. 12-2. Apparent OIP vs. N_p.

Considering Equations 12-10 and 12-11, it is easily seen that:

$$N_a = N + \frac{W_e}{D} \tag{12-12}$$

where:

D = denominator of the material balance equation, also called the "expansibility" of the reservoir system. Physically, it is the expansion associated with one STB of original oil in place (from the original pressure to the present static pressure).

For an oil reservoir without an initial gas cap,

$$D = B_g(R_{si} - R_s) - (B_{oi} - B_o) \tag{12-13}$$

AQUIFER MODELS

Normally, there are little data, if any, concerning the aquifer of a hydrocarbon reservoir. The usual approach to describe an aquifer is to use an implicit method or conceptual model.

Examining the type of trap is usually helpful. In blanket sands, with no indication of faulting, we would expect large (possibly capable of being regarded as infinite) aquifer sizes. On the other hand, many stratigraphic traps have limited water drives due to limited aquifer extent. Studying the type of trap should shed some light on what model to choose.

Water influx depends more on the properties of the aquifer than on the characteristics of the hydrocarbon reservoir. Of course, the aquifer model requires input of aquifer properties, which are usually not available. Thus, the needed aquifer parameters will have to be back-calculated on the basis of production history (performance data). During this "history-matching" stage, we are calibrating the model to be able to predict the performance that has already been obtained.

Three different aquifer models, will be considered: Schilthuis, van Everdingen and Hurst, and Fetkovich.

STEADY-STATE MODEL (SCHILTHUIS)

The simplest method for characterizing water influx is due to Schilthuis.[3] This model is often tried first since the calculations are considerably less involved than with either of the other two methods.

The assumptions inherent in the Schilthuis model are many. First, it is assumed that the aquifer is very large and highly permeable. In fact, the aquifer is taken to have permeability so high that the pressure gradient across the aquifer itself is negligible. And the aquifer is so huge that the pressure within the aquifer never declines; i.e., the initial pressure, p_i, always exists within the aquifer. The hydraulic analog to the Schilthuis steady-state model is shown in Figure 12-3.

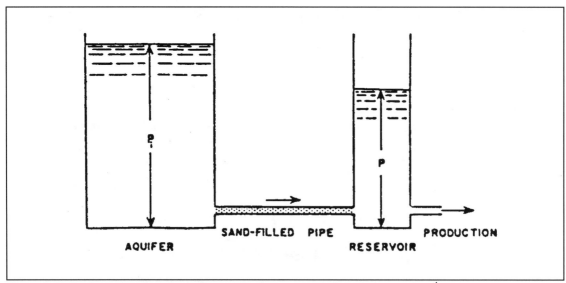

Fig. 12-3. Hydraulic analog of steady-state water influx into a reservoir (from Craft & Hawkins)[4]. Permission to publish by Prentice-Hall.

In the aquifer tank, the pressure surface is flat and remains constant at p_i. This could correspond to either an artesian aquifer being recharged with surface water or to an aquifer that is quite large compared to the size of the hydrocarbon reservoir.

The hydrocarbon trap of the aquifer-reservoir system is assumed to be small, at least as compared to the size of the aquifer. It is also assumed to be highly permeable such that at any particular time, there is a flat pressure surface across the reservoir as illustrated in the hydraulic analog.

A resistance to flow is assumed to be at the location of the original oil/water contact. This is shown in the hydraulic analog as a sand-filled pipe. In reality, there is a resistance to flow in the vicinity of the oil/water contact which is related to relative permeability effects.

The basic equation for the Schilthuis model is the integral of Darcy's law which is a steady-state relationship. The pressure in the aquifer is assumed to be unchanging, but with production, the pressure in the reservoir declines. This, of course, yields a larger and larger pressure differential across the original oil/water contact. Thus, the flow rate from the aquifer into the reservoir is not really steady state, but is treated as a "succession of steady states."

For this model to work, the reservoir/aquifer system must have a relatively high permeability: 50 md or more. The aquifer must, at the very least, be 10 to 20 times as large as the reservoir. Results will be much better if the aquifer is 100 times as large.

Schilthuis started with Darcy's law:

$$q_w = C_s \, (p_i - \bar{p})$$

(12-14)

where:

q_w = flow rate of water across the original oil/water contact,
C_s = aquifer constant (which contains the unchanging parts of Darcy's law: water viscosity, geometrical constants, etc.); relates to the deliverability of the aquifer, Bbls/time/psi,
p_i = the initial pressure, also the aquifer pressure for all time, psi,
$\bar{p}$ = static pressure of the reservoir, psi.

The material balance equation does not require water flow rate, rather cumulative water influx. Hence, Schilthuis integrated with respect to time:

$$W_e = C_s \int_o^t (p_i - \bar{p}) \, dt$$

For ease of application, this integral is represented as a sum. Thus, the pressure time history relationship must be broken into time intervals is illustrated in Figure 12-4.

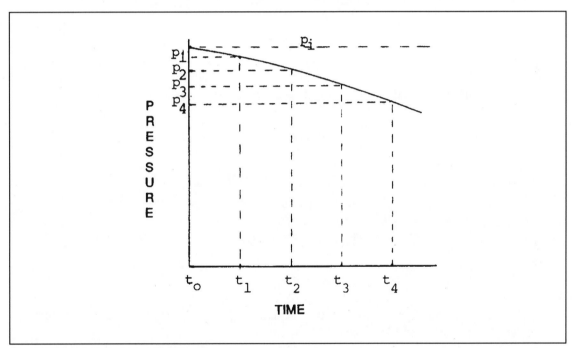

Fig. 12-4. Reservoir pressure with time.

If there are "n" time intervals out to the time of interest, then cumulative water influx can be represented as:

(12-15)

$$(W_e)_n = C_s \sum_{j=1}^{n} \left[p_i - 0.5 \, (\bar{p}_{j-1} + \bar{p}_j) \right] \Delta t_j$$

where:

$(W_e)_n$ = the cumulative water influx at time t_n, Bbls,

p_i = the initial pressure, psi,

$\bar{p}_j$ = the static reservoir pressure at t_j, psi,

Δt_j = the time interval between t_{j-1} and t_j.

This model contains no water influx rate dampening features. A change in rate in the field has an instantaneous effect on the aquifer influx. This means that there is no compressibility in the system between the reservoir and the aquifer. There is an immediate aquifer reaction to any pressure change. This is not entirely realistic, but it is an easy-to-use concept.

The Schilthuis model will be used with material balance to do two things:

(1) Calculate the original oil in place, N, and

(2) Solve for the steady-state aquifer constant (Schilthuis constant), Cs. This constant relates the rate of water influx per psi of pressure drop across the oil/water contact.

Recall material balance Equation 12-12:

$$N_a = N + \frac{W_e}{D} \tag{12-12}$$

Then, the Schilthuis equation can be substituted into this equation with the result:

$$N_a = N + C_s \left(\frac{\Sigma \Delta \bar{p} \, \Delta t}{D} \right) \tag{12-16}$$

where:

N_a = the apparent oil in place (calculated by assuming that W_e is zero), STB,

N = the actual original oil in place, STB,

$\Sigma \Delta \bar{p} \, \Delta t$ = the integral of pressure drop across the original oil/water contact with respect to time, and

D = the denominator of the material balance equation when solved for N; e.g., the denominator of Equation 12-10.

Assuming that N and C_s are constants, then Equation 12-16 is in the form of a straight line: y = mx + b. Thus, if the calculated N_a's are plotted versus:

$$\left(\frac{\Sigma \Delta \bar{p} \, \Delta t}{D} \right)$$

then a plot similar to that shown in Figure 12-5 should result.

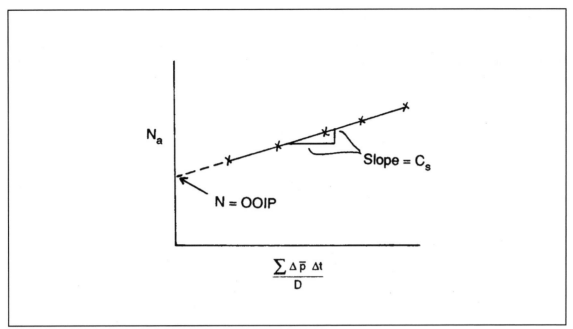

Fig. 12-5. Schilthuis material balance plot.

If the plotted points form a reasonably straight line, extrapolating the straight line to the "Y-intercept" yields N. The slope of the straight line is equal to C_s.

If the points do not plot as a straight line, but are nonlinear, then the aquifer- reservoir system is not behaving according to the Schilthuis assumptions, and this model should not be used. If the points are only slightly nonlinear, then the plot may be able to be extrapolated to obtain N without significant error. It would not be prudent to use the C_s to attempt to predict future aquifer performance.

The steady state model is used with the same logic for a gas reservoir as for an oil reservoir. A different material balance denominator, D, is used:

$$D = B_g - B_{gi} \tag{12-17}$$

The apparent gas in place (setting $W_e = 0$) at different times in the performance history is:

$$G_a = \frac{G_p B_g + W_p B_w}{B_g - B_{gi}} \tag{12-18}$$

Then, G_a is plotted versus $(\Sigma \Delta \bar{p} \Delta t) / D$. The plot (if linear) is extrapolated to obtain G, and the slope is measured for C_s.

Whether, the hydrocarbon is oil or gas, there are two basic procedures to obtain the "best" straight line through the plotted points:

(1) Visual inspection, and

(2) Method of least squares.

Development of Least-Squares Equations

If there are more data points than unknowns, then the least squares method may be used to determine the "best" values of the unknowns. Equation 12-16, a straight line equation, is the equation that will be considered.

$$Y_i = A + B X_i$$

where:

$Y_i = N_{ai}$
$X_i = [(\Sigma \Delta \bar{p} \, \Delta t) \, / D]_i$
$A = N$
$B = C_s$

To find the best constants A and B, the following residual function will be minimized:

$$R = \Sigma (Y_i - A - B X_i)^2$$

The conditions for minimizing this function with respect to A and B are:

$$\frac{\partial R}{\partial A} = 0 \quad \text{and} \quad \frac{\partial R}{\partial B} = 0$$

Evaluating these derivatives and simplifying:

$$\frac{\partial R}{\partial A} = 2\Sigma Y_i - 2nA - 2B \, \Sigma X_i = 0$$

So, $\Sigma Y_i = nA + B\Sigma X_i$

where: n = number of data points Y_i and X_i

$$\frac{\partial R}{\partial B} = 2 \, \Sigma Y_i X_i - 2A \, \Sigma X_i - 2B \, \Sigma X_i^2 = 0$$

So, $\Sigma Y_i X_i = A \, \Sigma X_i + B \, \Sigma X_i^2$

Solving for A:

$$A = \frac{(\Sigma Y_i X_i)(\Sigma X_i) - (\Sigma X_i^2)(\Sigma Y_i)}{(\Sigma X_i)(\Sigma X_i) - n \, \Sigma (X_i^2)} = N \tag{12-19}$$

Solving for B:

$$B = \frac{(\Sigma X_i)(\Sigma Y_i) - n (\Sigma Y_i X_i)}{(\Sigma X_i)(\Sigma X_i) - n \, \Sigma (X_i^2)} = C_s \tag{12-20}$$

Problem 12-1:

Calculate water influx, W_e, using the steady-state (Schilthuis) model. The basic data include:

C_s = 1000 Bbl/Month/psi

p_i = 2500 psia

Reservoir Static Pressure Data:

Time, Months	Pressure, psia
0	2500
12	2480
24	2470
36	2464
48	2460

Recall the steady-state aquifer equation:

$$W_e = C_s \, \Sigma \, \Delta \bar{p} \, \Delta t$$

where: $\Delta \bar{p} = p_i - \bar{p}_{avg}$

To evaluate $\bar{p}_{avg}$ and the $\Sigma \Delta \bar{p} \Delta t$ function, it is helpful to use a tabular calculation procedure:

Time Mo.	Δt Mo.	Pressure psia	Average Pressure	$\Delta \bar{p}$ psi	$\Delta \bar{p} \, \Delta t$	$\Sigma \Delta \bar{p} \, \Delta t$	W_e M Bbls
0		2500	2500	0	0	0	
12	12	2480	2490	10	120	120	120
24	12	2470	2475	25	300	420	420
36	12	2464	2467	33	396	816	816
48	12	2460	2462	38	456	1272	1272

Problem 12-2 — Steady State Model:

Calculations of OOIP (N) and Water Influx Constant (C_s)

Basic Data:

Porosity	16 %
Connate Water Saturation	25 %
Oil Compressibility	10×10^{-6} psi^{-1}
Water Compressibility	3×10^{-6} psi^{-1}
Formation Compressibility	4×10^{-6} psi^{-1}
Water Form. Vol. Factor	1.0 Bbl/STB
Bubblepoint Pressure	2150 psia
Initial Pressure	3000 psia

Pressure — Production Data:

Time Years	Oil Production MB	Water Production MB	Pressure psia	B_o
0	0	0	3000	1.3100
1	80.7	0	2870	1.3117
2	221.4	20	2810	1.3125
3	395.5	60	2760	1.3131
4	586.1	130	2720	1.3137
5	806.9	200	2690	1.3141

All of the static pressures are above the bubblepoint pressure, so this system can be analyzed as an undersaturated reservoir. Therefore, we need to calculate the effective oil compressibility:

$$c_{oe} = c_o + \frac{c_w\, S_w}{S_o} + \frac{c_f}{S_o}$$

$$= \left[10 + \frac{(3)(0.25)}{0.75} + \frac{4}{0.75} \right] \left[10^{-6} \right]$$

$$= 16.33 \times 10^{-6}\ psi^{-1}$$

Calculations of Expansibility, "D"

The expansibility (D) is the denominator of the material balance equation when solved for "N". Thus, for an undersaturated oil reservoir we have:

$$Expansibility = D = (c_{oe})(B_{oi})(p_i - \bar{p})$$

$$= (16.33)(10^{-6})(1.3100)(p_i - \bar{p})$$

$$= (21.392)(10^{-6})(p_i - \bar{p})$$

Thus, with each static pressure, an expansibility can be calculated:

Time Years	Pressure psia	$p_i - \bar{p}$ psi	Expansibility "D"
0	3000	0	—
1	2870	130	0.00278
2	2810	190	0.004064
3	2760	240	0.005134
4	2720	280	0.005990
5	2690	310	0.006632

Calculations of Withdrawals:

$$Withdrawals = N_p\, B_o + W_p\, B_w \qquad (B_w = 1.0)$$

$$N_a = \frac{Withdrawals}{Expansibility}$$

Time Years	N_p MB	B_o	W_p MB	$N_p B_o$	Withdrawals MB	N_a MMB
0	0	1.3100	0	0	0	
1	80.7	1.3117	0	105.85	105.85	38.076
2	221.4	1.3125	20	290.59	310.59	76.425
3	395.5	1.3131	60	519.33	579.33	112.842
4	586.1	1.3137	130	769.96	899.96	150.244
5	806.9	1.3141	200	1060.35	1260.35	190.407
						567.994

Calculation of $(\Sigma \Delta \bar{p} \, \Delta t) / D$:

Time Years	Pressure psia	$\bar{p}_{avg}$ psia	$p_i - \bar{p}_{avg}$ psi	Δt Mo.	$\Delta \bar{p} \, \Delta t$	$\Sigma \Delta \bar{p} \, \Delta t$	$(\Sigma \Delta \bar{p} \, \Delta t) / D$
0	3000	3000	0				
1	2870	2935	65	12	780	780	280.58×10^3
2	2810	2840	160	12	1920	2700	664.37×10^3
3	2760	2785	215	12	2580	5280	1028.44×10^3
4	2720	2740	260	12	3120	8400	1402.34×10^3
5	2690	2705	295	12	3540	11940	1800.36×10^3
							5.17609×10^6

Calculation of N and C_s (Least-Squares Method):

$$Y = N_a$$

$$\Sigma Y = 567.994 \text{ MMB}$$

$$X = \Sigma \Delta \bar{p} \, \Delta t / D$$

$$\Sigma X = 5.17609 \times 10^6$$

Recall the least-squares formulas:

$$N = \frac{(\Sigma Y_i \, X_i)(\Sigma \, X_i) - (\Sigma X_i^2)(\Sigma Y_i)}{(\Sigma \, X_i)(\Sigma \, X_i) - n \Sigma (X_i^2)} \qquad (12\text{-}19)$$

$$C_s = \frac{(\Sigma \, X_i)(\Sigma \, Y_i) - n (\Sigma Y_i \, X_i)}{(\Sigma \, X_i)(\Sigma \, X_i) - n \Sigma (X_i^2)} \qquad (12\text{-}20)$$

One more table is needed:

Time	X	X^2	Y,MMB	(Y)(X)
1	$280.58(10^3)$	$0.0787(10^{12})$	38.076	$10.683(10^6)$
2	$664.37(10^3)$	$0.4414(10^{12})$	76.425	$50.774(10^6)$
3	$1028.44(10^3)$	$1.0577(10^{12})$	112.842	$116.051(10^6)$
4	$1402.34(10^3)$	$1.9666(10^{12})$	150.244	$210.693(10^6)$
5	$1800.36(10^3)$	$3.2413(10^{12})$	190.407	$342.801(10^6)$
	$5.17609(10^6)$	$6.7857(10^{12})$	567.994	$731.002(10^6)$

Therefore,

$$N = \frac{(731.002)(5.17609)(10^{12}) - (6.7858)(567.994)(10^{12})}{(5.17609)(5.17609)(10^{12}) - (5)(6.7857)(10^{12})}$$

$$= 9.879 \ MMSTB$$

$$C_S = \frac{(5.17609)(567.994)(10^6) - (5)(731.002)(10^6)}{(5.17609)(5.17609)(10^{12}) - (5)(6.7858)(10^{12})}$$

$$= 100.2 \times 10^{-6} \ MMB \ / \ month \ / \ psi$$

$$C_S = 100.2 \ Barrels \ / \ month \ / \ psi$$

UNSTEADY-STATE FLOW AQUIFER MODEL — VAN EVERDINGEN AND HURST[6]

Figure 12-6 provides a hydraulic analog of the van Everdingen and Hurst aquifer model using a series of tanks interconnected with sand-filled pipes. Each tank has its own pressure (p_1, p_2, etc.), but all are less than the initial pressure p_i. Note that the hydrocarbon reservoir pressure, p, is actually that at the original oil/water contact which is the inner boundary of the aquifer.

Study of Figure 12-6 will show that, even with an infinite number of aquifer tanks, pressure can never fully stabilize as the water must come from ever increasing distances.

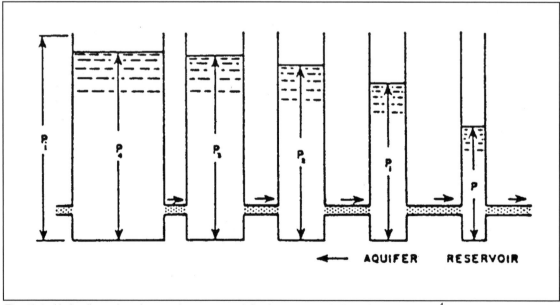

Fig. 12-6. *Hydraulic analog of unsteady-state water influx into a reservoir (from Craft & Hawkins[4]). Permission to publish by Prentice-Hall.*

The van Everdingen and Hurst mathematical model for flow of water through the aquifer involves the following assumptions.

(1) There is radial flow of water through the aquifer as shown here:

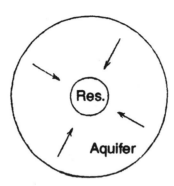

(2) A constant pressure drop across the aquifer exists throughout all flowing time.

(3) The aquifer properties are constant and uniform. Based on these assumptions, the van Everdingen and Hurst[6] aquifer equation is:

$$W_e = 2 \pi \alpha h \phi c_e \, r_f^2 \, [\, \Delta p \, Q \, (\, t_d \,) \,] \qquad (12\text{-}21)$$

where:

α = a fraction between 0 and 1 that represents the extent to which the aquifer surrounds the reservoir. In the sketch under assumption (1), $\alpha = 1.0$, because the aquifer completely surrounds the reservoir.)

W_e = cumulative water influx, cm^3,

h = the net aquifer thickness, cm,

ϕ = aquifer porosity, fraction,

c_e = effective compressibility of the aquifer, atm^{-1}

r_f = radius of the oil or gas reservoir, cm,

Δp = constant pressure drop across the aquifer, atm,

$Q(t_d)$ = cumulative influx function (developed by van Everdingen and Hurst[6]),

t_d = dimensionless time based on the reservoir radius; i.e.

$$t_d = \frac{k \, t}{\phi \, \mu_w \, c_e \, r_f^2} \qquad (12\text{-}22)$$

and

k = permeability of the aquifer, darcies,

t = time, seconds

μ_w = water viscosity, cp.

Equation 12-21 can be written as:

$$W_e = C_v \, \Delta p \, Q(t_d)$$

(12-23)

where:

$$C_v = 2 \pi \alpha h \phi c_e \, r_f^2$$

(12-24)

In so-called "practical oilfield" units, the water influx equation is:

$$W_e = \underbrace{1.119 \, \phi \, h \, c_e \, r_f^2 \, \alpha}_{C_v} \, \Delta p \, Q(t_d)$$

(12-25)

or

$$W_e = C_v \, \Delta p \, Q(t_d)$$

where:

$\quad \alpha$ = as defined with Equation 12-21,

$\quad W_e$ = cumulative water influx, Res. Bbls,

$\quad h$ = effective aquifer thickness, ft,

$\quad \phi$ = aquifer porosity, fraction,

$\quad c_e$ = effective compressibility of the aquifer, psi,$^{-1}$

$\quad r_f$ = radius of the oil or gas reservoir, ft,

$\quad \Delta p$ = constant pressure drop across the aquifer, psi,

$\quad Q(t_d)$ = cumulative influx function,

$\quad t_d$ = dimensionless time, i.e.

$$t_d = \frac{0.00633 \, k \, t}{\phi \, \mu_w \, c_e \, r_f^2} \quad \text{or} \quad t_d = (A)(t)$$

(12-26)

where:

$\quad k$ = aquifer permeability, md,

$\quad t$ = time, days, and

$\quad A$ = real time multiplier (a constant) to convert to dimensionless time.

Of the assumptions implicit in this model, only the one requiring a constant value of Δp needs to be modified to allow practical use of this theory. The superposition principle will be used to apply the theory to handle actual situations. Consider the basic differential equation that describes radial flow through an aquifer:

$$\frac{\partial^2 p}{\partial r^2} + \frac{1}{r} \frac{\partial p}{\partial r} = \frac{\phi \, \mu_w \, c_e}{k} \frac{\partial p}{\partial t}$$

(12-27)

where:

p = pressure at any radial position, r, and time, t,
r = radial position from center of reservoir, cm,
t = time, seconds

Application of Superposition Principle

Equation 12-27 is a linear partial differential equation. Hence, the principle of superposition can be used: the solutions are additive. To apply the superposition principle, consider Figure 12-7, a representative pressure-time profile at the water/oil contact.

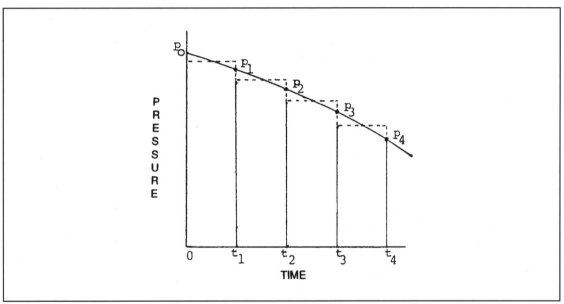

Fig. 12-7. Oil/water contact pressure vs. time.

To use superposition, the curvilinear pressure time shape must be approximated by a series of short-term constant pressures. Such an approximation is usually satisfactory as long as the time intervals are small. The time steps should be small enough that the actual pressure path is nearly linear across any of the time intervals.

The constant pressure to be used over any given interval is the average value of the beginning and ending interval pressures. Then, the difference between the constant pressure of the last time interval and that of the current time interval is assumed to be affecting the water influx behavior from the beginning of the current time interval throughout the remainder of the producing history.

These interval pressure changes are calculated as:

$$\Delta p_1 = p_0 - 0.5(p_0 + p_1) = 0.5(p_0 - p_1)$$

$$\Delta p_2 = 0.5(p_0 + p_1) - 0.5(p_1 + p_2) = 0.5(p_0 - p_2)$$

$$\Delta p_3 = 0.5(p_1 + p_2) - 0.5(p_2 + p_3) = 0.5(p_1 - p_3)$$

$$\Delta p_4 = 0.5(p_2 + p_3) - 0.5(p_3 + p_4) = 0.5(p_2 - p_4)$$

where:

$$p_o = p_i$$

By inspection of the equations for Δp_2, Δp_3, and Δp_4; a general recursion formula is evident. For $j > 1$,

$$\Delta p_j = 0.5\,(p_{j-2} - p_j)\qquad(12\text{-}28)$$

For $j = 1$:

$$\Delta p_1 = 0.5\,(p_o - p_1)\qquad(12\text{-}29)$$

When the basic van Everdingen and Hurst water influx equation is modified to allow pressure variations at the original water/oil contact, the resultant aquifer equation is:

$$W_e = C_v \sum_{j=1}^{n} \Delta p_j\, Q\,[\,A\,(t_n - t_{j-1})\,]\qquad(12\text{-}30)$$

where:

For Darcy units: $C_v = 2\pi h\,\alpha\,c_e\,r_f^2$

For practical oilfield units: $C_v = 1.119\,h\,\phi\,\alpha\,c_e\,r_f^2$

The constant "A" found in Equation 12-30 is a multiplier which converts real time into dimensionless time. Using practical oilfield units with time in days:

$$A = \frac{0.00633\,k}{\phi\,\mu_w\,c_e\,r_f^2}$$

where:

k = aquifer permeability, md,
μ_w = water viscosity, cp,
ϕ = porosity, fraction,
r_f = reservoir external radius or aquifer internal radius, ft,
c_e = effective aquifer compressibility, psi^{-1}
t = time, days

If the time steps are of equal length, then the practical evaluation of the water influx equation becomes easier. Let the time step size be "Δt". Then, the water influx equation can be written as follows:

$$W_e = C_v \sum_{j=1}^{n} \Delta p_j\, Q\,[\,A\,(n - j + 1)\,(\Delta t)\,]\qquad(12\text{-}31)$$

Recall the following material balance equation developed earlier in the chapter:

$$N_a = N + \frac{W_e}{D} \tag{12-12}$$

Substituting the equal time step van Everdingen and Hurst aquifer equation into the material balance equation:

$$N_a = N + C_v \frac{\Sigma \Delta p_j \; Q \, [\, A \, (\, n \, - \, j \, + \, 1 \,) \, (\, \Delta t \,) \,]}{D} \tag{12-32}$$

Equation 12-32 has three unknowns: (1) the original oil in place, N, (2) C_v, and (3) A. The determination of these three unknowns (most importantly, N) from an analysis of reservoir pressure production data is not straight forward.

Notice that "C_v" and "A" are the van Everdingen and Hurst aquifer model constants. "A" is the real time multiplier that converts to dimensionless time. Thus, it is found within the argument of the water influx function, $Q(t_d)$.

Cumulative Influx Function, $Q(t_d)$

Recall that the outer radius of the hydrocarbon (oil or gas) reservoir is r_f, which is the original oil/water contact. This is also the inner radius of the aquifer. Aquifers may be divided into two categories: finite and infinite. Of course, a finite aquifer will have an outer boundary or radius. There are no true infinite aquifers; however, there are some that are large enough to be "infinite-acting" during the producing life of the reservoir.

Figures 12-8 through 12-11 contain the graphical presentation of the influx function, $Q(t_d)$. Figures 12-8 and 12-9 are for a limited (finite) aquifer; while, Figures 12-10 and 12-11 describe an infinite aquifer. The value of $Q(t_d)$ is a function of dimensionless time.

Consider Figure 12-8 or 12-9. Notice that there is one curve for the infinite aquifer and several other curves relating to finite aquifers. The finite aquifers are characterized by their radius ratio: outside radius divided by inside radius. Note that every finite aquifer begins as an infinite-acting aquifer; i.e., its curve begins on the infinite aquifer curve. Then at a certain time, depending on the radius ratio, the finite aquifer behavior departs from the infinite aquifer response. At large time, each finite aquifer curve flattens and finally becomes a constant (unchanging) value. The maximum value (the flat part) of each finite aquifer curve is a function of its radius ratio:

$$[\, Q \, (\, t_d \,) \,]_{\max} = 0.5 \left[\left(\frac{r_a}{r_f} \right)^2 - 1 \right] \tag{12-33}$$

where:

r_a = outside radius of limited aquifer, ft,
r_f = radius of the hydrocarbon reservoir, or inside radius of the aquifer, ft

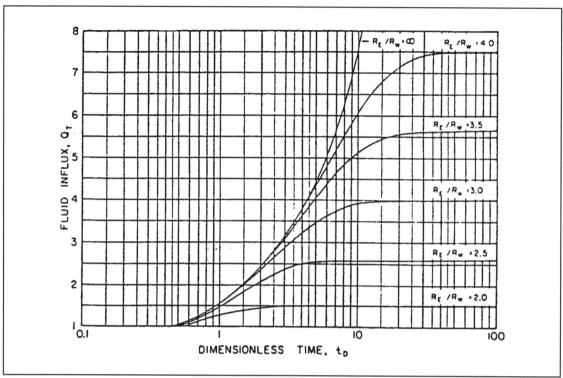

Fig. 12-8. Limited aquifer values of dimensionless influx Q(t) for values of dimensionless time t_D and aquifer limits given by the ratio r_e / r_w (from Craft & Hawkins[4]). (Note that $r_a / r_f = r_e / r_w$). Permission to publish by Prentice-Hall.

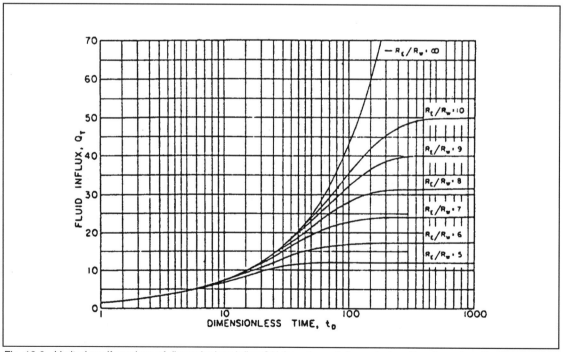

Fig. 12-9. Limited aquifer values of dimensionless influx Q(t) for values of dimensionless time t_D and aquifer limits given by the ratio r_e/r_w (from Craft & Hawkins[4]). (Note that $r_a / r_f = r_e / r_w$). Permission to publish by Prentice-Hall.

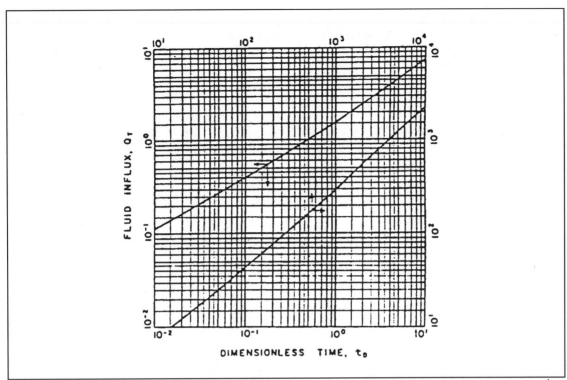

Fig. 12-10. *Infinite aquifer values of dimensionless influx Q(t) for values of dimensionless time t_D (from Craft & Hawkins[4]). Permission to publish by Prentice-Hall.*

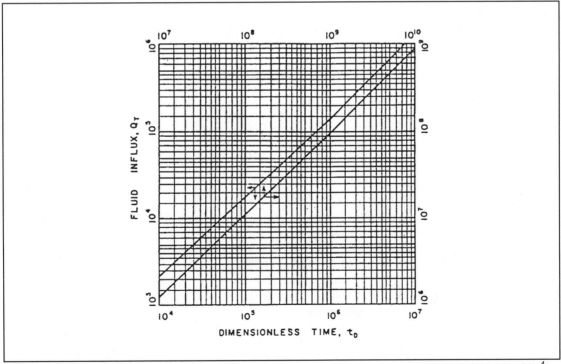

Fig. 12-11. *Infinite aquifer values of dimensionless influx Q(t) for values of dimensionless time t_D (from Craft & Hawkins[4]). Permission to publish by Prentice-Hall.*

For finite aquifers, Table 12-1 contains the numerical values of the $Q(t_d)$ function as a function of radius ratio (r_a/r_f) and dimensionless time, t_d. Radius ratios of 1.5 to 10 are considered. It has been found that finite aquifers with radius ratios larger than 10 usually can be treated as infinite aquifers.

The infinite aquifer cumulative-influx function values are given in Table 12-2. In the infinite aquifer influx function graph (Figures 12-10 and 12-11), the plot is reasonably linear as a log-log plot. For large values of dimensionless time, the slope approaches unity. Table 12-2 values of the influx function are given for dimensionless times up to 2×10.[12] This is generally adequate for any problem. However, if larger values are needed, then straight line extrapolation of Figure 12-11 will suffice for engineering applications.

Klins *et al.*[14] have published convenient equations for calculating values of $Q(t_d)$ for both finite and infinite aquifers.

Calculation of Original Hydrocarbon In Place and Aquifer Constants

The procedure for using reservoir history (performance data) to calculate the original hydrocarbon in place and the water influx constants (C_v and A) is summarized as follows:

1. Assume that the aquifer is infinite-acting over the time of the reservoir data used in the calculation. Therefore, the values of $Q(t_d)$ should be chosen from the infinite table: Table 12-2.

2. Calculate an approximate value of "A." To do this, values of permeability and porosity from the aquifer are needed. Such data are normally unavailable, so average values from the hydrocarbon reservoir are typically used. Aquifer water viscosity is also needed. This is usually determined on the basis of reservoir temperature. Also needed are the approximate radius of the reservoir and the effective compressibility of the aquifer. The effective compressibility of the aquifer is simply equal to the water compressibility plus the pore volume compressibility.

$$A = \frac{0.00633\, k}{\phi\, \mu_w\, c_e\, r_f^2}\ days^{-1}$$

where:

       k = aquifer permeability, md,
       ϕ = porosity, fraction,
   μ_w = aquifer water viscosity, cp,
       c_e = effective aquifer compressibility, psi^{-1}; i.e.,
       $c_e = c_w + c_f$
       r_f = radius of the hydrocarbon reservoir, ft.

Much of the data used in the previous equation are normally guessed or assumed, so the resulting "A" is only an approximation.

Choose a time step size, Δt. This is normally determined on the basis of the frequency of measurement of static pressure data for the reservoir. Typically, this is usually 365 days.

Table 12-1

Limited aquifer values of dimensionless water influx Q(t) for values of dimensionless time t_D and for several ratios of aquifer-reservoir radii r_e/r_w

(from van Everdingen and Hurst[6])

(Note that $r_v / r_w = r_a / r_i$)

$r_e/r_w = 1.5$

t_D	$Q(t)$
$5.0(10)^{-1}$	0.276
6.0	0.304
7.0	0.330
8.0	0.354
9.0	0.375
$1.0(10)^{-1}$	0.395
1.1	0.414
1.2	0.431
1.3	0.446
1.4	0.461
1.5	0.474
1.6	0.486
1.7	0.497
1.8	0.507
1.9	0.517
2.0	0.525
2.1	0.533
2.2	0.541
2.3	0.548
2.4	0.554
2.5	0.559
2.6	0.565
2.8	0.574
3.0	0.582
3.2	0.588
3.4	0.594
3.6	0.599
3.8	0.603
4.0	0.606
4.5	0.613
5.0	0.617
6.0	0.621
7.0	0.623
8.0	0.624

$r_e/r_w = 2.0$

t_D	$Q(t)$
$5.0(10)^{-1}$	0.278
7.5	0.345
$1.0(10)^{-1}$	0.404
1.25	0.458
1.50	0.507
1.75	0.553
2.00	0.597
2.25	0.638
2.50	0.678
2.75	0.715
3.00	0.761
3.25	0.783
3.50	0.817
3.75	0.848
4.00	0.877
4.25	0.905
4.50	0.932
4.75	0.958
5.00	0.983
5.50	1.028
6.00	1.070
6.50	1.108
7.00	1.143
7.50	1.174
8.00	1.203
9.00	1.253
1.0	1.295
1.1	1.330
1.2	1.358
1.3	1.382
1.4	1.402
1.6	1.432
1.7	1.444
1.8	1.453
2.0	1.468
2.5	1.487
3.0	1.495
4.0	1.499
5.0	1.500
8.0	
9.0	
10.0	

$r_e/r_w = 2.5$

t_D	$Q(t)$
$1.0(10)^{-1}$	0.408
1.5	0.509
2.0	0.599
2.5	0.681
3.0	0.758
3.5	0.829
4.0	0.897
4.5	0.962
5.0	1.024
5.5	1.083
6.0	1.140
6.5	1.195
7.0	1.248
7.5	1.299
8.0	1.348
8.5	1.395
9.0	1.440
9.5	1.484
1.0	1.526
1.1	1.605
1.2	1.679
1.3	1.747
1.4	1.811
1.5	1.870
1.6	1.924
1.7	1.975
1.8	2.022
2.0	2.106
2.2	2.178
2.4	2.241
2.6	2.294
2.8	2.340
3.0	2.380
3.2	2.444
3.4	2.491
4.2	2.525
4.6	2.551
5.0	2.570
6.0	2.599
7.0	2.613
8.0	2.619
9.0	2.622
10.0	2.624

$r_e/r_w = 3.0$

t_D	$Q(t)$
$3.0(10)^{-1}$	0.755
4.0	0.895
5.0	1.023
6.0	1.143
7.0	1.256
8.0	1.363
9.0	1.465
1.00	1.563
1.25	1.791
1.50	1.997
1.75	2.184
2.00	2.353
2.25	2.507
2.50	2.646
2.75	2.772
3.00	2.886
3.25	2.990
3.50	3.084
3.75	3.170
4.00	3.247
4.25	3.317
4.50	3.381
4.75	3.439
5.00	3.491
5.50	3.581
6.00	3.656
6.50	3.717
7.00	3.767
7.50	3.809
8.00	3.843
9.00	3.894
10.00	3.928
11.00	3.951
12.00	3.967
14.00	3.985
16.00	3.993
18.00	3.997
20.00	3.999
22.00	3.999
24.00	4.000

$r_e/r_w = 3.5$

t_D	$Q(t)$
1.00	1.571
1.20	1.761
1.40	1.940
1.60	2.111
1.80	2.273
2.00	2.427
2.20	2.574
2.40	2.715
2.60	2.849
2.80	2.976
3.00	3.098
3.25	3.242
3.50	3.379
3.75	3.507
4.00	3.628
4.25	3.742
4.50	3.850
4.75	3.951
5.00	4.047
5.50	4.222
6.00	4.378
6.50	4.516
7.00	4.639
7.50	4.749
8.00	4.846
9.00	4.932
9.50	5.009
10.00	5.078
	5.138
	5.241
12	5.321
13	5.385
14	5.435
15	5.476
16	5.506
17	5.531
18	5.551
20	5.579
25	5.611
30	5.621
35	5.624
40	5.625

$r_e/r_w = 4.0$

t_D	$Q(t)$
2.00	2.442
2.20	2.598
2.40	2.748
2.60	2.893
2.80	3.034
3.00	3.170
3.25	3.334
3.50	3.493
3.75	3.645
4.00	3.792
4.25	3.932
4.50	4.068
4.75	4.198
5.00	4.323
5.50	4.560
6.00	4.779
6.50	4.982
7.00	5.169
7.50	5.343
8.00	5.504
8.50	5.653
9.00	5.790
9.50	5.917
10	6.035
11	6.246
12	6.425
13	6.580
14	6.712
15	6.825
16	6.922
17	7.004
18	7.076
20	7.189
22	7.272
24	7.332
26	7.377
28	7.434
30	7.464
34	7.481
38	7.490
42	
46	7.494
50	7.497

$r_e/r_w = 4.5$

t_D	$Q(t)$
2.5	2.835
3.0	3.196
3.5	3.537
4.0	3.859
4.5	4.165
5.0	4.454
5.5	4.727
6.0	4.986
6.5	5.231
7.0	5.464
7.5	5.684
8.0	5.892
8.5	6.089
9.0	6.276
9.5	6.453
10	6.621
11	6.930
12	7.208
13	7.457
14	7.680
15	7.880
16	8.060
18	8.365
20	8.611
22	8.809
24	8.968
26	9.097
28	9.200
30	9.283
34	9.404
38	9.481
42	9.532
46	9.565
50	9.586
60	9.612
70	9.621
80	9.623
90	9.624
100	9.625

$r_e/r_w = 5.0$

t_D	$Q(t)$
3.0	3.195
3.5	3.542
4.0	3.875
4.5	4.193
5.0	4.499
5.5	4.792
6.0	5.074
6.5	5.345
7.0	5.605
7.5	5.854
8.0	6.094
8.5	6.325
9.0	6.547
9.5	6.760
10	6.965
11	7.350
12	7.706
13	8.035
14	8.339
15	8.620
16	8.879
18	9.338
20	9.731
22	10.07
24	10.35
26	10.59
28	10.80
30	10.98
34	11.26
38	11.46
42	11.61
46	11.71
50	11.79
60	11.91
70	11.96
80	11.98
90	11.99
100	12.00
120	12.00

$r_e/r_w = 6.0$

t_D	$Q(t)$
6.0	5.148
6.5	5.440
7.0	5.724
7.5	6.002
8.0	6.273
8.5	6.537
9.0	6.795
9.5	7.047
10.0	7.293
10.5	7.533
11	7.767
12	8.220
13	8.651
14	9.063
15	9.456
16	9.829
17	10.19
18	10.53
19	10.85
20	11.16
22	11.74
24	12.26
26	12.50
28	13.74
30	14.40
34	14.93
38	16.05
40	16.56
45	16.91
50	17.14
60	17.27
70	17.36
80	17.41
90	17.45
100	17.46
120	17.48
140	17.49
150	17.49
180	17.50
200	17.50
220	17.50

$r_e/r_w = 7.0$

t_D	$Q(t)$
9.00	0.861
9.50	7.127
10	7.389
11	7.902
12	8.397
13	8.876
14	9.341
15	9.791
16	10.23
17	10.65
18	11.06
20	11.46
22	11.85
24	12.58
26	13.27
28	13.92
30	14.53
35	15.11
40	16.39
45	17.49
50	18.43
60	19.24
70	20.51
80	21.45
90	22.13
100	22.63
120	23.00
140	23.47
160	23.71
180	23.85
200	23.92
500	23.96
	24.00

$r_e/r_w = 8.0$

t_D	$Q(t)$
9	7.861
10	7.398
11	7.920
12	8.431
13	8.930
14	9.418
15	9.895
16	10.361
17	10.82
18	11.26
19	11.70
20	12.13
22	12.95
24	13.74
26	14.50
28	15.23
30	15.92
34	17.22
38	18.41
40	18.97
45	20.26
50	21.42
55	22.46
60	23.40
70	24.98
80	26.26
90	27.28
100	28.11
120	29.31
140	30.08
160	30.58
180	30.91
200	31.12
240	31.34
280	31.43
320	31.47
360	31.49
400	31.50
500	31.50

$r_e/r_w = 9.0$

t_D	$Q(t)$
10	7.417
15	9.945
20	12.26
22	13.13
24	13.98
26	14.79
28	15.59
30	16.35
32	17.10
34	17.82
36	18.52
38	19.19
40	19.85
42	20.48
44	21.09
46	21.69
48	22.26
50	22.82
52	23.89
54	24.39
56	24.88
58	25.36
60	26.48
65	27.52
70	28.48
75	29.36
80	30.18
85	30.93
90	31.63
95	32.27
100	34.39
120	35.92
140	37.01
160	37.85
180	38.14
200	39.17
240	39.56
280	39.77
320	39.88
360	39.94
400	39.97
440	39.98
480	

$r_e/r_w = 10.0$

t_D	$Q(t)$
15	9.965
20	13.32
22	13.22
24	14.09
26	14.95
28	15.78
30	16.59
32	17.38
34	18.16
36	18.91
38	19.65
40	20.37
42	21.07
44	21.76
46	22.42
48	23.07
50	23.71
52	24.33
54	24.94
56	25.53
58	26.11
60	26.67
65	28.02
70	29.29
75	30.49
80	31.61
85	32.67
90	33.66
95	34.60
100	35.48
120	38.51
140	40.89
160	42.75
180	44.21
200	45.36
240	46.95
280	47.94
320	48.54
360	48.91
400	49.14
440	49.28
480	49.36

Table 12-2
Infinite aquifer values of dimensionless water influx Q(t) for values of dimensionless time t_D, radial flow
(from van Everdingen & Hurst[6])

Column pairs: Dimensionless time t_D | Fluid influx $Q(t)$ (repeated across the table)

t_D	$Q(t)$	t_D	$Q(t)$	t_D	$Q(t)$	t_D	$Q(t)$	t_D	$Q(t)$	t_D	$Q(t)$
0.00	0.000	79	35.697	455	150.249	1190	340.843	3250	816.090	35,000	6780.247
0.01	0.112	80	36.058	460	151.640	1200	343.308	3300	827.088	40,000	7650.096
0.05	0.278	81	36.418	465	153.029	1210	345.770	3350	838.067	50,000	8343.099
0.10	0.404	82	36.777	470	154.416	1220	348.230	3400	849.028	60,000	11,047.299
0.15	0.520	83	37.136	475	155.801	1225	349.400	3450	859.974	70,000	12,708.358
0.20	0.606	84	37.494	480	157.184	1230	350.688	3500	870.903	75,000	13,531.457
0.25	0.689	85	37.851	485	158.565	1240	353.144	3550	881.816	80,000	14,350.121
0.30	0.758	86	38.207	490	159.945	1250	355.597	3600	892.712	90,000	15,975.389
0.40	0.898	87	38.563	495	161.322	1260	358.048	3650	903.594	100,000	17,586.284
0.50	1.020	88	38.919	500	162.698	1270	360.496	3700	914.459	125,000	21,560.732
0.60	1.140	89	39.272	510	165.444	1275	361.720	3750	925.309	$1.5(10)^5$	$2.538(10)^4$
0.70	1.251	90	39.626	520	168.183	1280	362.942	3800	936.144	2.0	3.308 *
0.80	1.359	91	39.979	525	169.549	1290	365.386	3850	946.966	2.5	4.066 *
0.90	1.469	92	40.331	530	170.914	1300	367.828	3900	957.773	3.0	4.817 *
1	1.569	93	40.684	540	173.039	1310	370.267	3950	968.566	4.0	6.267 *
2	2.447	94	41.034	550	176.357	1320	372.704	4000	979.344	5.0	7.699 *
3	3.202	95	41.385	560	179.069	1325	373.922	4050	990.108	6.0	9.113 *
4	3.893	96	41.735	570	181.774	1330	375.139	4100	1000.858	7.0	$1.051(10)^5$ *
5	4.539	97	42.084	575	183.124	1340	377.572	4150	1011.595	8.0	1.180 *
6	5.153	98	42.433	580	184.473	1350	380.003	4200	1022.318	9.0	1.328 *
7	5.743	99	42.781	590	187.166	1360	382.432	4250	1033.028	$1.0(10)^6$	1.402 *
8	6.314	100	43.129	600	189.852	1370	384.859	4300	1043.724	1.5	2.126 *
9	6.869	105	44.858	610	192.533	1375	386.070	4350	1054.409	2.0	2.781 *
10	7.411	110	46.574	620	195.208	1380	387.283	4400	1065.082	2.5	3.427 *
11	7.940	115	48.277	625	196.544	1390	389.705	4450	1075.743	3.0	4.064 *
12	8.457	120	49.968	630	197.878	1400	392.125	4500	1086.390	4.0	5.313 *
13	8.964	125	51.648	640	200.542	1410	394.543	4550	1097.024	5.0	6.544 *
14	9.461	130	53.317	650	203.201	1420	396.959	4600	1107.646	6.0	7.761 *
15	9.949	135	54.976	660	205.854	1425	398.167	4650	1118.257	7.0	8.965 *
16		140	56.625	670	208.502	1430	399.373	4700	1128.854	8.0	$1.016(10)^6$ *
17	10.913	145	58.265	675	209.825	1440	401.786	4750	1139.439	9.0	1.134 *
18	11.386	150	59.895	680	211.145	1450	404.197	4800	1150.012	$1.0(10)^7$	1.252 *
19	11.855	155	61.517	690	213.784	1460	406.606	4850	1160.574	1.5	1.828 *
20	12.319	160	63.131	700	216.417	1470	409.013	4900	1171.125	2.0	2.398 *
21	12.778	165	64.737	710	219.046	1475	410.214	4950	1181.666	2.5	2.961 *
22	13.233	170	66.336	720	221.670	1480	411.418	5000	1192.198	3.0	3.517 *
23	13.684	175	67.928	725	222.980	1490	413.820	5100	1213.222	4.0	4.610 *
24	14.131	180	69.512	730	224.289	1500	416.220	5200	1234.203	5.0	5.689 *
25	14.573	185	71.090	740	226.904	1525	422.214	5300	1255.141	6.0	6.758 *
26	15.013	190	72.661	750	229.514	1550	428.196	5400	1276.037	7.0	7.816 *
27	15.450	195	74.226	760	232.120	1575	434.168	5500	1296.893	8.0	8.866 *
28	15.883	200	75.785	770	234.721	1600	440.128	5600	1317.709	9.0	9.911 *
29	16.313	205	77.338	775	236.020	1625	446.077	5700	1338.486	$1.0(10)^8$	$1.095(10)^7$ *
30	16.742	210	78.886	780	237.318	1650	452.016	5800	1359.225	1.5	1.604 *
31	17.167	215	80.428	790	239.912	1675	457.945	5900	1379.927	2.0	2.108 *
32	17.590	220	81.965	800	242.501	1700	463.863	6000	1400.593	2.5	2.607 *
33	18.011	225	83.497	810	245.086	1725	469.771	6100	1421.224	3.0	3.100 *
34	18.429	230	85.023	820	247.668	1750	475.669	6200	1441.820	4.0	4.071 *
35	18.845	235	86.545	825	248.957	1775	481.558	6300	1462.383	5.0	5.032 *
36	19.259	240	88.062	830	250.245	1800	487.437	6400	1482.912	6.0	5.984 *
37	19.671	245	89.575	840	252.819	1825	493.307	6500	1503.408	7.0	6.928 *
38	20.080	250	91.084	850	255.388	1850	499.167	6600	1523.872	8.0	7.865 *
39	20.488	255	92.589	860	257.953	1875	505.019	6700	1544.305	9.0	8.797 *
40	20.894	260	94.090	870	260.515	1900	510.861	6800	1564.706	$1.0(10)^9$	9.725 *
41	21.298	265	95.588	875	261.795	1925	516.695	6900	1585.077	1.5	$1.429(10)^8$ *
42	21.701	270	97.081	880	263.073	1950	522.520	7000	1605.418	2.0	1.880 *
43	22.101	275	98.571	890	265.629	1975	528.337	7100	1625.729	2.5	2.328 *
44	22.500	280	100.057	900	268.181	2000	534.145	7200	1646.011	3.0	2.771 *
45	22.897	285	101.540	910	270.729	2025	539.945	7300	1666.265	4.0	3.645 *
46	23.291	290	103.019	920	273.274	2050	545.737	7400	1686.490	5.0	4.510 *
47	23.684	295	104.495	925	274.545	2075	551.522	7500	1706.688	6.0	5.368 *
48	24.076	300	105.968	930	275.815	2100	557.299	7600	1726.859	7.0	6.220 *
49	24.466	305	107.437	940	278.353	2125	563.068	7700	1747.002	8.0	7.066 *
50	24.855	310	108.904	950	280.888	2150	568.830	7800	1767.120	9.0	7.909 *
51	25.244	315	110.367	960	283.420	2175	574.585	7900	1787.312	$1.0(10)^{10}$	8.747 *
52	25.633	320	111.827	970	285.948	2200	580.332	8000	1807.278	1.5	$1.288(10)^9$ *
53	26.020	325	113.284	975	287.211	2225	586.072	8100	1827.319	2.0	1.697 *
54	26.406	330	114.738	980	288.473	2250	591.806	8200	1847.336	2.5	2.103 *
55	26.791	335	116.189	990	290.995	2275	597.532	8300	1867.329	3.0	2.505 *
56	27.174	340	117.638	1000	293.514	2300	603.252	8400	1887.298	4.0	3.299 *
57	27.555	345	119.083	1010	296.030	2325	608.965	8500	1907.243	5.0	4.087 *
58	27.935	350	120.526	1020	298.543	2350	614.672	8600	1927.166	6.0	4.868 *
59	28.314	355	121.966	1025	299.799	2375	620.372	8700	1947.065	7.0	5.643 *
60	28.691	360	123.403	1030	301.053	2400	626.066	8800	1966.942	8.0	6.414 *
61	29.068	365	124.838	1040	303.560	2425	631.755	8900	1986.796	9.0	7.183 *
62	29.443	370	126.270	1050	306.065	2450	637.437	9000	2006.628	$1.0(10)^{11}$	7.948 *
63	29.818	375	127.699	1060	308.567	2475	643.113	9100	2026.438	1.5	$1.17(10)^{10}$ *
64	30.192	380	129.126	1070	311.066	2500	648.781	9200	2046.227	2.0	1.55 *
65	30.565	385	130.550	1075	312.314	2550	660.093	9300	2065.996	2.5	1.92 *
66	30.937	390	131.972	1080	313.562	2600	671.379	9400	2085.744	3.0	2.29 *
67	31.308	395	133.391	1090	316.055	2650	682.640	9500	2105.473	4.0	3.02 *
68	31.679	400	134.808	1100	318.545	2700	693.877	9600	2125.184	5.0	3.75 *
69	32.048	405	136.223	1110	321.032	2750	705.090	9700	2144.878	6.0	4.47 *
70	32.417	410	137.635	1120	323.517	2800	716.280	9800	2164.555	7.0	5.19 *
71	32.785	415	139.045	1125	324.760	2850	727.449	9900	2184.216	8.0	5.89 *
72	33.151	420	140.453	1130	326.000	2900	738.598	10,000	2203.861	9.0	6.58 *
73	33.517	425	141.859	1140	328.480	2950	749.725	12,500	2688.967	$1.0(10)^{12}$	7.28 *
74	33.883	430	143.262	1150	330.958	3000	760.833	15,000	3164.780	1.5	$1.08(10)^{11}$ *
75	34.247	435	144.664	1160	333.433	3050	771.922	17,500	3633.368	2.0	1.42 *
76	34.611	440	146.064	1170	335.906	3100	782.992	20,000	4095.800		
77	34.974	445	147.461	1175	337.142	3150	794.042	25,000	5005.726		
78	35.336	450	148.856	1180	338.376	3200	805.075	30,000	5899.508		

3. Evaluate (j)(A)(Δt) where the j's are successive integers from one to the number of time steps to be considered.

4. Evaluate the cumulative influx function, Q[(j)(A)(Δt)], for each of the values of (j)(A)(Δt) of Step 3.

5. Evaluate the values of Δp_j for each pressure change (each new step).

$$\Delta p_j = (p_{j-2} - p_j) / 2 \quad \text{for all } j > 1$$

$$\Delta p_1 = (p_o - p_1) / 2 \quad \text{for } j = 1$$

6. For each time step, evaluate the summation:

$$\Sigma \, \Delta p_j \, Q \, [\, (n - j + 1) (A) (\Delta t) \,]$$

which is conveniently done using a "Superposition Work Sheet" (Figure 12-14).

7. Divide each value of the summation (Step 6) by the expansibility factor "D". That is, calculate:

$$\left\{ \, \Sigma \, \Delta p_j \, Q \, [\, (n - j + 1) (A) (\Delta t) \,] \right\} / D$$

Recall that the expansibility was discussed earlier in this chapter. It is the denominator of the material balance equation. Because "D" is comprised of fluid properties, it is a strict function of pressure. Therefore, a new "D" must be calculated for each new time step or static pressure.

8a. Calculate the apparent hydrocarbon-in-place (N_a or G_a) at the end of each time step. Recall that this is calculated with the appropriate material balance equation and assuming that W_e = 0.

8b. Plot apparent hydrocarbon in place versus:

$$\left\{ \Sigma \, \Delta p_j \, Q \, [\, (n - j + 1) (A) (\Delta t) \,] \right\} / D$$

on Cartesian coordinate paper. This plot is illustrated in Figure 12-12 for an oil reservoir.

If the previously assumed value of "A" is correct, then the result of this plot will be a straight line. If the plotted points do not fall on a straight line, then the assumed value of "A" is not correct and should be modified. Figure 12-12 should assist in determining whether the next "A" should be either larger or smaller.

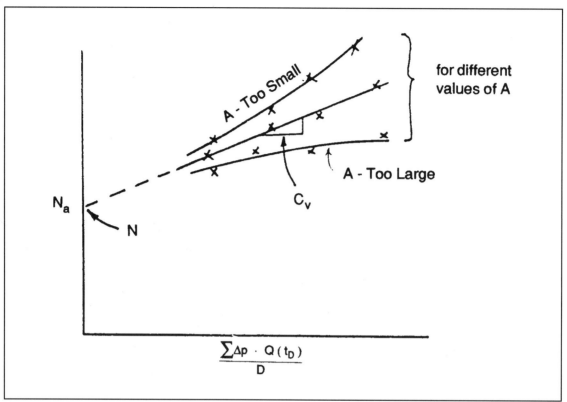

Fig. 12-12. van Everdingen and Hurst material balance plot.

9. If the "A" was too large or too small, estimate a new "A" and repeat Steps 3 through 8.

10. When an approximate straight line is obtained, then the value of "A" is nearly correct. The line may be extrapolated back to the Y-intercept to obtain N. The slope of the straight line is equal to C_v. Thus, the original hydrocarbon-in-place and the water influx constants have been determined.

These values of N, A, and C_v can be improved by using the method of least squares. An example of this procedure will be given later.

Problem 12-3:

The following data are available from a water drive oil reservoir:

Aquifer and reservoir area	28,850 acres
Oil reservoir area	451 acres
Porosity	0.22
Effective aquifer thickness	60 ft
Formation compressibility	$4.0 \times 10^{-6} \text{ psi}^{-1}$
Aquifer permeability	100 md
Water viscosity	0.3 cp
Water compressibility	$3.0 \times 10^{-6} \text{ psi}^{-1}$
Water saturation (oil reservoir)	0.26

The system is concentric and circular.

(a) Calculate the effective radii of the oil reservoir and the aquifer. Calculate the radius ratio, r_a/r_f:

$$r_a = \sqrt{[\,(28{,}850)\,(43{,}560)\,/\,\pi\,]} = 20{,}000 \; ft$$

$$r_f = \sqrt{[\,(451)\,(43{,}560)\,/\,\pi\,]} = 2{,}500 \; ft$$

$$r_a \,/\, r_f = 20{,}000 \,/\, 2{,}500 = 8$$

(b) Calculate the theoretical time conversion, A:

$$A = \frac{0.00633 \; k}{\phi \; \mu_w \; c_e \; r_f^2} = \frac{(0.00633)\,(100)}{(0.22)\,(0.3)\,(3+4)\,(10^{-6})\,(2500^2)}$$

$$A = 0.219 \; days^{-1}$$

So, $t_d = (A)(t)$ (with "t" in days)

(c) Calculate the theoretical value of C_v:

$$C_v = 1.119 \; \phi \; h \; c_e \; r_f^2 \; \alpha \qquad (\alpha = 1.0)$$
$$= (1.119)(0.22)(60)(3.0+4.0)(10^{-6})(2500)^2$$
$$C_v = 646.2 \; bbl/psi$$

(d) Calculate water influx at 100, 200, 400, and 800 days if the reservoir boundary pressure is lowered from the initial pressure of 3500 psia to 3450 psia and maintained:

t days	$t_d = 0.219t$	$Q(t_d)$*	Δp, psi	$Q(t_d) \times \Delta p$	W_e, bbl $C_v \times Q(t_d) \times \Delta p$
100	21.9	12.91	50	645.5	417,100
200	43.8	19.95	50	997.5	644,600
400	87.6	27.04	50	1352.0	873,700
800	175.2	30.83	50	1541.5	996,100

* from Table 12-1, using $r_a/r_f = 8$

(e) Given the data in columns (1) and (2) below, calculate the cumulative water influx at 500 days.

(1) t, days	(2) p, psia	(3) t_d	(4) $Q(t_d)$	(5) Δp	(6) $Q(t_d)$ Inverted	(7) $\Delta p \times Q(t_{dn} - t_{dj-1})$
0	3500	0		0		
100	3490	21.9	12.91	5	28.68	143.4
200	3476	43.8	19.95	12	27.04	324.5
300	3458	65.7	24.30	16	24.30	388.8
400	3444	87.6	27.04	16	19.95	319.2
500	3420	109.5	28.68	19	12.91	245.3
						Sum = 1421.2

Therefore,
$W_e = (646.2)(1421.2) = 918,400$ bbl

Discussion

The calculation of the radius ratio that was performed in part (a) could normally not be done in this manner. It is highly unlikely that the total area of aquifer and reservoir would be available. However, if this information were known, then the radius ratio could be calculated as shown.

Similarly, the aquifer properties are normally not known, so "A" and "C_v" cannot be calculated directly as was done in parts (b) and (c). If these data were available, the calculations could be made. Part (d) of this problem illustrates the use of the van Everdingen and Hurst aquifer equation when "A" and "C_v" are known and when the pressure at the oil/water contact is assumed to be unchanging.

Part (e) illustrates the use of the superposition principle which is needed when the pressure at the oil/water contact is changing with time. Columns (1) and (2) are given data of time and oil/water contact pressure. Columns (3) and (4) were determined exactly as in part (d). Column (5), Δp, was computed as:

$$\Delta p_1 = (p_0 - p_1)/2 \quad for \; j = 1$$

$$\Delta p_j = (p_{j-2} - p_j)/2 \quad for \; j > 1$$

The generation of column (6) can be confusing. This is where superposition is introduced. Notice that the numbers in column (6) are merely column (4) numbers turned upside down. Why? Notice that column (7) is equal to column (6) multiplied by column (5) or:

$$\Delta p \times [Q(t_d)]_{inverted}$$

This means that Δp_1 is associated with the longest acting cumulative influx function. And Δp_2 is associated with the next longest acting cumulative function, etc.

Consider the pressure-time diagram in Figure 12-13. Over each time interval, the pressure is represented as a constant value. For each time interval, the corresponding Δp is shown. Water influx is directly dependent on Δp. The superposition principle stipulates that each Δp is independent of all of the other Δp's in its effect on water influx. Thus, to find the total cumulative water influx, the solution for each Δp may be calculated and then all summed. Further, the length of time that a particular Δp is acting to cause water influx lasts from the beginning of its time interval out to t_n, or the time under consideration. For this problem, Δp_1 effects water influx from time 0 out to 500 days; Δp_2 lasts from the beginning of time interval 2 (100 days) out to 500 days; Δp_3 from 200 days to 500 days; etc.

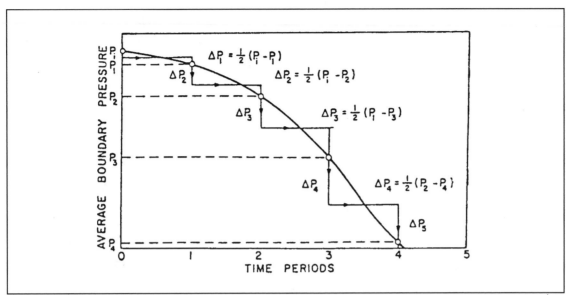

Fig. 12-13. *Sketch showing the use of step pressures to approximate the pressure-time curve (from Craft & Hawkins[4]). Permission to publish by Prentice-Hall.*

The only way to associate time with Δp is through the influx function, $Q(t_d)$. This is why the influx function column (4) was inverted: to associate the appropriate $Q(t_d)$ with each Δp.

Then, column (7) is summed and multiplied by C_v to obtain the water influx at 500 days.

The application of the superposition principle is easier if the time steps are all equal as in this problem where $\Delta t = 100$ days. The method illustrated here (inverting column 4 to create column 6) is correct only with equal time steps. There is no theoretical problem using superposition and unequal time steps. However, this computational technique would mismatch the Δp's and the $Q(t_d)$'s.

Figure 12-15 contains the generalized Superposition Worksheet; and this worksheet, applied to Example 12-3, is shown in Figure 12-14.

It helps the reservoir analyst go through the superposition steps by automatically matching the appropriate Δp with the proper $Q(t_d)$. However, it is only valid to use this sheet when the time steps are all equal.

The Δp's are entered into the left-hand column. Notice that there is a row associated with each Δp. Each row has an upper part and a lower part. In the upper part of each row, starting with the first block on the left that is not crosshatched, the $Q(t_d)$'s are written from left to right; i.e., $Q(A\Delta t)$, $Q(2A\Delta t)$, $Q(3A\Delta t)$, etc. After this is accomplished, then attention is turned to the lower part of each row. Into each of these blocks is entered the product of the Δp on the left and the $Q(t_d)$ just above it.

Then, sums are performed on each column: adding only the lower blocks in each row. This calculation yields the value of:

$$\Sigma \Delta p_j \, Q \,[\,(n - j + 1)(A)(\Delta t)\,]$$

associated with the end of each time step. Notice that the first column corresponds to the first time step, and the second column with the second time step, etc. Then to obtain W_e at the particular time under consideration, merely multiply the above sum by "C_v".

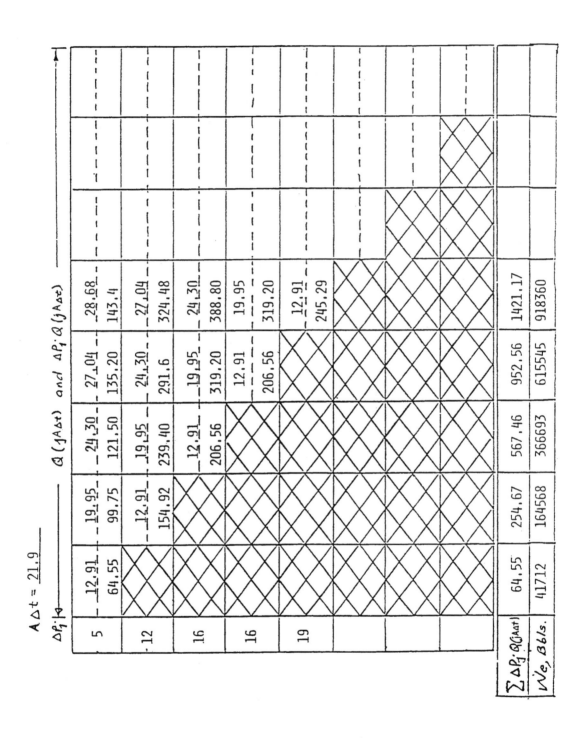

Fig. 12-14. Superposition Worksheet. for part (e), Problem 12-3.

$A\Delta t = \underline{\hspace{2cm}}$

$\Delta P_j \mid \quad Q(jA\Delta t) \quad$ and $\quad \Delta P_j \cdot Q(jA\Delta t)$

ΔP_j								
ΔP_1	$Q(1A\Delta t)$ $\Delta P_1 Q(1A\Delta t)$	$Q(2A\Delta t)$ $\Delta P_1 Q(2A\Delta t)$	$Q(3A\Delta t)$ $\Delta P_1 Q(3A\Delta t)$	$Q(4A\Delta t)$ $\Delta P_1 Q(4A\Delta t)$	$Q(5A\Delta t)$ $\Delta P_1 Q(5A\Delta t)$	$Q(6A\Delta t)$ $\Delta P_1 Q(6A\Delta t)$	$Q(7A\Delta t)$ $\Delta P_1 Q(7A\Delta t)$	$Q(8A\Delta t)$ $\Delta P_1 Q(8A\Delta t)$
ΔP_2	$Q(1A\Delta t)$ $\Delta P_2 Q(1A\Delta t)$	$Q(2A\Delta t)$ $\Delta P_2 Q(2A\Delta t)$	$Q(3A\Delta t)$ $\Delta P_2 Q(3A\Delta t)$	$Q(4A\Delta t)$ $\Delta P_2 Q(4A\Delta t)$	$Q(5A\Delta t)$ $\Delta P_2 Q(5A\Delta t)$	$Q(6A\Delta t)$ $\Delta P_2 Q(6A\Delta t)$	$Q(7A\Delta t)$ $\Delta P_2 Q(7A\Delta t)$	
ΔP_3	$Q(1A\Delta t)$ $\Delta P_3 Q(1A\Delta t)$	$Q(2A\Delta t)$ $\Delta P_3 Q(2A\Delta t)$	$Q(3A\Delta t)$ $\Delta P_3 Q(3A\Delta t)$	$Q(4A\Delta t)$ $\Delta P_3 Q(4A\Delta t)$	$Q(5A\Delta t)$ $\Delta P_3 Q(5A\Delta t)$	$Q(6A\Delta t)$ $\Delta P_3 Q(6A\Delta t)$		
ΔP_4	$Q(1A\Delta t)$ $\Delta P_4 Q(1A\Delta t)$	$Q(2A\Delta t)$ $\Delta P_4 Q(2A\Delta t)$	$Q(3A\Delta t)$ $\Delta P_4 Q(3A\Delta t)$	$Q(4A\Delta t)$ $\Delta P_4 Q(4A\Delta t)$	$Q(5A\Delta t)$ $\Delta P_4 Q(5A\Delta t)$			
ΔP_5	$Q(1A\Delta t)$ $\Delta P_5 Q(1A\Delta t)$	$Q(2A\Delta t)$ $\Delta P_5 Q(2A\Delta t)$	$Q(3A\Delta t)$ $\Delta P_5 Q(3A\Delta t)$	$Q(4A\Delta t)$ $\Delta P_5 Q(4A\Delta t)$				
ΔP_6	$Q(1A\Delta t)$ $\Delta P_6 Q(1A\Delta t)$	$Q(2A\Delta t)$ $\Delta P_6 Q(2A\Delta t)$	$Q(3A\Delta t)$ $\Delta P_6 Q(3A\Delta t)$					
ΔP_7	$Q(1A\Delta t)$ $\Delta P_7 Q(1A\Delta t)$	$Q(2A\Delta t)$ $\Delta P_7 Q(2A\Delta t)$						
ΔP_8	$Q(1A\Delta t)$ $\Delta P_8 Q(1A\Delta t)$							

$\sum \Delta P_j \cdot Q(jA\Delta t)$

Fig. 12-15. Superposition worksheet.

Problem 12-4:

This problem will illustrate the calculation of water influx, W_e, using the van Everdingen and Hurst radial aquifer model. An infinite aquifer is assumed.

Basic Data:

Porosity	0.21
Aquifer permeability	500 md
Aquifer effective thickness	50 ft
Reservoir radius	2000 ft
Water viscosity	0.3 cp
Water compressibility	3.0×10^{-6} psi^{-1}
Formation compressibility	3.6×10^{-6} psi^{-1}
α	1.0

$$c_e = c_w + c_f = (3.0 + 3.6)(10^{-6}) = 6.6 \times 10^{-6} \text{ psi}^{-1}$$

Calculation of the time coefficient, A, for time in months:

$$A = \frac{(0.00633)(30.42) \ k}{\phi \ \mu_w \ c_e \ r_f^2} = \frac{0.193 \ k}{\phi \ \mu_w \ c_e \ r_f^2}$$

$$A = \frac{(0.193)(500)}{(0.21)(0.3)(6.6)(10^{-6})(2000)^2} = 58.02 \ mo.^{-1}$$

Calculation of transmission constant, C_v:

$$C_v = 1.119 \ h \ \phi \ \alpha \ c_e \ r_f^2$$

$$C_v = (1.119)(50)(0.21)(6.6)(10^{-6})(2000)^2$$

$$C_v = 310.187 \ Barrels/psi$$

Field Pressure - Time Data:

Time, Months	Pressure at Original Oil/Water Contact, psi
0	2500
12	2480
24	2470
36	2464
48	2460

Calculation of Δp_j:

Time Step, j	p	p_{avg}	Δp_j
0	2500	2500	0
1	2480	2490	10
2	2470	2475	15
3	2464	2467	8
4	2460	2462	5

Example:

$$\text{For } j = 1, \Delta p_1 = (p_0 - p_1) / 2 = (2500 - 2480) / 2$$

$$\Delta p_1 = 10$$

$$\text{For } j = 2, \Delta p_j = (p_{j-2} - p_j) / 2 = (2500 - 2470) / 2$$

$$\Delta p_2 = 15$$

Calculation of the Cumulative Influx Function, $Q[(j)(A)(\Delta t)]$:

$$A = 58.02 \text{ mo.}^{-1}$$

$$\Delta t = 12 \text{ months}$$

$$(A)(\Delta t) = 696.24$$

j	(j)(A)(Δt)	Q[(j)(A)(Δt)]
1	696.24	215.427
2	1392.48	390.305
3	2088.72	554.692
4	2784.96	712.914

Table 12-2 was used to obtain the $Q(t_d)$ values in the third column based on the dimensionless time in the second column. Linear interpolation was used to determine the appropriate influx function value.

Example calculation:

For $(j)(A)(\Delta t) = 696.24$

$$Q(690) = 213.784 \text{ and } Q(700) = 216.417$$

Hence:

$$Q(696.24) = 213.784 + [(696.24 - 690)/(700 - 690)] [216.417 - 213.784]$$

$$= 215.427$$

Now that the Δp's, $Q(t_d)$'s, and C_v have been determined, W_e may be calculated at the end of each time step. This is accomplished with the worksheet in Figure 12-16. The use of this calculation sheet was explained in Example 12-3.

After the $\Sigma (\Delta p) [Q(t_d)]$ calculation is accomplished in the next to last row at the bottom of the worksheet, W_e is obtained by multiplying by C_v (310.187).

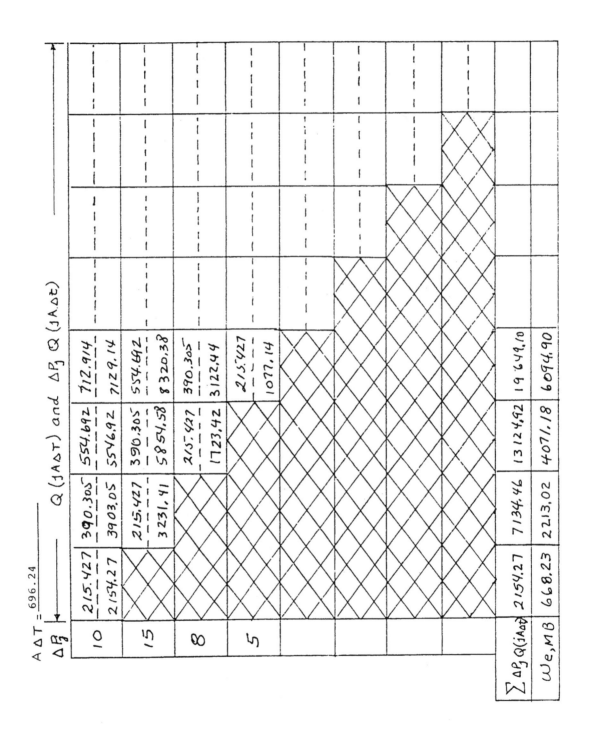

Fig. 12-16. Superposition worksheet, for Example 12-4.

Problem 12-5:

This problem illustrates the calculation of N and C_v. using the van Everdingen and Hurst model (assuming infinite radial aquifer). The least-squares method is used with a trial and error procedure.

Basic data:

Porosity	0.16
Connate water saturation	0.25
Oil compressibility	10×10^{-6} psi^{-1}
Water compressibility	3.0×10^{-6} psi^{-1}
Formation compressibility	4.0×10^{-6} psi^{-1}
Bubblepoint pressure	2100 psia
Initial pressure	3000 psia
Water formation vol. factor	1.0 Res. Bbls / STB

Pressure-Production Data:

Time Years	Pressure Psia	B_o	Cum. Oil Prod. N_p, MB	Water Prod. W_p, MB
0	3000	1.3100	0.0	0
1	2870	1.3117	952.92	10
2	2810	1.3125	2346.80	37
3	2760	1.3131	3919.25	153
4	2720	1.3137	5687.65	272

In this problem, the pressure and cumulative production data will be used with material balance and the van Everdingen and Hurst aquifer equation to calculate the original oil in place (N) and the aquifer constants (A and C_v). Note that all this pressure data is above the bubblepoint pressure. Therefore, the appropriate material balance formulation is that for an undersaturated oil reservoir.

Recall that the material balance equation can be written as:

$$N_a = N + W_e / D$$

For an undersaturated oil reservoir, the expansibility, "D" is:

$$D = c_{oe} B_{oi} (p_i - \bar{p})$$

Hence, the effective oil compressibility is needed:

$$c_{oe} = c_o + (c_w S_w / S_o) + (c_f / S_o)$$

$$= \left\{ 10 + [(3)(0.25)/0.75] + [4/0.75] \right\} (10^{-6})$$

$$= 16.33 \times 10^{-6} \ psi^{-1}$$

The effective compressibility of the aquifer is:

$$c_{we} = c_w + c_f = (3.0 + 4.0)(10^{-6})$$

$$= 7.0 \times 10^{-6} \ psi^{-1}$$

Calculations of expansibility, "D":

$$\text{Expansibility} = D = c_{oe} B_{oi} (p_i - \bar{p})$$

$$= (16.33)(10^{-6})(1.3100)(p_i - \bar{p})$$

$$= (21.392 \times 10^{-6})(p_i - \bar{p})$$

Time Years.	Pressure Psia	$p_i - \bar{p}$ Psi	Expansibility "D"
0	3000	—	—
1	2870	130	0.002781
2	2810	190	0.004064
3	2760	240	0.005134
4	2720	280	0.005990

The apparent oil in place for an oil reservoir is:

$$N_a = (\textit{cumulative withdrawals}) / (\textit{expansibility})$$

Of course, an undersaturated oil reservoir has no free gas; therefore, the expression for cumulative withdrawals is:

$$\textit{Cumulative withdrawals} = N_p B_o + W_p B_w \qquad (B_w = 1.0)$$

Time Years	N_p MB	$N_p B_o$ MB	W_p MB	Withdrawals MB	N_a MMB
0	0.0	0.0	0	0.0	—
1	952.92	1249.95	10	1259.95	453.056
2	2346.80	3080.18	37	3117.18	767.023
3	3919.25	5146.37	153	5299.37	1032.211
4	5687.65	7471.87	272	7743.87	1292.800

Time step size = 12 months

Calculation of Δp_j:

Time Step, j	Time, Mo.	p	p_{avg}	Δp_j
0	0	3000	3000	0
1	12	2870	2935	65
2	24	2810	2840	95
3	36	2760	2785	55
4	48	2720	2740	45

van Everdingen and Hurst Aquifer Model

Least-Squares Solution[15]

The material balance equation may be written as:

$$N = (Withdrawals - W_e) / D \tag{12-34}$$

Then,

$$Withdrawals = (N)(D) + W_e$$

Substituting in the van Everdingen and Hurst Equation:

$$Withdrawals = (N)(D) + C_v \Sigma \Delta pj \, Q(t_d) \tag{12-35}$$

Now, let:

$z_k = $ *cumulative withdrawals at* p_k

$x_k = $ *expansibility at* p_k, and

$$y_k = \sum_{j=1}^{k} \Delta p_j \, Q(t_d) \text{ at } p_k$$

Then, the least-squares equations are:

$$N = \frac{(\Sigma xz \, (\Sigma y^2) - (\Sigma xy)(\Sigma yz)}{(\Sigma x^2)(\Sigma y^2) - (\Sigma xy)(\Sigma xy)} \tag{12-36}$$

$$C_v = \frac{(\Sigma x^2)(\Sigma yz) - (\Sigma xy)(\Sigma xz)}{(\Sigma x^2)(\Sigma y^2) - (\Sigma xy)(\Sigma xy)} \tag{12-37}$$

A deviation function, Dev., is used as a measure of goodness of fit:

$$Dev. = \frac{1}{n} \sqrt{\sum_{j=1}^{n} (z_j - Nx_j - C_v y_j)^2} \tag{12-38}$$

This deviation function can be considered to be a function of "A," the time coefficient, or a function of (A)(Δt). Recall that Δt is the time step size, which is 12 months for this problem.

The least-squares procedure is to assume an "A;" and then calculate N, C_v, and Dev. Then, choose a different "A" and repeat the calculations. Continue investigating different "A's" until a minimum of the Dev. function is found.

Case 1:

To begin, let $(A)(\Delta t) = (10)(12) = 120$

Therefore,

Time, Years	Time Step, j	(j)(A)(Δt)	Q[(j)(A)(Δt)]
0	0	—	—
1	1	120	49.968
2	2	240	88.062
3	3	360	123.403
4	4	480	157.184

The $\Sigma \Delta p_j Q [(n - j + 1)(A)(\Delta t)]$ calculation is performed using the work sheet in Figure 12-17.

Recall that:

x_k = the expansibility at p_k (end of time step "k")

$y_k = \Sigma \Delta p_j Q(t_d)$ at the end of time step "k"

z_k = cumulative withdrawals at the end of time step "k"

k	x_k	x_k^2	y_k	y_k^2
0	—	—	....	—
1	0.002781	$(7.734)(10^{-6})$	3,247.9	10,548,854.4
2	0.004064	$(16.516)(10^{-6})$	10,471.0	109,641,841.0
3	0.005134	$(26.358)(10^{-6})$	19,135.3	366,159,706.1
4	0.005990	$(35.880)(10^{-6})$	29,032.3	842,874,443.3
		$(86.488)(10^{-6})$		1,329,244,844.8

k	z_k	$(x_k)(z_k)$	$(y_k)(z_k)$	$(x_k)(y_k)$
0	—	—	—	—
1	1259.95	3.5039	4,092,191.6	9.0324
2	3117.18	12.6682	32,639,991.8	42.5541
3	5299.37	27.2070	101,405,034.8	98.2406
4	7743.87	46.3858	224,822,357.0	173.9035
		89.7649	362,959,575.2	323.7306

Therefore,

$$N = \frac{(\Sigma xz)(\Sigma y^2) - (\Sigma xy)(\Sigma yz)}{(\Sigma x^2)(\Sigma y^2) - (\Sigma xy)(\Sigma xy)}$$

$$= \frac{(89.7649)(1.32922)(10^9) - (323.731)(3.62960)(10^8)}{(86.488)(10^{-6})(1.32922)(10^9) - (323.731)(323.731)}$$

$$N = 178,733 \ MSTB$$

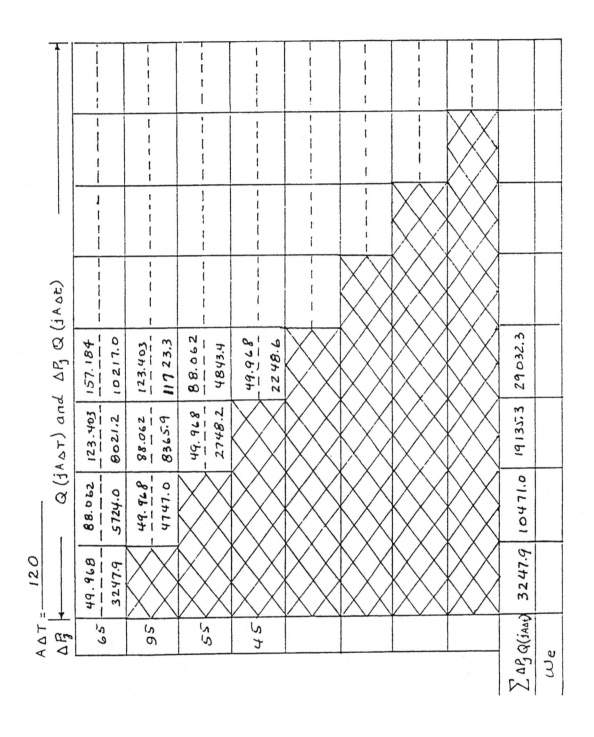

Fig. 12-17. Superposition worksheet, Example 12-5.

$$C_v = \frac{(\Sigma x^2)(\Sigma yz) - (\Sigma xy)(\Sigma xz)}{(\Sigma x^2)(\Sigma y^2) - (\Sigma xy)(\Sigma xy)}$$

$$= \frac{(86.488)(10^{-6})(3.62960)(10^8) - (323.731)(89.7649)}{(86.488)(10^{-6})(1.32922)(10^9) - (323.731)(323.731)}$$

$$C_v = 0.2295 \ MB \ / \ psi$$

$$Dev. = \frac{1}{n} \sqrt{\Sigma (z_k - Nx_k - C_v y_k)^2}$$

k	$[z_k - Nx_k - C_v y_k]$	$[z_k - Nx_k - C_v y_k]^2$
0	—	—
1	17.5005	306.267
2	-12.2854	150.931
3	-9.7966	95.973
4	10.3465	107.050
		660.221

$$Dev. = \frac{1}{4} \sqrt{660.22}$$

$$Dev. = 6.424$$

The calculation details have all been reported for the first case where $(A)(\Delta t) = 120$. For the other cases investigated, only the superposition work sheets and major results will be given.

Case II:

        $(A)(\Delta t) = 600$ (See Figure 12-18)

Results:

$$\Sigma(x)(z) = 89.765$$
$$\Sigma(y)(z) = 14.235 \times 10^8$$
$$\Sigma(x)(y) = 1267.42$$
$$\Sigma x^2 = 86.488 \times 10^{-6}$$
$$\Sigma y^2 = 20.460 \times 10^9$$

Then,

    $N = 198,303 \ MSTB$

    $C_v = 0.05729 \ MB/psi$

   Dev. $= 0.298$

So, $(A)(\Delta t) = 600$ provides a better fit than does $(A)(\Delta t) = 120$.

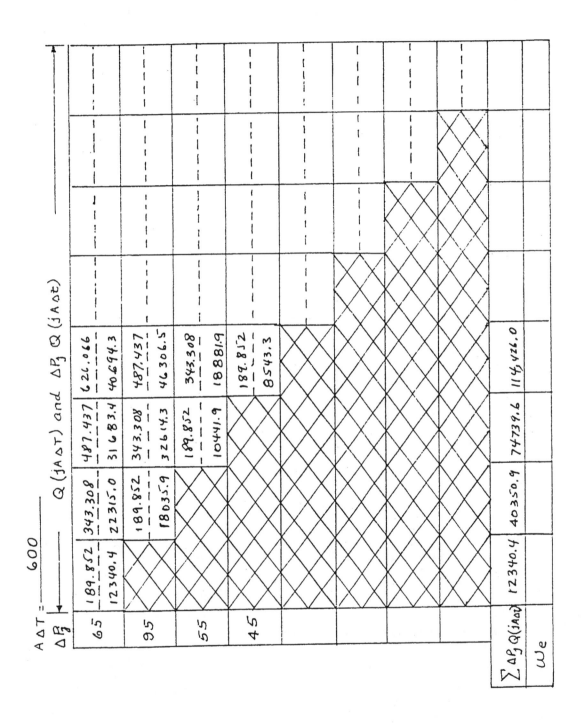

Fig. 12-18. Superposition worksheet, Example 12-5.

Case III:

$$(A)(\Delta t) = 1200 \qquad \text{(See Figure 12-19)}$$

Results:

$$\Sigma(x)(z) = 89.765$$

$$\Sigma(y)(z) = 26.037 \times 10^8$$

$$\Sigma(x)(y) = 2317.08$$

$$\Sigma x^2 = 86.488 \times 10^{-6}$$

$$\Sigma y^2 = 68.464 \times 10^9$$

Then,

$$N = 203,723 \text{ MSTB}$$

$$C_v = 0.03114 \text{ MB/psi}$$

$$\text{Dev.} = 2.615$$

Unfortunately, the deviation function is larger than that of Case II. Therefore, a smaller $(A)(\Delta t)$ will be investigated next.

Case IV:

$$(A)(\Delta t) = 700 \qquad \text{(See Figure 12-20)}$$

Results:

$$\Sigma(x)(z) = 89.765$$

$$\Sigma(y)(z) = 16.267 \times 10^8$$

$$\Sigma(x)(y) = 1448.12$$

$$\Sigma x^2 = 86.488 \times 10^{-6}$$

$$\Sigma y^2 = 26.718 \times 10^9$$

Then,

$$N = 199,782 \text{ MSTB}$$

$$C_v = 0.05005 \text{ MB/psi}$$

$$\text{Dev.} = 0.1635$$

Based on the Dev. values, the best case appears to be between:

$$600 < (A)(\Delta t) < 1200$$

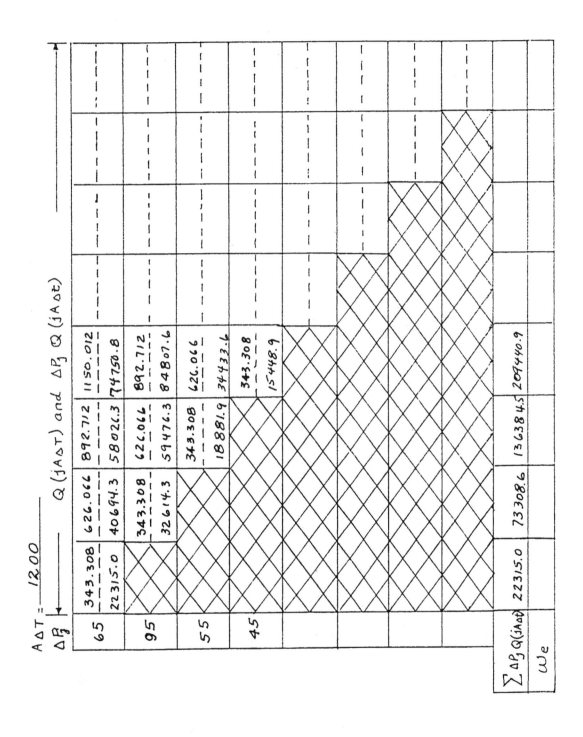

Fig. 12-19. Superposition worksheet, Example 12-5.

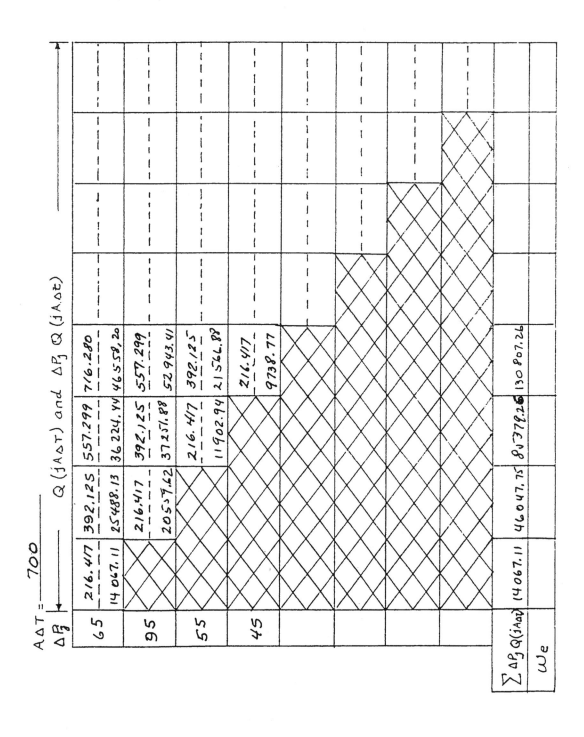

Fig. 12-20. Superposition worksheet, Example 12-5.

Case V:

$$(A)(\Delta t) = 720 \qquad \text{(See Figure 12-21)}$$

Results:

$$\Sigma(x)(z) = 89.765$$
$$\Sigma(y)(z) = 16.169 \times 10^8$$
$$\Sigma(x)(y) = 1483.88$$
$$\Sigma x^2 = 86.488 \times 10^{-6}$$
$$\Sigma y^2 = 28.056 \times 10^9$$

Then,

$$N = 200{,}043 \text{ MSTB}$$
$$C_v = 0.04883 \text{ MB/psi}$$
$$\text{Dev.} = 0.2406$$

Since the Dev. function increased in value over that computed in Case IV, the best case should be in the interval:

$$600 < [\,(A)(\Delta t)\,]_{best} < 700$$

Case VI:

$$(A)(\Delta t) = 680 \qquad \text{(See Figure 12-22)}$$

Results:

$$\Sigma(x)(z) = 89.765$$
$$\Sigma(y)(z) = 15.863 \times 10^8$$
$$\Sigma(x)(y) = 1412.23$$
$$\Sigma x^2 = 86.488 \times 10^{-6}$$
$$\Sigma y^2 = 25.409 \times 10^9$$

Then,

$$N = 199{,}510 \text{ MSTB}$$
$$C_v = 0.05134 \text{ MB/psi}$$
$$\text{Dev.} = 0.0835$$

Case VI yielded the smallest deviation of all the values of $(A)(\Delta t)$ considered. Additional cases could be investigated, but notice that N and C_v did not change appreciably from the values calculated in Case IV. So, we have convergence, and

$$N = 199.5 \text{ MMSTBO}$$
$$C_v = 0.05134 \text{ MB/psi}$$
$$A = 680 / 12 = 56.67 \text{ mo.}^{-1}$$

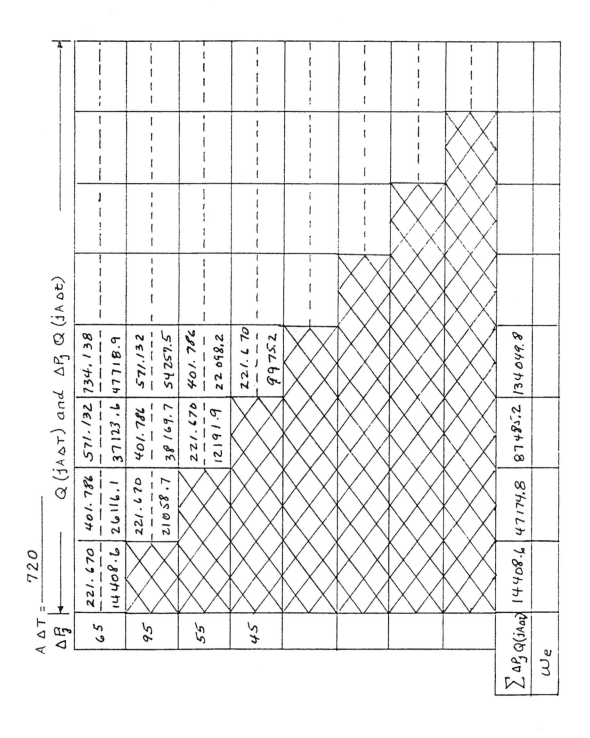

Fig. 12-21. Superposition worksheet, Example 12-5.

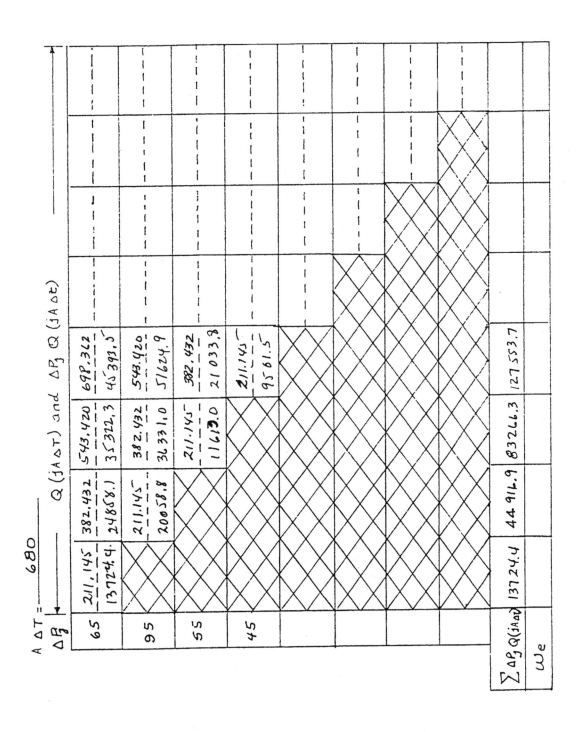

Figure 12-22. Superposition worksheet, Example 12-5.

Problem 12-6:

This example illustrates the procedure for using the van Everdingen and Hurst aquifer model to calculate the original gas in place, G, and the aquifer constants, C_v and A, in a water drive gas reservoir.

I. Theory:

The material balance equation for a gas reservoir may be written as:

$$G = \frac{G_p B_g + W_p B_w - W_e}{B_g - B_{gi}} \qquad (12\text{-}39)$$

where:

G = original gas in place, scf,
G_p = cumulative gas production, scf,
W_p = cumulative water production, STB,
W_e = cumulative water encroachment, reservoir barrels,
B_w = water formation volume factor, BBL/STB,
i = subscript indicating initial conditions,
B_g = gas formation volume factor, res. barrels / scf,
B_g = 0.005034 T z / p
T = formation temperature, deg. R,
z = gas deviation factor, and
p = reservoir pressure, psia.

Recall that the van Everdingen and Hurst aquifer equation with equal time steps is:

$$W_e = C_v \sum_{j=1}^{n} \Delta p_j \left\{ Q \left[A \left(n - j + 1 \right) \left(\Delta t \right) \right] \right\}$$

where:

$$C_v = 1.119 \ h \ \phi \ \alpha \ c_e \ r_f^2$$

and

$$A = \frac{0.00633 \ k}{\phi \ \mu_w \ c_e \ r_f^2}$$

with permeability in md. and time in days.

Combining the gas material balance equation with the van Everdingen and Hurst equation and rearranging:

$$G_p B_g + W_p B_w = G (B_g - B_{gi}) + C_v \sum_{j=1}^{n} \Delta p_j Q [(n-j+1)(A)(\Delta t)] \tag{12-40}$$

The unknowns are G, C_v, and A. The solution procedure is analogous to that used for an oil reservoir (Problem 12-5).

Let:

$$x = B_g - B_{gi} \quad (expansibility)$$

$$y = \sum_{j=1}^{n} \Delta p_j Q [(A)(n - j + 1)(\Delta t)]$$

$$z = G_p B_g + W_p B_w \quad (cumulative\ withdrawals)$$

The least-squares equation for G and C_v are:

$$G = \frac{(\Sigma xz)(\Sigma y^2) - (\Sigma xy)(\Sigma yz)}{(\Sigma x^2)(\Sigma y^2) - (\Sigma xy)(\Sigma xy)} \tag{12-41}$$

$$C_v = \frac{(\Sigma x^2)(\Sigma yz) - (\Sigma xy)(\Sigma xz)}{(\Sigma x^2)(\Sigma y^2) - (\Sigma xy)(\Sigma xy)} \tag{12-42}$$

It is assumed that the pressure-production data used are taken while the aquifer is "infinite-acting." As long as the aquifer is infinite acting during the production history for the reservoir, past performance can be matched with the G, C_v, and A determined in the solution process. A value of "A" is assumed to start the procedure. Then, after G and C_v are calculated, the deviation of A is calculated as a measure of goodness of fit as in the following.

$$dev(A) = \frac{1}{n} \sum_{k=1}^{n} Abs \left[\left(\frac{G x_k + C_v y_k}{z_k} \right) - 1 \right] \tag{12-43}$$

Different A's are investigated until the minimum of the deviation function is found.

II. Problem Statement and Solution:

Basic Data

Permeability	100	md
Porosity	20	%
Connate water saturation	30	%
Pay thickness	50	ft
Productive area	2000	acres
Effective aquifer compressibility	7×10^{-6}	psi-1
Water viscosity	0.4	cp
Reservoir temperature	160	°F
Water formation volume factor	1.02	BBl/STB
Pore volume calculated GIP	134,188	MMSCF

Pressure - Production Data:

Time Years	Pressure Psia	Gas Prod. MMSCF	Water Prod. MSTB
0	3000	0.0	0.0
1	2963	1,628.25	0.0
2	2944	3,201.13	0.0
3	2930	4,881.13	0.0
4	2919	6,547.37	0.0
5	2910	8,290.96	1.0
6	2901	10,114.06	3.0
7	2894	11,916.60	7.0
8	2887	13,773.81	10.0
9	2880	15,692.67	15.0

PVT Data:

Pressure Psia	z-factor	$B_g \times 10^3$ BBL/SCF	$(B_g - B_{gi})(10^3)$ BBL/SCF
3000	0.7780	0.80940	0.00000
2963	0.7750	0.81635	0.00695
2944	0.7730	0.81950	0.01010
2930	0.7720	0.82235	0.01295
2919	0.7710	0.82438	0.01498
2910	0.7705	0.82639	0.01699
2901	0.7700	0.82842	0.01902
2894	0.7695	0.82988	0.02048
2887	0.7690	0.83135	0.02195
2880	0.7685	0.83283	0.02343

Recalling that $z = G_p B_g + W_p B_w$ = cumulative production

Time Years	G$_p$ MMSCF	G$_p$B$_g$ MMBBLS	W$_p$ MSTB	W$_p$B$_w$ MMBBLS	z MMBBLS
0	0.00	0.0000	0.0	0.00000	0.0000
1	1,628.25	1.3292	0.0	0.00000	1.3292
2	3,201.13	2.6233	0.0	0.00000	2.6233
3	4,881.83	4.0146	0.0	0.00000	4.0146
4	6,547.37	5.3975	0.0	0.00000	5.3975
5	8,290.96	6.8516	1.0	0.00102	6.8526
6	10,114.06	8.3787	3.0	0.00306	8.3818
7	11,916.60	9.8893	7.0	0.00714	9.8964
8	13,773.81	11.4509	10.0	0.01020	11.4611
9	15,692.67	13.0693	15.0	0.01530	13.0846

The pressure drop associated with time step, j, is calculated using Equations 12-28 and 12-29.

Time Step, j	p	p$_{avg}$	Δp$_j$
0	3000	3000.0	—
1	2963	2981.5	18.5
2	2944	2953.5	28.0
3	2930	2937.0	16.5
4	2919	2924.5	12.5
5	2910	2914.5	10.0
6	2901	2905.5	9.0
7	2894	2897.5	8.0
8	2887	2890.5	7.0
9	2880	2883.5	7.0

Calculate an approximate "A".

(a) *Effective reservoir radius:*

$$r_f = \sqrt{(2000)(43,560)/\pi} = 5266.04 \; feet$$

(b) *A convenient time step size for this problem is one year.*

Therefore, for time in years:

$$A = \frac{(0.00633)(365)k}{\phi \; \mu_w \; c_e \; r_f^2} = \frac{(0.00633)(365)(100)}{(0.2)(0.4)(7 \times 10^{-6})(5266.04)^2} = 14.88 \; yr^{-1}$$

Recall that the previous equation for A requires aquifer data which are usually not available. Therefore, reservoir data are normally used, yielding only an approximate value of A. Thus, the value of A will also have to be determined.

Based on the approximate A calculation, the first A to be investigated is 10. So, since the time step size is one year.

$$(A)(\Delta t) = 10$$

Time, Years	Time Step, j	(j)(A)(Δt)	Q[(j)(A)(Δt)]**
0	0	—	—
1	1	10	7.402
2	2	20	12.320
3	3	30	16.741
4	4	40	20.884
5	5	50	24.840
6	6	60	28.658
7	7	70	32.369
8	8	80	35.991
9	9	90	39.539

** These values of Q(t_d) were not obtained from Table 12-2. Quite accurate values of the infinite aquifer influx function, Q(t_d), may be obtained with the equations of Edwardson *et al.*[13].

For $0.01 < t_d < 200$,

$$Q(t_d) = \frac{C_1 \sqrt{t_d} + C_2 t_d + C_3 t_d \sqrt{t_d} + C_4 t_d^2}{1 + C_5 \sqrt{t_d} + C_6 t_d}$$

(12-44)

where:

$C_1 = 1.12838$
$C_2 = 1.19328$
$C_3 = 0.269872$
$C_4 = 0.00855294$
$C_5 = 0.616599$
$C_6 = 0.0413008$

For $t_d \geq 200$,

$$Q(t_d) = \frac{(2.02566)(t_d) - 4.29881}{\ln(t_d)}$$

(12-45)

The $\Sigma \Delta p_j \{Q[(n-j+1)(A)(\Delta t)]\}$ calculation was performed using a superposition work sheet as was illustrated in Problem 12-5.

Recall that:

$x_k = B_g - B_{gi}$ (the expansibility at p_k),

$y_k = \Sigma \Delta p_j Q(t_d)$ at the end of time step "k,"

$z_k =$ cumulative withdrawals at the end of time step "k"

k	$(x_k)(10^3)$	x_k^2	y_k	y_k^2
0	—	—	—	—
1	0.00695	$0.48303(10^{-10})$	136.93	18,749.8
2	0.01010	$1.02010(10^{-10})$	435.17	189,372.9
3	0.01295	$1.67703(10^{-10})$	776.80	603,418.2
4	0.01498	$2.24400(10^{-10})$	1150.90	1,324,570.8
5	0.01699	$2.88660(10^{-10})$	1548.53	2,397,945.2
6	0.01902	$3.61760(10^{-10})$	1969.37	3,878,418.2
7	0.02048	$4.19430(10^{-10})$	2409.67	5,806,509.5
8	0.02195	$4.81802(10^{-10})$	2865.40	8,210,517.2
9	0.02343	$5.48965(10^{-10})$	3339.88	11,154,798.4
		$26.43033(10^{-10})$		33,584,300.2

k	z_k, MMB	$(x_k)(z_k)$	$(y_k)(z_k)$	$(x_k)(y_k)$
0	—	—	—	—
1	1.3292	$0.92379(10^{-5})$	182.0074	$0.9517(10^{-3})$
2	2.6233	$2.64953(10^{-5})$	1,141.5815	$4.3952(10^{-3})$
3	4.0146	$5.19891(10^{-5})$	3,118.5413	$10.0596(10^{-3})$
4	5.3975	$8.08546(10^{-5})$	6,211.9828	$17.2405(10^{-3})$
5	6.8526	$11.64257(10^{-5})$	10,611.4567	$26.3095(10^{-3})$
6	8.3818	$15.94218(10^{-5})$	16,506.8655	$37.4574(10^{-3})$
7	9.8964	$20.26783(10^{-5})$	23,847.0582	$49.3500(10^{-3})$
8	11.4611	$25.15711(10^{-5})$	32,840.6359	$62.8955(10^{-3})$
9	13.0846	$30.65722(10^{-5})$	43,700.9939	$78.2534(10^{-3})$
		$120.52460(10^{-5})$	138,161.1232	$286.9128(10^{-3})$

Then,

$$G = \frac{(\Sigma\, xz)(\Sigma\, y^2) - (\Sigma\, xy)(\Sigma\, yz)}{(\Sigma\, x^2)(\Sigma\, y^2) - (\Sigma\, xy)(\Sigma\, xy)}$$

$$= \frac{(120.52460)(10^{-5})(33,584,300.2) - (286.9128)(10^{-3})(138,161.1232)}{(26.43033)(10^{-10})(33,584,300.2) - [(286.9128)(10^{-3})]^2}$$

$$G = 129,882 \ MMscf$$

$$C_v = \frac{(\Sigma\, x^2)(\Sigma\, yz) - (\Sigma\, xy)(\Sigma\, xz)}{(\Sigma\, x^2)(\Sigma\, y^2) - (\Sigma\, xy)(\Sigma\, xy)}$$

$$= \frac{(26.43033)(10^{-10})(138,161.1232) - (286.9128)(10^{-3})(120.5246)(10^{-5})}{(26.43033)(10^{-10})(33,584,300.2) - [(286.9128)(10^{-3})]^2}$$

$$C_v = 0.0030043 \ MMB\,/\,psi = 3004.3 \ BBL\,/\,psi$$

Recall that these results were obtained with an assumed A = 10. To consider the goodness of fit with this A, the deviation function is used:

$$Dev\,(A) = \frac{1}{n}\sum_{k=1}^{n} Abs\left[\left(\frac{G\,x_k + C_v\,y_k}{z_k}\right) - 1\right]$$

k	$(Gx_k + C_v y_k)/z_k$	$\lvert[(Gx_k + C_v y_k)/z_k] - 1\rvert$
1	0.98861	0.01139
2	0.99843	0.00157
3	1.00027	0.00027
4	1.00106	0.00106
5	1.00092	0.00092
6	1.00061	0.00061
7	1.00029	0.00029
8	0.99985	0.00015
9	0.99942	0.00058
		0.01684

$$Dev = (1/9)(0.01684) = 0.00187$$

For the other A's tested, the following results were obtained:

A	G, MMSCF	C_v, MMB/psi	Dev(A)
18.0	136,131	0.0018537	0.00087
13.0	132,789	0.0024279	0.00062
15.0	134,285	0.0021355	0.00004
14.9	134,215	0.0021675	0.00001

So, the correct A is near 14.9.

Therefore,

$$G = 134\ \text{BSCF}$$

$$C_v = 0.00217\ \text{MMB/psi}$$

van Everdingen and Hurst Review

Problems 12-5 and 12-6 illustrated the method for calculating original hydrocarbon in place and the aquifer constants from reservoir performance data. It was assumed that the aquifer was infinite acting during the reservoir history used to "calibrate" the aquifer model. The N (or G) calculated will be correct even if in the future, the aquifer reaches finite behavior. In this case, the infinite aquifer model will no longer be correct for the purpose of predicting future aquifer behavior. On the other hand, if the aquifer is quite large, then the calculated aquifer constants (A and C_v), used with the infinite aquifer model, should describe future aquifer behavior for the economic life of the reservoir.

Even if finite aquifer behavior begins during reservoir history, it is still possible to use the van Everdingen and Hurst model (and material balance) to calculate N (or G) and the aquifer model constants: A, C_v, and (r_a/r_f). Unfortunately, the procedure is cumbersome. With a limited aquifer, the authors suggest the use of the Fetkovich model which is simpler to use and gives answers well within reservoir engineering accuracy.

FETKOVICH FINITE AQUIFER MODEL

In 1971, Fetkovich[11] developed a method of describing the approximate water influx behavior of a finite aquifer. The results of this model closely match those of the van Everdingen and Hurst method for the finite aquifer case. However, the Fetkovich theory is simpler, and the application is much easier. A trial and error procedure is used to evaluate the water influx constants using production data.

BASIC THEORY

Fetkovich began with the premise that the productivity index concept is adequate for describing water influx from a finite aquifer into a hydrocarbon reservoir. That is, the water influx rate is directly proportional to the pressure drop between the average aquifer pressure and the pressure at the oil/water contact.

$$q_w = J(\bar{p}_a - p_f) = \frac{\partial W_e}{\partial t} \tag{12-46}$$

where:

q_w = water influx rate, Bbl/day,
J = productivity index of the aquifer, Bbl/day/psi,
$\bar{p}_a$ = average aquifer pressure, psia,
p_f = pressure at the oil/water contact, psia,
W_e = cumulative water influx volume, Bbls, and
t = time, days,

From compressibility considerations, the cumulative water influx is proportional to the total pressure drop in the aquifer.

$$W_e = V_a c_e (p_i - \bar{p}_a)$$ (12-47)

where:

V_a = the initial volume of water in the aquifer, Bbl,
c_e = effective aquifer compressibility (c_w + c_f), psi^{-1},
c_w = water compressibility, psi^{-1}, and
c_f = formation compressibility, psi^{-1}
p_i = the initial pressure in the aquifer, psia

Differentiating Equation 12-47 with respect to time and combining with Equation 12-46 results in:

$$-V_a c_e \frac{\partial \bar{p}_a}{\partial t} = J(\bar{p}_a - p_f).$$ (12-48)

Assuming that p_f remains constant over the period of interest, separating variables, and integrating; the following equation is obtained:

$$\bar{p}_a = p_f + (p_i - p_f) e^{-\left(\frac{Jt}{V_a c_e}\right)}$$ (12-49)

where:

$\bar{p}_a = p_i$ at $t = 0$

Solving Equation 12-46 for $\bar{p}_a$, and then substituting the results into Equation 12-49,

$$p_f + \frac{1}{J} \frac{\partial W_e}{\partial t} = p_f + (p_i - p_f) e^{-\left(\frac{Jt}{V_a c_e}\right)}$$ (12-50)

Now, separating variables and integrating:

$$W_e = (p_i - p_f) V_a c_e \left[1 - e^{-\left(\frac{Jt}{V_a c_e}\right)} \right]$$ (12-51)

Because p_f was assumed to be constant over the interval of interest, as Equation 12-51 stands, the principle of superposition must be used to handle changes in the oil/water contact pressure. However, Fetkovich has shown that an incremental form of Equation 12-51 can be used without superposition. This modification is described as follows:

$$\Delta W_{e,n} = (\bar{p}_{a,n-1} - \bar{p}_{f,n}) V_a c_e \left[1 - e^{-\left(\frac{J\Delta t_n}{V_a c_e}\right)} \right]$$ (12-52)

where:

$\Delta W_{e,n}$ = incremental water influx during time step n, (time t_{n-1} to t_n), bbls

$\bar{p}_{a,n-1}$ = average aquifer pressure at the end of time step (n-1) which is also the start of time step (n), psia,

$\bar{p}_{f,n}$ = average water/oil contact pressure during time step (n), psia; i.e.,

$$\bar{p}_{f,n} = \frac{p_{f,n-1} + p_{f,n}}{2}$$

Δt_n = size of time step (n), days.

If $\Delta t_n = \Delta t$ is a constant for all time steps, then the expression:

$$V_a c_e \left[1 - e^{-\left(\frac{J \Delta t_n}{V_a c_e}\right)} \right] = V_a c_e \left[1 - e^{-\left(\frac{J \Delta t}{V_a c_e}\right)} \right]$$

is a constant. Hence, Equation 12-52 can be written as:

$$\Delta W_{e,n} = A \left(\bar{p}_{a,n-1} - \bar{p}_{f,n} \right) \tag{12-53}$$

where:

$$A = V_a c_e \left[1 - e^{-\left(\frac{J \Delta t}{V_a c_e}\right)} \right]$$

Equation 12-47 can be rewritten such that $\bar{p}_{a,n-1}$ can be evaluated:

$$\bar{p}_{a,n-1} = p_i \left[1 - \frac{\sum\limits_{j=1}^{n-1} \Delta W_{e,j}}{V_a c_e p_i} \right] \tag{12-54}$$

The average water/oil contact pressure during time step (n) is:

$$\bar{p}_{f,n} = \frac{(p_{f,n-1} + p_{f,n})}{2} \tag{12-55}$$

Equation 12-54 yields the average aquifer pressure at the end of time step (n-1). Writing a similar equation for the average aquifer pressure at the end of time step (n):

$$\bar{p}_{a,n} = p_i \left[1 - \frac{\sum\limits_{j=1}^{n} \Delta W_{e,j}}{V_a c_e p_i} \right] \tag{12-56}$$

By substituting Equation 12-53 into Equation 12-56, an expression for $\bar{p}_{a,n}$ in terms of $\bar{p}_{a,n-1}$ and $\bar{p}_{f,n}$ can be written:

$$\bar{p}_{a,n} = p_i - \frac{1}{V_a\, c_e}\, \sum_{j=1}^{n} \Delta W_{e,j}$$

$$\bar{p}_{a,n} = p_i - \frac{1}{V_a\, c_e}\, \sum_{j=1}^{n} [\,A\,(\bar{p}_{a,j-1} - \bar{p}_{f,j})\,]$$

$$\bar{p}_{a,n} = p_i - B\left[\, \sum_{j=1}^{n} \bar{p}_{a,j-1} - \sum_{j=1}^{n} \bar{p}_{f,j} \,\right] \qquad (12\text{-}57)$$

where:

$$B = \frac{A}{V_a\, c_e}$$

$$B = 1 - e^{-\left(\frac{J\,\Delta t}{V_a\, c_e}\right)} \qquad (12\text{-}58)$$

Rewriting Equation 12-47 at time level (n):

$$W_{e,n} = V_a\, c_e\,(p_i - \bar{p}_{a,n}) \qquad (12\text{-}59)$$

Then, incremental water influx can be determined by:

$$\Delta W_{e,n} = W_{e,n} - W_{e,n-1}$$

Problem 12-7:

The objective of this problem is to demonstrate that for a finite aquifer, the Fetkovich method produces results that match closely those of the van Everdingen and Hurst procedure.

Basic Data:

Permeability of the aquifer	500 md.
Porosity of the aquifer	20 %
Effective aquifer compressibility ($c_w + c_f$)	7.0×10^{-6} psi^{-1}
Pay thickness	20 ft
Water viscosity	0.8 cp.
Field radius	5000 ft
Aquifer radius	40,000 ft
Radius ratio	8

Water/Oil Contact Pressure Data:

Time, Years	Pressure, psia	Time, Years	Pressure, psia
0	3000		
1	2923	11	2698
2	2880	12	2680
3	2848	13	2665
4	2821	14	2650
5	2800	15	2633
6	2780	16	2620
7	2762	17	2607
8	2744	18	2593
9	2730	19	2580
10	2713	20	2568

Calculation of the van Everdingen and Hurst water influx constants:

$$C_v = 1.119 \; \phi \; h \; c_e \; r_f^2 \; \alpha \qquad (\alpha = 1.0)$$

$$= (1.119) (0.2) (20) (7) (10^{-6}) (5000)^2$$

$$= 783.3$$

$$A = \frac{0.00633 \; k}{\phi \; \mu_w \; c_e \; r_f^2} = \frac{(0.00633) (500)}{(0.2) (0.8) (7) (10^{-6}) (5000)^2}$$

$$A = 0.113 \; days^{-1}$$

As annual pressure data is available, the time step size is chosen to be 365 days.

Hence,

$$(A) (\Delta t) = (0.113) (365) = 41.2$$

Using these water influx constants, the van Everdingen and Hurst method, and the pressure history at the water/oil contact; water influx as a function of time was predicted. The method used (although not shown here) was similar to that of Problem 12-3. The following table contains the results of the calculations.

Water influx volumes predicted by van Everdingen and Hurst theory:

Time Years	Field Pressure	Water Influx Thousand Bbls.
0	3000	0
1	2923	581.4
2	2880	1705.2
3	2848	2699.9
4	2821	3532.3
5	2800	4219.7
6	2780	4798.0
7	2762	5310.6
8	2744	5782.1
9	2730	6207.9
10	2713	6608.2
11	2698	7005.3
12	2680	7408.6
13	2665	7814.1
14	2650	8197.9
15	2633	8588.7
16	2620	8966.8
17	2607	9309.9
18	2593	9647.3
19	2580	9982.2
20	2568	10301.0

As will be illustrated in Problem 12-8, if production data and fluid properties are available along with the field pressure data, then the original oil-in-place and the Fetkovich aquifer model constants can be determined simultaneously. The purpose here is to demonstrate how closely the Fetkovich model can come to the van Everdingen and Hurst results. The Fetkovich method needs two influx constants: $(V_a c_e)$ and B. The first will be calculated by using van Everdingen and Hurst influx results and Equation 12-59. "B" will be determined by trial and error.

The calculation procedure is:

1. Assume a value of $B = 1 - e^{-\left(\dfrac{J\,\Delta t}{V_a\,c_e}\right)}$

   The value of B can only vary from 0 to 1.

2. For each time step, calculate the sum:

$$\sum_{j=1}^{n} \bar{p}_{f,j} = 0.5 \sum_{j=1}^{n} (p_{f,j-1} + p_{f,j})$$

   where:
   $$p_{f,0} = p_i$$

3. For each time step, calculate the sum:

$$\sum_{j=1}^{n} \bar{p}_{a,j-1}$$

   This sum can be evaluated since only previous time steps are involved. (For time step n $= 1, \bar{p}_{a,0} = p_i$).

4. For each time step, calculate $\bar{p}_{a,n}$:

$$\bar{p}_{a,n} = p_i - B\left[\left(\sum_{j=1}^{n} \bar{p}_{a,j-1} - \sum_{j=1}^{n} \bar{p}_{f,j}\right)\right]$$

5. Calculate the coefficient, $(V_a\,c_e)$:

$$(V_a\,c_e)_n = \frac{(W_e)_{v.E\&H.}}{p_i - \bar{p}_{a,n}}$$

6. Determine the average value of $(V_a\,c_e)_n$. This "constant" will vary, which is a result of (1) uncertainty of theory, and (2) inaccuracies in the field data. For this problem, there should be a minimum of data inaccuracies. Therefore, the differences in $(V_a\,c_e)$ would be primarily due to the theory being unreliable. It will be shown that with the correct "B" the variations in $(V_a\,c_e)$ are minimal.

7. As a measure of the variations of results between the Fetkovich and van Everdingen and Hurst models, the following deviation function was used.

$$Dev(B) = \frac{1}{n}\sum_{j=1}^{n}\left\{ Abs\left[1 - \frac{(W_e)_{Fetk}}{(W_e)_{vE\&H}} \right]\right\}_j$$

   As indicated, this function is dependent on the assumed value of B.

8. Repeat Steps 1 through 7 using different assumed values of B. The correct B is the one which minimizes the deviation function defined in Step 7.

Calculation of water influx by the Fetkovich method:

The calculation details of the previous procedure will be illustrated for the best "B" found:

$$B = 1 - e^{-\left(\frac{J\,\Delta t}{V_a\,c_e}\right)} = 0.6075$$

For the other "B's," only the results will be given.

Calculation of: $\sum_{j=1}^{n} \bar{p}_{f,j}$, $\sum_{j=1}^{n} \bar{p}_{a,j-1}$, and $\bar{p}_{a,n}$ for $B = 0.6075$

(1) n	(2) $p_{f,n}$	(3) $\bar{p}_{f,n}$	(4) $\sum_{j=1}^{n} \bar{p}_{f,j}$	(5) $\sum_{j=1}^{n} \bar{p}_{a,j-1}$	(6) $(5)-(4)$	(7) $B*(6)$	(8) $\bar{p}_{a,n}$
0	3000.0						
1	2923.0	2961.5	2961.5	3000.0	38.5	23.4	2976.6
2	2880.0	2901.5	5863.0	5976.6	113.6	69.0	2931.0
3	2848.0	2864.0	8727.0	8907.6	180.6	109.7	2890.3
4	2821.0	2834.5	11561.5	11797.9	236.4	143.6	2856.4
5	2800.0	2810.5	14372.0	14654.3	282.3	171.5	2828.5
6	2780.0	2790.0	17162.0	17482.8	320.8	194.9	2805.1
7	2762.0	2771.0	19933.0	20287.9	354.9	215.6	2784.4
8	2744.0	2753.0	22686.0	23072.3	386.3	234.7	2765.3
9	2730.0	2737.0	25423.0	25837.6	414.6	251.9	2748.1
10	2713.0	2721.5	28144.5	28585.7	441.2	268.1	2731.9
11	2698.0	2705.5	30850.0	31317.7	467.7	284.1	2715.9
12	2680.0	2689.0	33539.0	34033.6	494.6	300.4	2699.6
13	2665.0	2672.5	36211.5	36733.1	521.6	316.9	2683.1
14	2650.0	2657.5	38869.0	39416.2	547.2	332.4	2667.6
15	2633.0	2641.5	41510.5	42083.8	573.3	348.3	2651.7
16	2620.0	2626.5	44137.0	44735.5	598.5	363.6	2636.4
17	2607.0	2613.5	47650.5	47371.9	621.4	377.5	2622.5
18	2593.0	2600.0	49350.5	49994.4	643.9	391.2	2608.8
19	2580.0	2586.5	51937.0	52603.2	666.2	404.7	2595.3
20	2568.0	2574.0	54511.0	55198.5	687.5	417.7	2582.3

Now, that $\bar{p}_{a,n}$ has been calculated, $(V_a\,c_e)_n$ can be determined by using the van Everdingen and Hurst water influx values and the following manipulation of Equation 12-59.

$$(V_a\,c_e)_n = \frac{(W_{e,n})_{vE\&H}}{p_i - \bar{p}_{a,n}}$$

Another table is needed:

n	$(W_e)_{vE\&H}$ MBBLS	$p_i - \bar{p}_{a,n}$ psi	$(V_a c_e)_n$ MBBL/psi
1	581.4	23.4	24.846
2	1705.2	69.0	24.708
3	2699.9	109.7	24.610
4	3532.3	143.6	24.598
5	4219.7	171.5	24.607
6	4798.0	194.9	24.620
7	5310.6	215.6	24.631
8	5782.1	234.7	24.639
9	6207.9	251.9	24.646
10	6608.2	268.1	24.653
11	7005.3	284.1	24.656
12	7408.6	300.4	24.658
13	7814.1	316.9	24.659
14	8197.9	332.4	24.660
15	8588.7	348.3	24.661
16	8966.8	363.6	24.661
17	9309.9	377.5	24.661
18	9647.3	391.2	24.663
19	9982.2	404.7	24.664
20	10301.0	417.7	24.664
		Sum =	493.165

The average value of $(V_a c_e)$ may be calculated as:

$$[(V_a c_e)]_{avg} = \frac{1}{n} \sum_{j=1}^{n} (V_a c_e)_j = 493.165 / 20 = 24.658$$

Using this average value of $(V_a c_e)$, water influx will be calculated by the Fetkovich method. The appropriate equation is:

$$[(W_{e,n})]_{Fetk} = [(V_a c_e)_{avg}](p_i - \bar{p}_{a,n}) = (24.658)(p_i - \bar{p}_{a,n})$$

Therefore:

n	$p_i - \bar{p}_{a,n}$ psi	$(W_e)_{Fetk}$ MBBLS
1	23.4	577.0
2	69.0	1701.4
3	109.7	2705.0
4	143.6	3540.9
5	171.5	4228.8
6	194.9	4805.8
7	215.6	5316.3
8	234.7	5787.2
9	251.9	6211.4
10	268.1	6610.8
11	284.1	7005.3
12	300.4	7407.3
13	316.9	7814.1
14	332.4	8196.3
15	348.3	8588.4
16	363.6	8965.6
17	377.5	9308.4
18	391.2	9646.2
19	404.7	9979.1
20	417.7	10299.6

To check the goodness of fit, the deviation function is calculated:

$$Dev\,(B) = \frac{1}{n} \sum_{j=1}^{n} \left\{ Abs \left[1 - \frac{(W_{e,j})_{Fetk}}{(W_{e,j})_{vE\&H}} \right] \right\}$$

Using this relationship, for:

$$B = 0.6075$$
$$Dev = 0.00110$$

For other "B's" that were tested, the calculation details have been omitted. However, some results are given as follows:

Assumed B	Dev (B)
0.5900	0.00600
0.5925	0.00521
0.5950	0.00443
0.5975	0.00365
0.6000	0.00289
0.6025	0.00219
0.6050	0.00156
0.6075	0.00110
0.6100	0.00124
0.6150	0.00163

Therefore, the best value of B is: B = 0.6075

and the value of $(V_a c_e)_{avg}$ = 24.658 MBBL / psi

Problem 12-8:

This problem will illustrate how the Fetkovich limited aquifer model can be used simultaneously to determine the original hydrocarbon in place and the water influx constants.

First, a review of some material balance concepts will be presented. The general material balance equation for an oil reservoir, neglecting rock and water expansion, can be written as:

$$N = \frac{N_p (B_o - R_s B_g) + G_p B_g + W_p B_w - W_e}{B_o - B_{oi} + (R_{si} - R_s) B_g + \frac{m B_{oi}}{B_{gi}} (B_g - B_{gi})} \tag{12-60}$$

where:

N = original oil in place, STB
N_p = cumulative oil production, STB,
G_p = cumulative gas production, SCF or MSCF (depending on the units of Bg; the product of $G_p B_g$ must be equal to barrels)
W_e = cumulative water influx from the aquifer into the reservoir, reservoir barrels,
W_p = cumulative water production, STB
m = ratio of the initial hydrocarbon volume of the gas cap to the initial hydrocarbon volume of the oil zone (m = 0 if no initial gas cap), dimensionless.

Other terms were defined in the "Oil Reservoir Drive Mechanism" chapter. Let "D" equal the denominator of the previous equation.

$$D = B_o - B_{oi} + (R_{si} - R_s) B_g + \frac{m B_{oi}}{B_{gi}} (B_g - B_{gi}) \tag{12-61}$$

Let the oil pressure factor be defined to be the coefficient of the cumulative oil production term (N_p) in Equation 12-60.

$$F_o = (B_o - R_s B_g) / D \tag{12-62}$$

Similarly, the gas pressure factor and the water pressure factor are:

$$F_g = B_g / D \tag{12-63}$$

$$F_w = 1 / D \tag{12-64}$$

Therefore, the general oil reservoir material balance equation can be written as:

$$N = N_p F_o + G_p F_g + B_w W_p F_w - W_e F_w \tag{12-65}$$

Notice that each pressure factor is comprised entirely of fluid properties (except possibly for "m," which is a constant). This is useful because for a given reservoir, the pressure factors are functions of pressure only. If fluid property data is available, then the pressure factors can be calculated. Recall that the apparent oil in place is the oil in place that is calculated by assuming that $W_e = 0$. That is:

$$N_a = N_p F_o + G_p F_g + B_w W_p F_w \tag{12-66}$$

Thus, we can write:

$$N_a = N + W_e F_w \tag{12-67}$$

This is the form of the material balance equation that will be used with the Fetkovich aquifer model.

Recall the Fetkovich aquifer Equation (12-59):

$$W_e = (V_a \, c_e) (p_i - \overline{p}_a)$$

where:

V_a = original volume of water in the aquifer, Bbl,
c_e = effective aquifer compressibility ($c_w + c_f$), psi^{-1}
p_i = initial pressure, psia,
$\overline{p}_a$ = average pressure in the aquifer, psia

Substituting the Fetkovich aquifer model into the previous material balance equation:

$$N_a = N + (V_a \, c_e) (F_w) (p_i - \overline{p}_a) \tag{12-68}$$

Given production history and static pressure data, the values of N and $(V_a c_e)$ can be determined by least-squares calculations. A trial and error procedure will be used.

Considering an oil reservoir, the procedure is:

1. For each time step: *(Equal time steps are assumed here.)*

   (a) Calculate the denominator of the material balance equation (expansibility), "D":

   $$D = B_o - B_{oi} + (R_{si} - R_s)B_g + (mB_{oi})(B_g - B_{gi})/B_{gi}$$

   (b) Calculate the pressure factors:

   $$F_o = (B_o - R_s B_g) / D$$

   $$F_g = B_g / D$$

   $$F_w = 1 / D$$

2. For each time step, calculate the apparent oil in place, N_a.

   $$N_a = N_p F_o + G_p F_g + B_w W_p F_w = (N_a)_{mat.bal.}$$

3. Assume a value of B where:

   $$B = 1 - e^{-\left(\dfrac{J \Delta t}{V_a c_e}\right)}$$

   Recall that B can only vary from 0 to 1.

4. For each time step, calculate the sum:

   $$\sum_{j=1}^{n} \bar{p}_{f,j} = 0.5 \sum_{j=1}^{n} (p_{f,j-1} + p_{f,j})$$

   where:

   > $p_{f,j}$ = the reservoir static at the oil/water contact at the end of time step "j," and
   > $\bar{p}_{f,j}$ = the average static pressure over time step "j."

   Also, note that: $p_{f,0} = p_i$

5. For each time step, calculate the sum:

   $$\sum_{j=1}^{n} \bar{p}_{a,j-1}$$

   This sum can be calculated since only previous time steps are involved.

   (For time step $n = 1$, $\bar{p}_{a,0} = p_i$).

6. For each time step, calculate $\bar{p}_{a,n}$:

$$\bar{p}_{a,n} = p_i - B \left[\sum_{j=1}^{n} \bar{p}_{a,j-1} - \sum_{j=1}^{n} \bar{p}_{f,j} \right]$$

7. For each time step, calculate the quantity:

$$F_w (p_i - \bar{p}_{a,n})$$

8. Using the following least-squares equations, calculate N and $(V_a c_e)$.

   Let $Y_j = N_{a,j}$

   and $X_j = F_{wj}(p_i - \bar{p}_{a,j})$

   Then:

$$N = \frac{(\Sigma Y_j X_j)(\Sigma X_j) - (\Sigma X_j^2)(\Sigma Y_j)}{(\Sigma X_j)(\Sigma X_j) - n \Sigma (X_j^2)} \tag{12-69}$$

   and

$$(V_a c_e) = \frac{(\Sigma X_j)(\Sigma Y_j) - n (\Sigma Y_j X_j)}{(\Sigma X_j)(\Sigma X_j) - n \Sigma (X_j^2)} \tag{12-70}$$

9. For each time step, calculate the apparent oil-in-place using the combined material balance and Fetkovich aquifer equation:

$$(N_{a,n})_{model} = N + (V_a c_e)(F_w)(p_i - \bar{p}_{a,n})$$

10. Calculate the following deviation function, which is a measure of goodness of fit.

$$Dev (B) = \frac{1}{n} \sum_{j=1}^{n} Abs \left[1 - \frac{(N_{a,j})_{model}}{(N_{a,j})_{mat.bal.}} \right] \tag{12-71}$$

    As shown, this function is dependent on the assumed value of B.

11. Repeat Steps 3 through 10 using different assumed values of B. The best B minimizes the deviation function of Step 10.

Problem Statement:

Using the Fetkovich limited-aquifer model together with the following basic data, calculate the original oil in place, N, and the Fetkovich aquifer constants: B and $(V_a c_e)$.

Time Years	Oil/Water Contact Pressure Psia	Cum. Oil Prod. MSTB	Cum. Gas Prod. MMSCF	Cum. Water Prod. MSTB
0	3000	0	0	0
1	2923	564.2	395.8	3.3
2	2880	1418.8	996.0	15.7
3	2848	2155.3	1513.9	32.7
4	2821	2766.9	1944.2	52.3
5	2800	3265.5	2295.2	72.2
6	2780	3686.7	2591.9	92.6
7	2762	4058.0	2853.5	113.6
8	2744	4400.1	3094.8	136.0
9	2730	4702.4	3308.0	158.1
10	2713	4992.3	3512.6	182.0
11	2698	5274.3	3711.7	207.7
12	2680	5563.7	3916.2	237.3
13	2665	5846.5	4116.0	269.1
14	2650	6113.6	4304.8	301.8
15	2633	6386.7	4498.0	338.6
16	2620	6642.3	4678.9	375.5
17	2607	6874.9	4843.6	411.4
18	2593	7104.3	5006.0	449.4
19	2580	7328.5	5164.8	488.9
20	2568	7539.7	5314.4	528.4

Using the reservoir PVT data, the pressure functions were calculated for each of the previously reported static pressures:

Pressure	D	F_o	F_g, Bbl/MCF	F_w
3000				
2923	0.00979755	79.0749	99.5162	102.0663
2880	0.01546637	50.1392	63.3602	64.6564
2848	0.01978181	39.2288	49.7287	50.5515
2821	0.02348956	33.0562	42.0175	42.5721
2800	0.02641652	29.4071	37.4592	37.8551
2780	0.02924028	26.5788	33.9266	34.1994
2762	0.03181248	24.4394	31.2547	31.4342
2744	0.03441440	22.6006	28.9583	29.0576
2730	0.03645896	21.3397	27.3840	27.4281
2713	0.03896691	19.9737	25.6785	25.6628
2698	0.04120279	18.8960	24.3331	24.2702
2680	0.04391531	17.7358	22.8849	22.7711
2665	0.04620047	16.8641	21.7969	21.6448
2650	0.04850860	16.0669	20.8020	20.6149
2633	0.05115273	15.2421	19.7728	19.5493
2620	0.05319573	14.6609	19.0477	18.7985
2607	0.05525656	14.1181	18.3706	18.0974
2593	0.05749638	13.5722	17.6898	17.3924
2580	0.05959582	13.0978	17.0981	16.7797
2568	0.06155058	12.6852	16.5835	16.2468

Calculate the apparent oil in place with the material balance equation:

$$(N_a)_{mat.bal.} = N_p F_o + G_p F_g + B_w W_p F_w \quad (B_w = 1.0)$$

Time Years	(N_a) MMSTB		Time Years	(N_a) MMSTB
0	—			
1	84.34		11	195.02
2	135.26		12	193.70
3	161.48		13	194.13
4	175.38		14	194.00
5	184.74		15	192.90
6	189.09		16	193.56
7	191.93		17	193.48
8	193.02		18	192.79
9	195.27		19	192.49
10	194.58		20	192.36

Assume "B." Let B = 0.6075. For this "B," the details of the calculation procedure will be illustrated. For the other "B's" investigated, only the results will be given. So,

Calculation of $\sum\limits_{j=1}^{n} \bar{p}_{f,j}$, $\sum\limits_{j=1}^{n} \bar{p}_{a,j-1}$, and $\bar{p}_{a,n}$ for B = 0.6075

(1)	(2)	(3)	(4)	(5)	(6)	(7)	(8)
n	$p_{f,n}$	$\bar{p}_{f,n}$	$\sum\limits_{j=1}^{n} \bar{p}_{f,j}$	$\sum\limits_{j=1}^{n} \bar{p}_{a,j-1}$	$(5)-(4)$	$B*(6)$	$\bar{p}_{a,n}$
0	3000.0						
1	2923.0	2961.5	2961.5	3000.0	38.5	23.4	2976.6
2	2880.0	2901.5	5863.0	5976.6	113.6	69.0	2931.0
3	2848.0	2864.0	8727.0	8907.6	180.6	109.7	2890.3
4	2821.0	2834.5	11561.5	11797.9	236.4	143.6	2856.4
5	2800.0	2810.5	14372.0	14654.3	282.3	171.5	2828.5
6	2780.0	2790.0	17162.0	17482.8	320.8	194.9	2805.1
7	2762.0	2771.0	19933.0	20287.9	354.9	215.6	2784.4
8	2744.0	2753.0	22686.0	23072.3	386.3	234.7	2765.3
9	2730.0	2737.0	25423.0	25837.6	414.6	251.9	2748.1
10	2713.0	2721.5	28144.5	28585.7	441.2	268.1	2731.9
11	2698.0	2705.5	30850.0	31317.7	467.7	284.1	2715.9
12	2680.0	2689.0	33539.0	34033.6	494.6	300.4	2699.6
13	2665.0	2672.5	36211.5	36733.1	521.6	316.9	2683.1
14	2650.0	2657.5	38869.0	39416.2	547.2	332.4	2667.6
15	2633.0	2641.5	41510.5	42083.8	573.3	348.3	2651.7
16	2620.0	2626.5	44137.0	44735.5	598.5	363.6	2636.4
17	2607.0	2613.5	46750.5	47371.9	621.4	377.5	2622.5
18	2593.0	2600.0	49350.5	49994.4	643.9	391.2	2608.8
19	2580.0	2586.5	51937.0	52603.2	666.2	404.7	2595.3
20	2568.0	2574.0	54511.0	55198.5	687.5	417.7	2582.3

Let $X = F_w(p_i - \bar{p}_a) = F_w(3000 - \bar{p}_a)$

and $\qquad\qquad Y = (N_a)_{mat.bal.}$

j	X_j	x_j^2	Y_j	$(X_j)(Y_j)$
1	2388.361	5,704,268.27	84.34	201,434.37
2	4461.292	19,903,126.31	135.26	603,434.36
3	5545.500	30,752,570.25	161.48	895,487.34
4	6113.354	37,373,097.13	175.38	1,072,160.03
5	6492.150	42,148,011.62	184.74	1,199,359.79
6	6665.463	44,428,397.00	189.09	1,260,372.40
7	6777.214	45,930,629.60	191.93	1,300,750.68
8	6819.819	46,509,931.19	193.02	1,316,361.46
9	6909.138	47,736,187.90	195.27	1,349,147.38
10	6880.197	47,337,110.76	194.58	1,338,748.73
11	6895.164	47,543,286.59	195.02	1,344,694.88
12	6840.438	46,791,592.03	193.70	1,324,992.84
13	6859.237	47,049,132.22	194.13	1,331,583.68
14	6852.393	46,955,289.83	194.00	1,329,364.24
15	6809.021	46,362,766.98	192.90	1,313,460.15
16	6835.135	46,719,070.47	193.56	1,323,008.73
17	6831.768	46,673,054.01	193.48	1,321,810.47
18	6803.907	46,293,150.46	192.79	1,311,725.23
19	6790.745	46,114,217.66	192.49	1,307,150.51
20	6786.288	46,053,704.82	192.36	1,305,410.36
	127,356.584	834,378,595.20	3639.52	23,750,457.63

Substituting into the least-squares formula for N (Equation 12-69):

$$N = \frac{(23,750,457.63)(127,356.584) - (834,378,595.2)(3,639.52)}{(127,356.584)(127,356.584) - (20)(834,378,595.2)}$$

$$= 25.56 \ MMSTB$$

Substituting into Equation 12-70:

$$(V_a \ c_e) = \frac{(127,356.584)(3,639.52) - (20)(23,750,457.63)}{(127,356.584)(127,356.584) - (20)(834,378,595.2)}$$

$$= 0.024563 \ MMBBL/psi \ = 24.563 \ MBBL/psi$$

Now that N and $(V_a \ c_e)$ have been computed, the apparent oil in place can be calculated using the combined material balance—Fetkovich aquifer equation (Equation 12-68):

$$(N_{a,n})_{model} = N + (V_a \ c_e)(F_w)(p_i - \bar{P}_{a,n})$$

Then, the deviation can be determined:

$$Dev(B) = \frac{1}{n} \sum_{j=1}^{n} Abs\left[1 - \frac{(N_{a,j})_{model}}{(N_{a,j})_{mat.bal.}}\right]$$

n	$(N_a)_{model}$	$(N_a)_{mat.bal.}$	$\left\| 1 - \dfrac{(N_a)_{model}}{(N_a)_{mat.bal.}} \right\|$
1	84.23	84.34	0.00130
2	135.14	135.26	0.00089
3	161.77	161.48	0.00179
4	175.72	175.38	0.00193
5	185.03	184.74	0.00157
6	189.28	189.09	0.00100
7	192.03	191.93	0.00052
8	193.08	193.02	0.00031
9	195.27	195.27	0.00000
10	194.56	194.58	0.00010
11	194.93	195.02	0.00046
12	193.58	193.70	0.00062
13	194.04	194.13	0.00046
14	193.88	194.00	0.00062
15	192.81	192.90	0.00047
16	193.45	193.56	0.00057
17	193.37	193.48	0.00057
18	192.68	192.79	0.00057
19	192.36	192.49	0.00068
20	192.25	192.36	0.00057
			0.01500

$$B = 0.6075$$

$$Dev(B) = \frac{1}{n} \sum_{j=1}^{n} Abs \left[1 - \frac{(N_{a,j})_{model}}{(N_{a,j})_{mat.bal.}} \right] = (1/20)(0.01500)$$

$$Dev(B) = 0.00075$$

Thus far, the details of the calculation procedure have been illustrated for B = 0.6075. The results of other "B" trials are given as follows:

Computed results:

Assumed B	Computed OOIP, MMB	Computed $(V_a\,c_e)$, MB/psi	Dev(B)
0.5900	29.23	24.130	0.00179
0.5925	28.71	24.182	0.00143
0.5950	27.66	24.246	0.00108
0.5975	27.14	24.312	0.00073
0.6000	26.62	24.372	0.00045
0.6025	26.61	24.435	0.00037
0.6050	26.09	24.498	0.00048
0.6075	25.56	24.563	0.00075

From the above tabulation of results, the minimum deviation occurs with B = 0.6025. Hence the following comprise the computed results for this problem:

$$N = 26.61 \ \text{MMSTBO}$$
$$(V_a \, c_e) = 24.435 \ \text{MBBL/psi}$$
$$B = 0.6025$$

Problem 12-9:

The Fetkovich model will be used with a water-drive gas reservoir to obtain G and the aquifer constants.

Basic Data:

Permeability	100 md
Porosity	20%
Connate water saturation	30%
Pay thickness	50 ft
Productive area	2000 acres
Effective aquifer compressibility	$7 \times 10^{-6} \ \text{psi}^{-1}$
Water viscosity	0.4 cp
Reservoir temperature	160 °F
Water formation volume factor	1.02 BBL/STB
Pore volume calculated GIP	134, 188 MMscf
Effective reservoir radius	5266 ft
Time interval	1.0 year

Pressure - Production Data:

Time Years	Pressure Psia	Gas Prod. MMSCF	Water Prod. MSTB
0	3000	0.0	0.0
1	2963	1,626.58	0.0
2	2944	3,158.58	0.0
3	2930	4,695.97	0.0
4	2919	6,071.69	0.0
5	2910	7,364.37	1.0
6	2901	8,583.27	3.0
7	2894	9,638.48	7.0
8	2887	10,616.99	10.0
9	2880	11,539.62	15.0

PVT Data:

Pressure Psia	z-factor	$B_g \times 10^3$ BBL/scf	$(B_g - B_{gi})(10^3)$ BBL/scf
3000	0.7780	0.80940	0.00000
2963	0.7750	0.81635	0.00695
2944	0.7730	0.81950	0.01010
2930	0.7720	0.82235	0.01295
2919	0.7710	0.82438	0.01498
2910	0.7705	0.82639	0.01699
2901	0.7700	0.82842	0.01902
2894	0.7695	0.82988	0.02048
2887	0.7690	0.83135	0.02195
2880	0.7685	0.83283	0.02343

We begin with the assumption that the aquifer is infinite acting over the time period of the production data. Therefore, a calculation procedure will be attempted using an infinite van Everdingen and Hurst model as was illustrated in Problem 12-6.

I. Calculate an initial guess for "A," the time multiplier to convert to dimensionless time. For time in years:

$$A = \frac{(0.00633)(365)k}{\phi\,\mu_w\,c_e\,r_f^2} = \frac{(0.00633)(365)(100)}{(0.2)(0.4)(7 \times 10^{-6})(5266)^2} = 14.88 \; yr^{-1}$$

II. Select a range of values of A and solve for G and C_v by least squares (as illustrated in Example 12-6). For each A, also calculate dev(A), and find the A that minimizes this deviation function. The results of these investigations are presented below:

A	C_v, MB/psi	G, MMscf	Dev(A)
100.	0.2123	221,721	0.05701
10.	1.4784	211,327	0.05228
1.	8.9169	194,161	0.04463
0.1	43.8604	173,270	0.03538
0.01	175.0706	157,298	0.02855
0.001	608.9513	149,440	0.02471

Notice that the dev(A) is continuing to decrease as A becomes smaller. The smallest value of A gave the smallest deviation. Since A = 0.001 is too small for serious consideration, the assumed infinite aquifer model is likely incorrect. Therefore, the Fetkovich model will be used.

Theory — Fetkovich Model, Water Drive Gas Reservoir:

The material balance equation:

$$G = \frac{G_p B_g + W_p B_w - W_e}{B_g - B_{gi}}$$

The Fetkovich aquifer equation:

$$W_e = (V_a c_e)(p_i - \bar{p}_a)$$

Combining these two equations:

$$G_p B_g + W_p B_w = G(B_g - B_{gi}) + (V_a c_e)(p_i - \bar{p}_a) \qquad (12\text{-}72)$$

or,

$$z = Gx + (V_a c_e)y$$

where:

     z = cum. production = $G_p B_g + W_p B_w$
     x = expansibility = $B_g - B_{gi}$
     y = cum. aquifer pressure drop = $p_i - \bar{p}_a$.

Gas Reservoir Solution Procedure (using equal time steps):

1. For each time step, calculate the reservoir expansibility, $(B_g - B_{gi})$.

2. At each point in history, calculate the cumulative production: $(G_p B_g + W_p B_w)$.

3. Assume a value of "B" (must be between 0 and 1.0). It is often desirable to make the first three attempts as B = 0.4, 0.5, and 0.6. Based on these results, other values of B can be investigated.

4. For each time step, calculate the sum:

$$\sum_{j=1}^{n} \bar{p}_{f,j} = 0.5 \sum_{j=1}^{n} (p_{f,j-1} + p_{f,j})$$

where:

     $p_{f,j}$ = the reservoir static pressure at the oil/water contact at the end of time step "j", and
     $\bar{p}_{f,j}$ = the average static pressure over time step "j."

Also, note that: $p_{f,0} = p_i$

5. For each time step, calculate the sum:

$$\sum_{j=1}^{n} \bar{p}_{a,j-1}$$

This sum can be calculated since only previous time steps are involved. (For time step n = 1, $\bar{p}_{a,0} = p_i$.)

6. For each time step, calculate $\bar{p}_{a,n}$:

$$\bar{p}_{a,n} = p_i - B \left[\sum_{j=1}^{n} \bar{p}_{a,j-1} - \sum_{j=1}^{n} \bar{p}_{f,j} \right]$$

7. For each time step, calculate the quantity:

$$(p_i - \bar{p}_{a,n})$$

8. Let:

$$z_j = (G_p B_g + W_p B_w)_j$$
$$x_j = (B_g - B_{gi})_j$$
$$y_j = (p_i - \bar{p}_{a,j})$$

9. Calculate G and $(V_a c_e)$ with the least-squares formulas:

$$G = \frac{(\Sigma xy)(\Sigma z) - (\Sigma y)(\Sigma xz)}{(\Sigma x)(\Sigma xy) - (\Sigma y)(\Sigma x^2)} \qquad (12\text{-}73)$$

and

$$(V_a c_e) = \frac{(\Sigma x)(\Sigma xz) - (\Sigma x^2)(\Sigma z)}{(\Sigma x)(\Sigma xy) - (\Sigma y)(\Sigma x^2)} \qquad (12\text{-}74)$$

10. Calculate the deviation function as a measure of goodness of fit:

$$Dev\,(B) = \frac{1}{n} \sum_{j=1}^{n} Abs \left[\left(\frac{G x_j + (V_a c_e)\, y_j}{z_j} \right) - 1 \right] \qquad (12\text{-}75)$$

11. Repeat Steps 3 through 10 using different assumed values of B. The best B minimizes the deviation function of Step 10.

Solution:

Calculate cumulative production: $z = G_pB_g + W_pB_w$

Time Years	G_p MMscf	G_pB_g MMBBLS	W_p MSTB	W_pB_w MMBBLS	z MMBBLS
0	0.0	0.0	0.0	0.00000	0.00000
1	1,626.58	1.32786	0.0	0.00000	1.32786
2	3,158.58	2.58846	0.0	0.00000	2.58846
3	4,695.97	3.86173	0.0	0.00000	3.86173
4	6,071.69	5.00538	0.0	0.00000	5.00538
5	7,364.37	6.08584	1.0	0.00102	6.08686
6	8,583.27	7.11055	3.0	0.00306	7.11361
7	9,638.48	7.99878	7.0	0.00714	8.00592
8	10,616.99	8.82643	10.0	0.01020	8.83663
9	11,539.62	9.61054	15.0	0.01530	9.62584

Assume a value of B between 0 and 1.0.

The calculation details will be illustrated with the best B found: $B = 0.2975$.

Calculation of

$$\sum_{j=1}^{n} \bar{p}_{f,j}, \ \sum_{j=1}^{n} \bar{p}_{a,j-1} \text{ ,and } \bar{p}_{a,n} \text{ for } B = 0.2975$$

(1) n	(2) $p_{f,n}$	(3) $\bar{p}_{f,n}$	(4) $\sum_{j=1}^{n} \bar{p}_{f,j}$	(5) $\sum_{j=1}^{n} \bar{p}_{a,j-1}$	(6) (5) − (4)	(7) B * (6)	(8) $\bar{p}_{a,n}$
0	3000						
1	2963	2981.5	2981.5	3000.00	18.50	5.50	2994.5
2	2944	2953.5	5935.0	5994.50	59.50	17.70	2982.3
3	2930	2937.0	8872.0	8976.80	104.80	31.18	2968.8
4	2919	2924.5	11796.5	11945.62	149.12	44.36	2955.6
5	2910	2914.5	14711.0	14901.26	190.26	56.60	2943.4
6	2901	2905.5	17616.5	17844.66	228.16	67.88	2932.1
7	2894	2897.5	20514.0	20776.78	262.78	78.18	2921.8
8	2887	2890.5	23404.5	23698.60	294.10	87.50	2912.5
9	2880	2883.5	26288.0	26611.10	323.10	96.12	2903.9

Notice that column (7) is equal to: $y_n = p_i - \bar{p}_{a,n}$

Recall that: $x_n = (B_{gn} - B_{gi})$

and $z_n = (G_pB_g + W_pB_w)_n$

Summarizing:

n	$(x_n)(10^3)$	x_n^2	y_n	z_n
0	—	—	—	—
1	0.00695	$0.48303(10^{-10})$	5.50	1.32786
2	0.01010	$1.02010(10^{-10})$	17.70	2.58846
3	0.01295	$1.67703(10^{-10})$	31.18	3.86173
4	0.01498	$2.24400(10^{-10})$	44.36	5.00538
5	0.01699	$2.88660(10^{-10})$	56.80	6.08686
6	0.01902	$3.61760(10^{-10})$	67.88	7.11361
7	0.02048	$4.19430(10^{-10})$	78.18	8.00592
8	0.02195	$4.81802(10^{-10})$	87.50	8.83663
9	0.02343	$5.48965(10^{-10})$	96.12	9.62584
	0.14685	$26.43033(10^{-10})$	485.02	52.45229

Then:

n	$(x_n)(z_n)$	$(x_n)(y_n)$
0		
1	$0.00923(10^{-3})$	$0.03823(10^{-3})$
2	$0.02614(10^{-3})$	$0.17877(10^{-3})$
3	$0.05001(10^{-3})$	$0.40378(10^{-3})$
4	$0.07498(10^{-3})$	$0.66451(10^{-3})$
5	$0.10342(10^{-3})$	$0.96163(10^{-3})$
6	$0.13530(10^{-3})$	$1.29108(10^{-3})$
7	$0.16396(10^{-3})$	$1.60113(10^{-3})$
8	$0.19396(10^{-3})$	$1.92063(10^{-3})$
9	$0.22553(10^{-3})$	$2.25209(10^{-3})$
	$0.98253(10^{-3})$	$9.31185(10^{-3})$

Substituting into the least-squares formulas (Equations 12-73 and 12-74):

$$G = \frac{(\Sigma xy)(\Sigma z) - (\Sigma y)(\Sigma xz)}{(\Sigma x)(\Sigma xy) - (\Sigma y)(\Sigma x^2)}$$

$$= \frac{(9.31185)(10^{-3})(52.45229) - (485.02)(0.98253)(10^{-3})}{(0.14685)(10^{-3})(9.31185)(10^{-3}) - (485.02)(26.43033)(10^{-10})}$$

$$G = 138,900 \ MMSCF$$

$$(V_a c_e) = \frac{(\Sigma x)(\Sigma xz) - (\Sigma x^2)(\Sigma z)}{(\Sigma x)(\Sigma xy) - (\Sigma y)(\Sigma x^2)}$$

$$= \frac{(0.14685)(10^{-3})(0.98253)(10^{-3}) - (26.43033)(10^{-10})(52.45229)}{(0.14685)(10^{-3})(9.31185)(10^{-3}) - (485.02)(26.43033)(10^{-10})}$$

$$(V_a c_e) = 0.06608 \ MMB/psi$$

Recall that the previous results were obtained with an assumed B = 0.2975. To consider the goodness of fit with this B, the deviation function is used:

$$Dev\,(B) = \frac{1}{n} \sum_{j=1}^{n} Abs\left[\left(\frac{G\,x_j + (V_a\,c_e)\,y_j}{z_j}\right) - 1\right]$$

(12-75)

j	$\dfrac{G\,x_j + (V_a\,c_e)\,y_j}{z_j}$	$\left\|\dfrac{G\,x_j + (V_a\,c_e)\,y_j}{z_j} - 1\right\|$
1	1.00085	0.00085
2	0.99395	0.00605
3	0.99943	0.00057
4	1.00142	0.00142
5	1.00225	0.00225
6	1.00202	0.00202
7	1.00069	0.00069
8	0.99943	0.00057
9	0.99802	0.00198
		0.01640

$$Dev = (1/9)(0.01640) = 0.00182$$

The results obtained with some of the other B's investigated are:

B	G, MMscf	Dev(B)
0.2960	139,500	0.00188
0.2980	138,600	0.00184
0.2990	138,200	0.00207
0.3000	137,700	0.00229

Hence, the best results are:

$$B = 0.2975$$

$$G = 138,900 \text{ MMscf}$$

$$(V_a\,c_e) = 0.06608 \text{ MMB/psi}$$

As a practical note, using a combined aquifer material balance equation to analyze production history yields good results when the time interval between static pressure points is reasonably small, such as no more than one year. The analysis of data with a substantially larger time interval (2 years or more) will likely yield an original hydrocarbon in place value that is only approximate. Later time N_a's (or G_a's) are typically not changing as much as those at early time. Thus, when the analysis is performed on later time data and the resulting straight line relationship is extrapolated to obtain N or G, the value will likely be too large. For this example, using only the data points every two years yields G = 178.8 BCF, which is 29% too high.

WATER DRIVE PREDICTION

The previous sections have discussed the use of different aquifer models with material balance and production data to calculate the original hydrocarbon in place (N or G). In the process the particular aquifer model constants were determined. For example, when using the Schilthuis model, C_s is calculated. If the model is that of van Everdingen and Hurst, than two constants are computed: A and C_v.

If the original hydrocarbon in place and the aquifer model constants have been determined successfully, then the model can be used together with the material balance equation to predict future performance. There is an important rule of thumb for such predictions, however: projecting into the future further than the amount of time (history) used to calibrate the aquifer model will likely result in considerable error. If the aquifer model was history matched (or calibrated) with five years of production data, it would be unwise to put much faith in a 10 year prediction.

Another problem with predicting future performance using an aquifer model is that there is no quantitative way to correlate water/oil ratio or water cut as a function of production or time. This relationship is individual reservoir dependent. However, sometimes future water production is projected with the use of decline curve relationships.

Gas/oil ratio usually can be reasonably correlated by considering past performance, the instantaneous GOR equation, and a saturation (material balance) equation. This procedure was illustrated in the "Solution Gas Drive" chapter. Sometimes, the GOR relationship can be developed even more simply by looking for a correlation between producing GOR versus solution GOR.

Recovery efficiency in a water drive reservoir is usually estimated on the basis of relative permeability characteristics, Buckley-Leverett frontal displacement calculations, and sweep efficiency approximations.

To predict future performance with an aquifer - material balance model, the suggested technique is to step forward in time using discrete time steps. To begin the calculations for a new time step, estimate the static pressure that will exist in the reservoir at the end of the time step. Then calculate the water influx using the aquifer model (Schilthuis, van Everdingen and Hurst, or Fetkovich) that was calibrated with the reservoir history. Next, using the material balance equations, solve for water influx at the end of the time step. If the two influx calculations do not agree, then the estimated reservoir static pressure must be changed and the calculations repeated. When the cumulative water influx agrees by both the aquifer and material balance equations then the correct pressure was estimated, and computations for the current time step are completed. Now, another time step can be considered.

The procedure just outlined is simple, but some of the calculation details need to be considered more closely. There are a number of places where engineering judgment is required to permit a successful solution to this problem. One of the main areas of concern is the water/oil ratio relationship. This will be discussed separately in the next section.

It is important to ensure that the aquifer model constants are correct and applicable throughout the prediction interval. (Do not predict too far into the future.)

The following stepwise procedure, although modified, came from Slider.[7] Recall that the technique is to step forward in time using discrete time steps.

WATER DRIVE PREDICTION PROCEDURE

1. Estimate the average reservoir pressure in the uninvaded part of the reservoir at the end of the current time step. This estimation can be performed by a plot of average reservoir pressure versus time. This plot contains reservoir history as well as results of previous time steps. Extrapolate the pressure versus time plot to the end of the current time step. Next, on the basis of the average uninvaded reservoir pressure just estimated, determine the corresponding pressure at the original oil/water contact. This is often done on the basis of the rate of water influx and Darcy's law:

$$q_w = \frac{\Delta W_e}{\Delta t} = 0.00708 \; \frac{k_w \, h \; (p_e - \bar{p})}{\mu_w \; \ln \, (r_e / r_w)}$$

or

$$p_e = \bar{p} + \frac{\Delta W_e}{\Delta t} \; \frac{(\mu_w) \, [\ln \, (r_e / r_w)]}{(0.00708) \, (k_w) \, (h)} \qquad (12\text{-}76)$$

where:

    ΔW_e = incremental water influx during this time step, reservoir barrels,
    Δt = time step size, days,
    k_w = effective permeability to water in invaded zone, millidarcies,
    h = effective reservoir thickness, ft.
    μ_w = water viscosity, cp,
    $\bar{p}$ = average reservoir pressure in the uninvaded part, at the end of the current time step, psia,
    p_e = pressure at the original oil/water contact at the end of the current time step, psia,
    r_e = distance from the middle of the uninvaded zone to the original oil/water contact (using a concentric, radial aquifer reservoir representation), ft, and
    r_w = distance from the middle of the uninvaded zone to the current oil/water contact, ft.

As stated earlier, this procedure is an iterative process. On the first pass through a new time step, the incremental water influx from the last time step is used for the ΔW_e called for in the above equation.

2. Calculate cumulative water influx at the end of the current time step with the aquifer equation (Schilthuis, van Everdingen and Hurst, or Fetkovich). The aquifer constants should have been calculated previously from production history. Since the pressure at the end of the time step was estimated in Step 1, this computation involves entering the new pressure data into the aquifer equation.

3. Decide which wells will be producing during the current time step. This should be based on the position of the oil/water contact at the beginning of the time step. Often, deciding which wells will be producing requires engineering judgment. For thin zones with edgewater drive, a conservative approach is to assume that the well is shut in as soon as the advancing oil/water contact reaches the wellbore.

In thick formations with bottomwater drive, this approach would not be realistic. This situation is sometimes handled by allowing a producer to remain on until there is only an arbitrary uninvaded producing interval that remains, say five feet.

In this step, engineering judgment is also necessary to decide if coning will be a mechanism in the reservoir, how it should be handled, and the amount of bypassed oil. There is a discussion of coning methods in a later chapter.

4. Estimate the cumulative oil production at the end of the current time step. Initially, this can be done by extrapolation of a plot of cumulative oil production versus time.

5. Estimate the cumulative water production at the end of the current time step.

   Initially, this can be done by extrapolation of a plot of water/oil ratio versus time.

   The reservoir water/oil ratio may be estimated at the end of the time step using the method discussed in the next section, "Prediction of Future Water/Oil Ratios."

   Having estimates of the reservoir water/oil ratio at the beginning and end of the time step, convert these to surface water/oil ratios by multiplying by (B_o/B_w); and then calculate the average surface water/oil ratio for the time step. Multiply the average surface water/oil ratio for the time step times the incremental cumulative oil for the time step to obtain the incremental cumulative water over the time step. This is added to the cumulative water production at the end of the last time step to obtain the estimate of the cumulative water production at the end of the current time step.

6. Evaluate the oil saturation in the uninvaded part of the reservoir at the end of the time step using the following equation:

$$S_{oUN} = \frac{Reservoir\ oil\ volume\ in\ the\ uninvaded\ zone}{Pore\ volume\ in\ the\ uninvaded\ zone} \tag{12-77}$$

$$S_{oUN} = \frac{(N - N_p)\,B_o - S_{oBY}\,(W_e - B_w W_p)\,/\,(1 - S_{oBY} - S_{wc})}{[NB_{oi}\,/\,(1 - S_{wc})] - [(W_e - B_w W_p)\,/\,(1 - S_{oBY} - S_{wc})]}$$

where:
   S_{oBY} = the average oil saturation behind the advancing water front (usually determined by Buckley-Leverett frontal displacement calculations), fraction, and
   S_{wc} = average water saturation ahead of the water/oil contact, fraction.

This equation does not account for a gas cap. For an oil reservoir with an initial gas cap, the uninvaded zone oil saturation equation may be found in the "Combination Drive" chapter (Equation 15-24).

7. Determine the average oil production capability of the active producing wells for the current time step. Based on this capability, calculate the incremental oil production for the current time step. There are several approaches that can be used at this point. Two of the common techniques will be mentioned.

(a) One popular method is to calculate an average productivity index of the producing wells. This approach assumes that an earlier average producing well P.I., J_i, (before any water production) is known as well as the static pressure and k_{ro} at that time. Using the oil saturation of the uninvaded zone (Step 6) at the end of the time step, k_{ro} can be determined from the relative permeability relationship. Similarly, as the pressure has been estimated at the end of the time step (Step 1), B_o and μ_o can be determined as well.

Therefore,

$$J = J_i \left[\frac{k_{ro}}{\mu_o B_o} \right] / \left[\frac{k_{ro}}{\mu_o B_o} \right]_i \qquad (12\text{-}78)$$

Then, from the water/oil ratio prediction scheme, an estimate of h_o/h_t is available for conditions at the end of the time step. Therefore, the corrected productivity index at the end of the time step is:

$$J_{cor} = (J)(h_o / h_t) \qquad (12\text{-}79)$$

where:

h_o = total completion interval footage (considering all active wells) that is above the present water/oil contact, ft, and

h_t = total completion interval footage of all active wells, ft.

Considering the current time step to be time step "n," the average single well productivity index over the time step is:

$$J_{avg} = (J_{cor,n} + J_{cor,n-1}) / 2 \qquad (12\text{-}80)$$

Then, incremental oil production over the current time step is:

$$\Delta N_p = (J_{avg}) \left[\left(\frac{\bar{p}_n + \bar{p}_{n-1}}{2} - p_w \right) \right] (\Delta t)(\#\ active\ wells) \qquad (12\text{-}81)$$

where:

ΔN_p = the incremental oil production over the current time step, STB,

$\bar{p}_n$ = the estimated reservoir static pressure at the end of the time step, psia,

$\bar{p}_{n-1}$ = reservoir static pressure at the beginning of the time step, psia,

p_w = the average bottomhole well pressure, psia, and

Δt = the time step size, days

(b) Another method that is sometimes used is to assume that the total reservoir fluid rate from each producing well remains constant even though the percentages of water and oil are changing. This would simulate a constant volume pump being used in each well.

From Step 5, an average water/oil ratio for the time step was estimated. Using the *reservoir* average WOR for the time step, then the incremental oil production for the time step is:

$$\Delta N_p = \left(\frac{1}{WOR + 1} \right) (q_{oi} B_{oi}) \left(\frac{1}{B_o} \right) (\Delta t) (\# \ active \ wells) \tag{12-82}$$

where:

ΔN_p = the incremental oil production over the current time step, STB,

WOR = the average reservoir water/oil ratio for the time step,

q_{oi} = the initial average oil rate from one well, assuming no water production, ST-BOPD,

B_{oi} = the oil formation volume factor existing at the time of the initial oil rate, BBL/STB,

B_o = the oil formation volume factor associated with average pressure of the time step, BBL/STB,

Δt = time step size, days.

This formula assumes that the constant reservoir volumetric rate from a single well is equal to $(q_{oi})(B_{oi})$.

8. Calculate cumulative oil production at the end of the current time step: add the incremental oil production for the time step (Step 7) to the cumulative oil production from the last time step:

$$N_{p,n} = N_{p,n-1} + \Delta N_{p,n} \tag{12-83}$$

Compare this value with that of step (4). Repeat Steps 4 through 7 until acceptable agreement is found.

9. Calculate the cumulative gas production, G_p, to the end of the current time step. This is generally done with the instantaneous gas/oil equation:

$$R = R_s + \frac{k_g}{k_o} \frac{\mu_o}{\mu_g} \frac{B_o}{B_g} \tag{12-84}$$

where:

R = surface producing GOR, made up of free gas from the reservoir and gas that was in solution in the oil while in the reservoir, scf/STB,

R_s = solution-gas/oil ratio, scf/STB

Also, the results of the oil saturation equation, Step 5, are needed. The oil saturation in the uninvaded zone is known at the beginning and at the end of the current time step. Therefore, the gas saturation at these points can be calculated. (The water saturation in the uninvaded zone is assumed to be unchanging.)

$$S_g = 1 - S_{wc} - S_{oUN} \tag{12-85}$$

Hence, k_g/k_o can be determined at the beginning and end of the current time step. The other parameters in the instantaneous GOR equation are fluid properties, all functions of pressure. Hence, since the pressures are known at the beginning and end of the time step, R can be calculated at these times as well. Therefore, the cumulative gas production at the end of the time step is:

$$G_{p,n} = G_{p,n-1} + \Delta N_{p,n} \left(\frac{R_{n-1} + R_n}{2} \right) \tag{12-86}$$

10. Calculate water influx at the end of the current time step using the appropriate material balance equation. This assumes that the original oil in place, N, is known. This should have been obtained through analysis of the reservoir history production data. Recall that the use of these equations requires that the assumptions made in the development of the equations are true. These conditions can be found in the "Oil Reservoir Drive Mechanisms" chapter.

Oil reservoir with an initial gas cap:

$$W_e = N_p B_o + B_g (G_p - N_p R_s) + W_p B_w$$

$$- N [B_g (R_{si} - R_s) + (B_o - B_{oi}) + (mB_{oi} / B_{gi}) (B_g - B_{gi})] \tag{12-87}$$

Undersaturated oil reservoir (pressure above bubblepoint):

$$W_e = N_p B_o + W_p B_w - N c_{oe} B_{oi} (p_i - p) \tag{12-88}$$

where:
$$c_{oe} = c_o + [S_w c_w / (1 - S_w)] + [c_f / (1 - S_w)]$$

Performance below the bubblepoint (no initial gas cap):

$$W_e = N_p B_o + B_g (G_p - N_p R_s) + W_p B_w - N [B_g (R_{si} - R_s) + (B_o - B_{oi})] \tag{12-89}$$

11. Compare W_e calculated by material balance (step 10) with W_e calculated by the aquifer equation (Step 2).

   (a) If there is agreement, then convergence is achieved for this time step. Proceed to the next time step.

   (b) If W_e by material balance is larger than by fluid flow, choose a lower value for reservoir static pressure, and repeat Steps 1 through 11.

   (c) If W_e by material balance is less than by fluid flow, choose a higher value for reservoir static pressure, and repeat Steps 1 through 11.

Discussion

It should be evident that modifications can be introduced to the previous procedure. No two water drive reservoirs are alike, and no two calculation procedures should be the same. Refinements can be made for wells with differing productivity indices. Different methods for the estimation of water production can be incorporated into the procedure. Downdip reinjection of produced water could easily be introduced into the calculation. Where the mobility ratio is unfavorable, considering saturation distributions along streamlines would improve the prediction since displacements would not be piston-like.

Pirson[8] and Slider[7] have presented theory and procedures devoted to water drive prediction, and these sources are recommended reading.

In general, the reservoir engineer should model water drive reservoirs with the simplest available and applicable system. When the calculation techniques become so complex that the analyst does not understand the procedures involved, then a serious question arises as to the value of the answers. The fact that large sums of time and money have been spent does not guarantee that the answer is better than that obtained by a simpler method.

As mentioned earlier, a significant problem with aquifer model-material balance prediction techniques is the lack of a quantitative way to correlate water production with time. The method given in the next section does not consider coning. If coning is thought to be a mechanism in the reservoir, then this effect will have to be added. Several different coning calculation techniques will be presented in a later chapter.

Prediction of Future Water/Oil Ratio (WOR)

This section contains a method for predicting approximate future water/oil ratios from a water drive reservoir. It is assumed that this procedure is being used with a future water encroachment prediction such as discussed in the last section. The method will be presented in two parts with the division at a reservoir WOR = 5.0. Experience has shown that at low values, WOR has a dependence on the structural relationship of the current oil/water contact with respect to the completion interval. However, for high water/oil ratios (WOR > 5), there is little effect of structure remaining in the WOR relationship.

The effect of coning is not considered here.

$WOR \leq 5.0$

For reservoir water/oil ratios less than or equal to 5.0, the water-invaded thickness is assumed to flow 100% water, while the uninvaded pay is assumed to flow only oil. Hence, the reservoir water/oil ratio is:

$$WOR = \left(\frac{k_{rw}}{k_{ro}} \right) \left(\frac{h_w}{h_o} \right) \left(\frac{\mu_o}{\mu_w} \right) \qquad (12\text{-}90)$$

where:

k_{rw} = the relative permeability to water at bypassed oil saturation, fraction,
k_{ro} = the relative permeability to oil in the uninvaded oil zone, fraction,
h_w = pay thickness flowing water, ft,

h_o = pay thickness flowing oil, ft,
μ_o = oil viscosity, cp,
μ_w = water viscosity, cp

It should be apparent that all factors on the right hand side of the previous equation are relatively constant except h_w and h_o. Of course, μ_o is changing slightly with pressure change. The determination of the ratio h_w/h_o is done with three steps.

1. Calculate the pore volume for the uninvaded oil zone. This is based on the amount of water encroachment, water production, bypassed oil saturation, and original pore volume of the oil zone.

$$(PV)_{above} = (PV)_t - \frac{(W_e - W_p B_w)}{(1 - S_{wc} - S_{oBY})} \tag{12-91}$$

where:

$(PV)_{above}$ = pore volume of the oil zone above the present oil/water contact, barrels,
$(PV)_t$ = the original oil zone pore volume, barrels,
W_e = predicted cumulative water encroachment at the time of the desired WOR, barrels,
W_p = cumulative water production at the time of the desired WOR, STB,
B_w = water formation volume factor, BBL/STB,
S_{wc} = water saturation in the uninvaded oil zone, fraction, and
S_{oBY} = the bypassed oil saturation in the invaded zone (S_{or} is often used), fraction.

2. After the uninvaded oil zone pore volume has been calculated, the height of the current oil/water contact above the original oil/water contact is determined. This is done through the use of a pre-constructed diagram such as that illustrated in Figure 12-23 for the subject reservoir. In order to prepare this plot of height above the original oil/water contact versus pore volume above the height considered, structure and isopach maps are needed.

   To use this plot of height versus pore volume, the plot is entered on the horizontal pore volume axis at the value calculated in part 1 (uninvaded oil zone pore volume), and the present height of the oil/water contact above the original oil/water contact is determined.

3. Determine the ratio: h_w / h_o. Once again a pre-constructed plot is needed. Figure 12-24 illustrates the needed diagram of the reservoir completion intervals versus height above the original oil/water contact.

   This plot is not made for a single well, but for the total reservoir. First, the total feet of perforations (or completion intervals) for all of the wells in the reservoir is determined. Then, various heights above the original oil/water are considered. At each height, the total feet of well completions below the level under consideration is calculated and then divided by the total feet of completion intervals in the reservoir. These fractional "watered-out" completion intervals are plotted versus height above the original oil/water contact as shown in Figure 12-24.

   With the present height of the oil/water contact, calculated in Step 2, enter the plot and determine h_w. Then, $h_o = 1 - h_w$. Thus, the ratio h_w/h_o may be calculated, and the reservoir WOR determined with Equation 12-90.

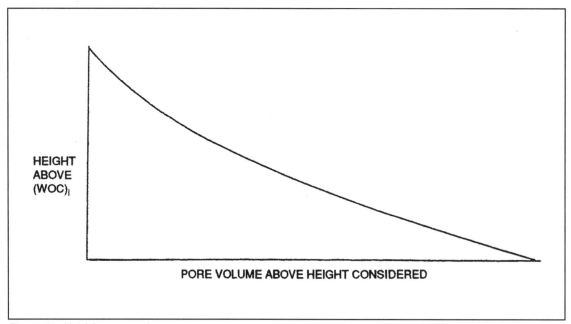

Fig. 12-23. Height vs. pore volume.

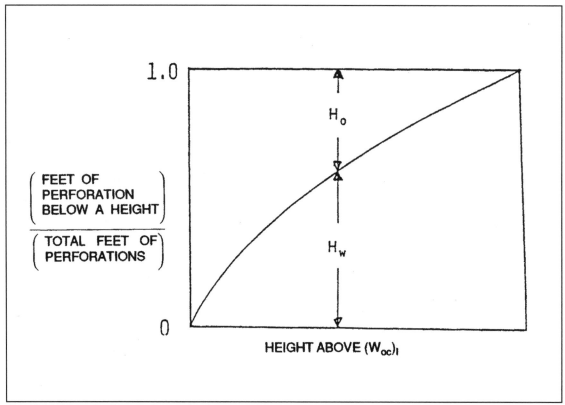

Fig. 12-24. Normalized watered-out perforations vs. height.

WOR>5.0 :

For water/oil ratios exceeding 5.0, there is little effect of structure remaining in the WOR relationship. The authors suggest using an empirical method that has its roots in decline curve analysis. Not only good for water drive reservoirs, the relationship is often successful in predicting waterflooding behavior.

For high WOR's, there is normally a semi-logarithmic relationship between WOR and flooded-out pore volume. Flooded-out pore volume is calculated as:

$$(PV)_{flooded} = \frac{W_e - W_p B_w}{1 - S_{wc} - S_{oBY}}$$

(12-92)

where these terms on the right were all defined and discussed under Equation 12-91.

The plot shown in Figure 12-25 is constructed quite simply. Make sure that semilog paper is used with the WOR axis on the log scale. Using the method for WOR ≤ 5.0, W_e and W_p are known when WOR = 5.0. Then, using the above equation, the flooded-out pore volume for WOR = 5.0 can be calculated. Thus, the first end point for the plot is in hand. The second end point comes from a rule of thumb. "Flood out" is reached in a water drive reservoir at a WOR approximately equal to 100. The entire oil zone is assumed to be flooded at this point. Thus, the second end point is plotted at a WOR = 100 and PV equal to the entire original oil-zone pore volume. Then, a straight line is drawn between the two end points.

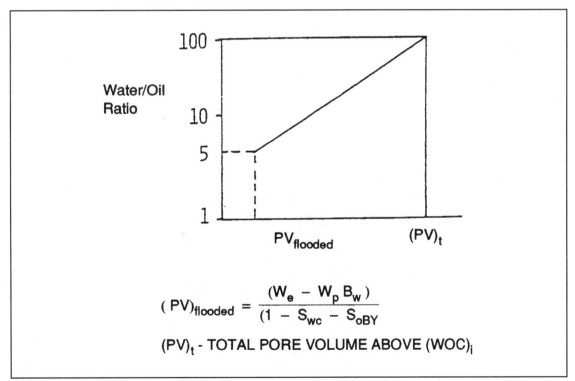

Fig. 12-25. WOR vs. invaded pore volume.

FIELD CASE HISTORIES

A. Field with a Strong Water drive

Willmon[9] has studied the displacement efficiency of a strong water drive in Redwater Field, Alberta, Canada. Figures 12-26 and 12-27 show the location of the field and a schematic cross section, respectively. Table 12-3 provides a summary of the reservoir data.

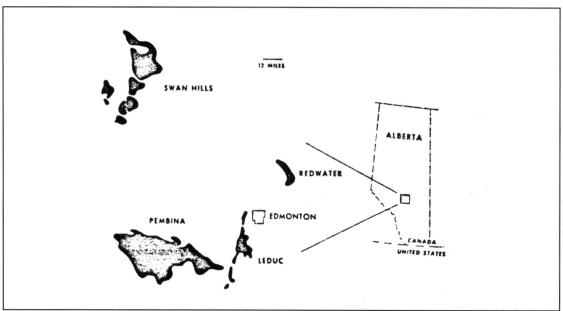

Fig. 12-26. Location of Redwater Field in relation to other fields in Central Alberta, Canada (from Willmon[9]). Permission to publish by JPT.

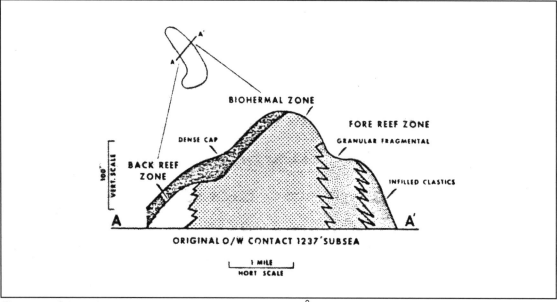

Fig. 12-27. Schematic cross section, Redwater Field (from Willmon[9]). Permission to publish by JPT.

Table 12-3
Reservoir Data — Redwater Field
(from Willmon[9]):

Average porosity	6.5	percent
Horizontal permeability	500	md
Average connate water saturation	25	percent
Average oil zone thickness	101	ft.
Areal extent	37,833	acres
Oil formation volume factor	1.133	B/STB
Original oil in place	1,277	MMSTB
Original reservoir pressure	1,050	psi
Bubblepoint pressure	485	psi
Original gas in solution	195	scf/STB
Crude oil viscosity	2.7	cp
Reservoir temperature	94	°F

The Redwater Field was discovered in 1948 and by 1967 had produced approximately 25 percent of the original oil in place. This was felt to be more than adequate to determine the original oil in place to be 1,277 million STB. Over 900 wells have been drilled on 40-acre spacing. Figure 12-28 provides the pressure-production history through 1966. Notice that during the last few years shown, reservoir pressure was nearly stabilized indicating that withdrawals were approximately balanced by water encroachment. Material balance calculations, producing well performance, and water/oil contact measurements in observation wells indicate that the displacement efficiency is 64 percent of the original oil in place. Extensive coring with a pressure-retaining core barrel and low-invasion, oil-base coring fluid have confirmed the connate water saturations and residual oil saturations for this displacement efficiency.

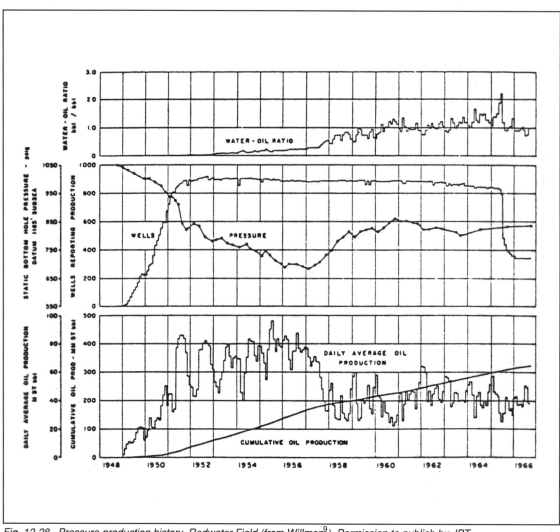

Fig. 12-28. Pressure-production history, Redwater Field (from Willmon[9]). Permission to publish by JPT.

B. Field with a Partial Water drive

Willingham and Howald[10] have reported on the performance of the Tensleep sandstone reservoir in the Torchlight Field, Wyoming. Figure 12-29 shows the location of the field in relation to other fields on the east side of the Big Horn Basin, Big Horn County, Wyoming. Figure 12-30 provides a structure map and a cross section of the Tensleep reservoir. Figure 12-31 contains an isopachous map of the Tensleep sandstone above the original water/oil contact. Table 12-4 provides a summary of the reservoir and fluid properties.

Field performance indicated that the reservoir was producing by a limited water drive and that 25 years would be needed for depletion. Oil production is from a uniform sand, 20 to 30 feet in thickness located near the top of the 220 foot-thick Tensleep Sandstone. Notice the reported tilting of the original water/oil contact shown in Figure 12-30. In 1957, the limited water drive was supplemented by outside water injection. Predictions coupled with past performance indicate that ultimate recovery by water displacement would be 43 percent of the original oil in place. The injection of water was a rate acceleration project.

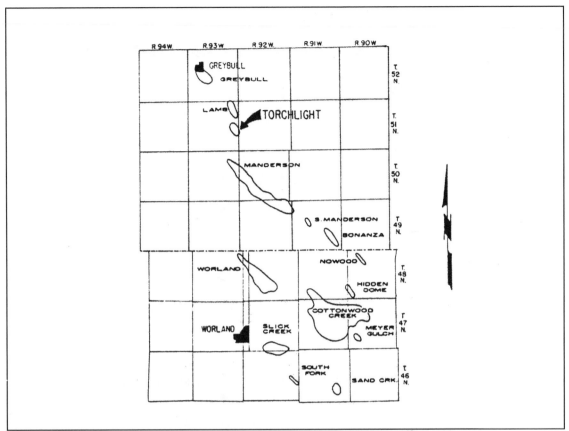

Fig. 12-29. General location map, Torchlight Field, Wyoming (from Willingham & Howald[10]). Permission to publish by JPT.

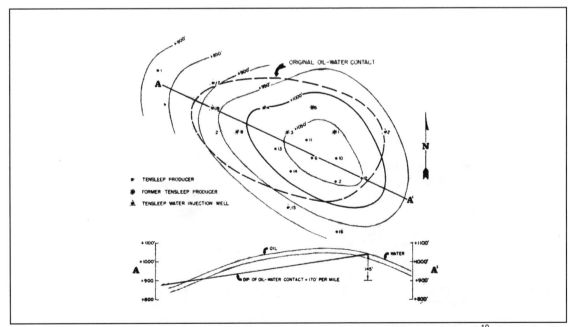

Fig. 12-30. Structure and cross section of the Torchlight Tensleep reservoir (from Willingham & Howald[10]).Permission to publish by JPT.

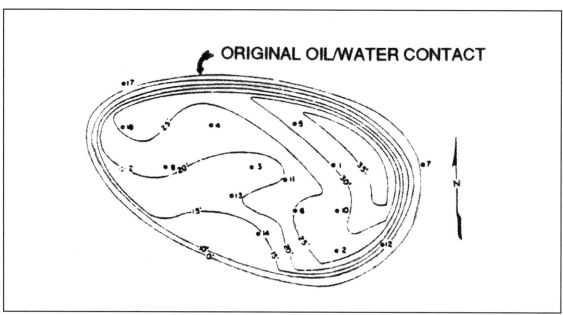

Fig. 12-31. *Sandstone isopachous map above the original oil/water contact, Tensleep sand (from Willingham & Howald[10]). Permission to publish by JPT.*

Table 12-4
Reservoir and fluid characteristics, Tensleep Torchlight Reservoir
(from Willingham & Howald[10]):

Discovery date	Oct. 31, 1947
Type structure	Anticline
Primary producing mechanism	Water Drive
Type rock	Sandstone
Crude gravity	33.7 °API
Original solution GOR	2 SCF/STB
Bubblepoint pressure	59 psia
Oil formation volume factor @ bubblepoint	1.0225
Reservoir temperature	100 °F
Oil viscosity at 100 °F and 59 psia	3.5 cp
Water viscosity at 100 °F	0.68 cp
Compressibility of reservoir oil	7.7×10^{-6} psi^{-1}
Compressibility of reservoir water	3.1×10^{-6} psi^{-1}
Compressibility of pore volume	3.9×10^{-6} psi^{-1}
Average depth to tensleep pay	3,090 ft
Average net pay thickness	20.7 ft
Average porosity, percent	16.9
Average permeability (core data)	297 md
Average connate water saturation, percent	13.1
Productive acres	291
Original oil in place	6,710,000 STB
Cumulative oil production to 1/1/65	2,774,000 STB
Cumulative water production to 1/1/65	2,424,000 STB

REFERENCES

1. Merle, H. A., Kentie, J. P., van Opstal, G and Schneider, G. M. G.: "The Bachaquero Study — A Composite Analysis of the Behavior of a Compaction Drive/Solution-gas Drive Reservoir", paper 5529 presented at 1975 SPE Annual Technical Conference and Exhibition, Dallas (Sept. 28 - Oct. 1).

2. Smith, Charles R.: *Mechanics of Secondary Oil Recovery*, Reinhold Publ. Corp., New York City (1966) 19ff.

3. Schilthuis, R. J.: "Active Oil and Reservoir Energy," *Trans.*, AIME, **188**, 33ff.

4. Craft, B. C. and Hawkins, M. F.: *Applied Petroleum Reservoir Engineering*, Prentice-Hall, Inc., Englewood Cliffs, N.J. (1959) 205ff.

5. Hurst, W.: "Water Influx Into a Reservoir and Its Application to the Equation of Volumetric Balance," *Trans.*, AIME (1943) **151**, 57ff.

6. van Everdingen, A. F. and Hurst, W.: "The Application of the Laplace Transformation to Flow Problems in Reservoirs," *Trans.*, AIME (1949) **186**, 305ff.

7. Slider, H. C.: *Practical Petroleum Reservoir Engineering Methods*, Petroleum Publishing Co., Tulsa (1976) 353 ff.

8. Pirson, S. J.: *Oil Reservoir Engineering*, McGraw-Hill Book Co., New York City (1958) Ch. 12.

9. Willmon, G. J.: "A Study of Displacement Efficiency in the Redwater Field," *JPT* (April 1967) 449-456.

10. Willingham, R. W. and Howald, C. D.: "Case History of the Tensleep Reservoir, Torchlight Field, Wyoming," *JPT* (October 1965) 1159-1163.

11. Fetkovich, M. J.: "A Simplified Approach to Water Influx Calculations — Finite Aquifer Systems," *JPT* (July 1971) 814 - 828.

12. Dake, L. P.: *Fundamentals of Reservoir Engineering*, Elsevier Scientific Publ. Co., Amsterdam (1978).

13. Edwardson, J. J., *et al*.: "Calculation of Formation Temperature Disturbances Caused by Mud Circulation," *Trans.*, AIME (1962) **226**, 416-426.

14. Klins, M. A., Bouchard, A. J. and Cable, C. L.: "A Polynomial Approach to Determining the van Everdingen-Hurst Dimensionless Variables," *SPE Reservoir Engineer* (February 1988).

15. Tracy, G. W.: "Water Drive," *Applied Reservoir Engineering* course manual, Oil & Gas Consultants International, Inc., Tulsa (1986).

13

WATER CONING AND FINGERING

INTRODUCTION

Premature water production is often a result of water coning and/or fingering. "Coning" results near a producing well when water flows from the free water level in a generally vertical direction. Hence, breakthrough of the coned water occurs at the lowest part of the well's completion interval. Production from a well causes a pressure sink at that location in the reservoir. If the wellbore pressure is low enough, the well is directly above the oil/water contact and there are no permeability barriers to vertical flow of the water into the wellbore, then water coning occurs.

On the other hand, if the early water production occurs in a reservoir with a nonzero dip angle where the water/oil contact becomes unstable and the water underruns the oil using the "horizontal" permeability, this is called water "fingering" or "tonguing."

This chapter discusses these two processes that lead to premature water production. These phenomena are important because coning and fingering of water decrease project profitability in several different ways. First, oil productivity is reduced because of relative permeability effects. Second, lifting costs rise resulting from the heavier wellbore fluid, and water disposal/reinjection costs can be substantial. Third, recovery efficiency is reduced because the economic limit water cut is reached with producible oil remaining in the drainage area of the well.

WATER CONING

The presence of water coning into a well results from the pressure drawdown being too great near the well. If the pressure drawdown, Δp, is larger than the gravity pressure differential, which tends to keep the oil on top of the water, then coning can occur. So, it is possible to have coning when:

$$\Delta p > 0.433 \left(\gamma_w - \gamma_o \right) h_c \qquad (13\text{-}1)$$

where:

$\Delta p = \bar{p} - p_{well}$ = pressure drawdown at the well, psi,

γ_w = formation water specific gravity,

γ_o = reservoir oil specific gravity,

h_c = vertical distance from the bottom of the well's completion interval to the oil/water contact, ft.

There is no reference to time in Equation 13-1. The inference is that whenever this inequality is satisfied, coning will occur instantaneously; any time that the inequality is not satisfied, there will be no coning. Because p_{well} may be controlled by regulating the rate of fluid withdrawal from the well, water coning is a rate-sensitive process.

Historically, the initial research and engineering efforts in the area of water coning were concentrated on its elimination. Howard and Fast[10] (Amoco) had the idea of injecting a "pancake" of cement just below the completion interval to act as a barrier to the vertical movement of water. In the field, preliminary results were disappointing as the formation broke down. Although a coning panacea had not been discovered, hydraulic fracturing was introduced.

Many other approaches were tested, all of which failed. So, for the elimination of coning, oil industry analysts had to return to the implications of equation 13-1, i.e., reduce the pressure drawdown by reducing producing rate. To calculate "critical rate," which is the maximum rate that precludes coning, the methods of Meyer and Garder[1] and Chaney *et al.*[2] will be discussed. Unfortunately, to completely eliminate coning by reducing the drawdown, the permissible well rate may be quite low. It is normally more economical to handle water production rather than restrict the oil production.

The early water coning studies and methods did not consider the time to breakthrough. Faced with the prospect of zero profit (because the calculated critical rate was so low), many operators chose to produce their wells at higher rates and solve water problems later. It was observed that water production often does not begin until considerable production time has elapsed, sometimes months or even years. Sobocinski and Cornelius[3] used a physical sand model to correlate the time needed for water cone development. Later, Bournazel and Jeanson[5] used additional laboratory studies and actual field data to modify the Sobocinski and Cornelius breakthrough-time method.

More recently, Kuo and DesBrisay[4] published a review of coning methods. They also presented a method for after-breakthrough prediction of water production due to coning as a function of time in a bottomwater drive reservoir.

MEYER AND GARDER[1]

Meyer and Garder directed their studies toward the calculation of the critical oil rate. They assumed that:

1. The flow of oil and/or gas to the wellbore is strictly radial.

2. The flow of water from the water/oil contact to the bottom of the wellbore is strictly vertical.

3. The pressure drawdown controlling the flow of oil and/or gas is restricted to the gravitational pressure difference. The gravitational pressure difference is equal to the gravitational gradient (due to the density difference between the flowing hydrocarbon phase and formation water) times the vertical distance between the lowest point of the completion interval and the water/oil contact.

With these assumptions, the following equation for critical rate was developed:

$$q_c = \frac{0.001535 \ (\rho_w - \rho_o) \ k \ (h^2 - D^2)}{\mu_o B_o \ \ln(r_e / r_w)} \tag{13-2}$$

where:

q_c = critical oil rate or the maximum oil rate that precludes coning water, STB/D,

ρ_w = formation water density, gm/cc,

ρ_o = formation oil density, gm/cc,

k = formation permeability, md,

h = oil zone thickness, ft.,

D = completion interval thickness, ft.,

μ_o = reservoir oil viscosity, cp,

B_o = oil formation volume factor, RES BBL/STB, .

r_e = external drainage radius, ft., and

r_w = wellbore radius, ft.

An example is provided to illustrate the magnitude of the Meyer and Garder critical oil rate.

Problem 13-1:

Basic data:

k = 100 md

h = 50 ft

D = 10 ft

ρ_w = 1.05 gm/cc

ρ_o = 0.8 gm/cc

μ_o = 1.0 cp

B_o = 1.2 Res Bbl/STB

r_e = 745 ft

r_w = 0.25 ft

$$q_c = \frac{(0.001535)(1.05 - 0.8)(100)(50^2 - 10^2)}{(1.0)(1.2) \ \ln(745/0.25)} = 9.6 \ STB/D$$

Although there was an attempt by many to use the critical rate calculated with equation 13-2 as a guide, it became evident that these rates were generally too low to be economically feasible. There was sentiment in the industry that the assumptions used by Meyer and Garder were too restrictive. Chaney *et al.*[2] have provided additional research in this area.

CHANEY, NOBLE, HENSON, AND RICE[2]

Using the Muskat[6] analysis of the potential distribution around a partially penetrating well, Chaney *et al.*[2] pursued the coning critical rate problem both analytically and experimentally. The experimental work was performed with an electrical-resistance-capacitance-network simulator constructed to describe flow into a well. Once the potential distribution was established, the electrical "flow rate" was measured for various conditions of pay thickness and well penetration. As flow in porous media is analogous to flow of electricity, Chaney *et al.* were able to relate the simulator results to the critical flow rate of an oil well.

These studies included not only water coning up into the well, but also the coning of gas from a gas/oil contact (above the well completion) down into the upper perforations. Thus, the Chaney *et al.* method provides a method to calculate the critical flow rate (maximum rate) that precludes coning of either water or gas.

The Chaney *et al.* results are presented in Figures 13-1 through 13-5. These figures are each for a different zone thickness, ranging from 12.5 ft (See Figure 13-1) to 100 ft (See Figure 13-5). Each figure presents critical rate in reservoir barrels per day versus distance from the top perforation to the top of the formation or to the gas/oil contact. There are two sets of curves in each figure: one set for gas coning and the other set for water coning. The water coning curves are spaced farther apart and decrease in value as the horizontal axis increases. The gas coning curves are closer together and have positive slopes.

The following values were used in the preparation of these curves:

$$k = 1000 \text{ md}$$
$$\mu_o = 1.0 \text{ cp}$$
$$\Delta\rho = 0.30 \text{ gm/cc}$$

Therefore, to convert to surface-oil rate and to conditions other than those used in the preparation of the charts, the flow rate read from the appropriate chart should be used with the following equation:

$$q_c = \frac{q_{curve} \, (k) \, (\rho_w - \rho_o) \, (0.00333)}{\mu_o \, B_o} \quad STB/D \tag{13-3}$$

where each of the terms in this equation (except q_{curve}) was defined under Equation 13-2.

Problem 13-2:

The Chaney *et al*[2]. procedure will be illustrated by considering the same basic data used in Problem 13-1. As it is 50 ft from the top of the oil zone down to the oil/water contact, Figure 13-3 will be entered to find q_{curve}. Since the perforated interval is 10 ft, Curve B is used. It is assumed that the completion interval is at the top of the oil zone (as far away from the underlying water as possible); therefore, the horizontal axis value of Figure 13-3 is zero. Hence, $q_{curve} = 280$ RB/D. Inserting this value and the appropriate data from Problem 13-1 into Equation 13-3 yields:

$$q_c = \frac{(280)(100)(0.25)(0.00333)}{(1)(1.2)} = 19.4 \ STB \ / \ D$$

For water coning situations, *if the completion interval is at the top of the oil zone,* an alternative to using the graphical representations for determining the values of q_{curve} is to use Equation 13-4 developed by the authors. This equation generally yields results that are within 5% of the values read from the curves. For h = 25 ft, the errors can be slightly larger, but always less than 10%.

$$q_{curve} = 0.1313 \ (h^2 - D^2) + 34.0 \left(\frac{57-h}{44}\right)^2 - 250 \left(\frac{D}{h} - 0.30\right)\left(\frac{h-10}{90}\right)^2 - 40.0$$

(13-4)

where:
$$10 \ feet \leq h \leq 100 \ feet; \text{ and } 0.1 \leq D/h \leq 0.50$$

Applying Equation 13-4 to Example Problem 13-2:

$$q_{curve} = (0.1313)(50^2 - 10^2) + (34.0)\left(\frac{57-50}{44}\right)^2$$

$$- (250)\left(\frac{10}{50} - 0.30\right)\left(\frac{50-10}{90}\right)^2 - 40.0 = 280.9 \ B/D$$

Therefore,

$$q_c = (19.4)(280.9/280) = 19.5 \ STB/D$$

Notice that the Chaney *et al.*[2] method yields critical rate values about twice that of Meyer and Garder. Even the critical rate of Chaney *et al.*[2] is still usually too low for serious economic consideration. There are few operators who restrict oil rates to levels low enough to preclude coning. More often, the surface facilities are designed to handle the water production. When designing the lift equipment (pumps or gas lift), sufficient capacity is included to handle the volumes of water to be encountered.

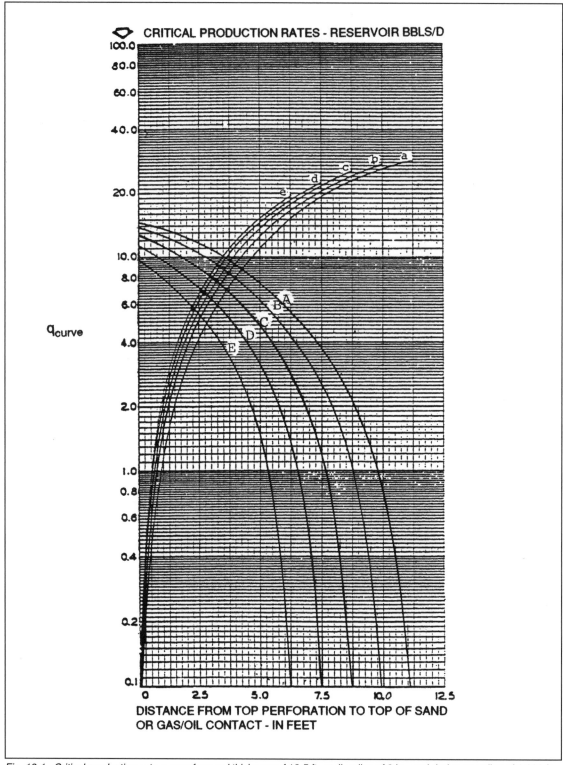

Fig. 13-1. *Critical-production-rate curves for sand thickness of 12.5 ft., well radius of 3 in., and drainage radius of 1,000 ft. Water coning curves: A, 1.25 ft. perforated interval; B, 2.5 ft.; C, 3.75 ft.; D, 5.00 ft.; and E, 6.25 ft. Gas coning curves: a, 1.25 ft perforated interval; b, 2.5 ft.; c, 3.75 ft.; d, 5.00 ft., and e, 6.25 ft. (From Chaney et al.[2]). Permission to publish by Oil & Gas Journal.*

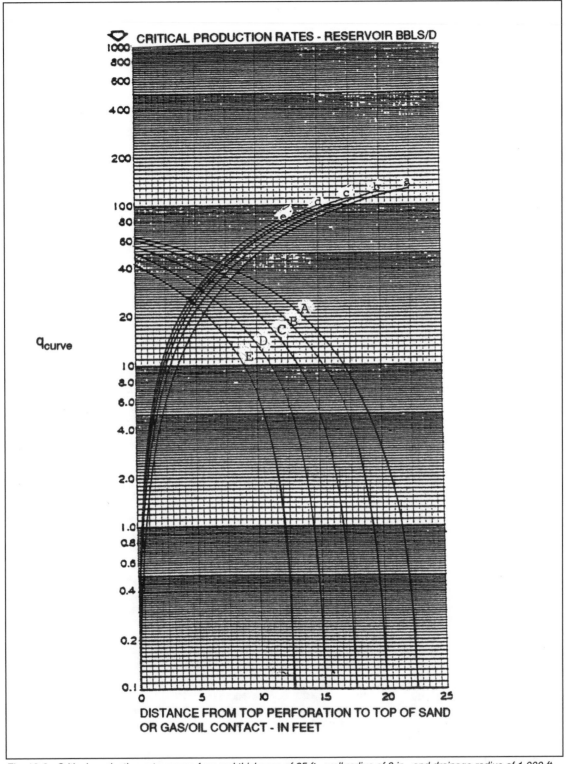

Fig. 13-2. *Critical-production-rate curves for sand thickness of 25 ft., well radius of 3 in., and drainage radius of 1,000 ft. Water coning curves: A, 2.5 ft. perforated interval, B, 5 ft.; C, 7.5 ft.; D, 10 ft.; and E, 12.5 ft. Gas coning curves: a, 2.5 ft perforated interval; b, 5 ft.; c, 7.5 ft.; d, 10 ft.; and e, 12.5 ft. (From Chaney et al.[2]). Permission to publish by Oil & Gas Journal.*

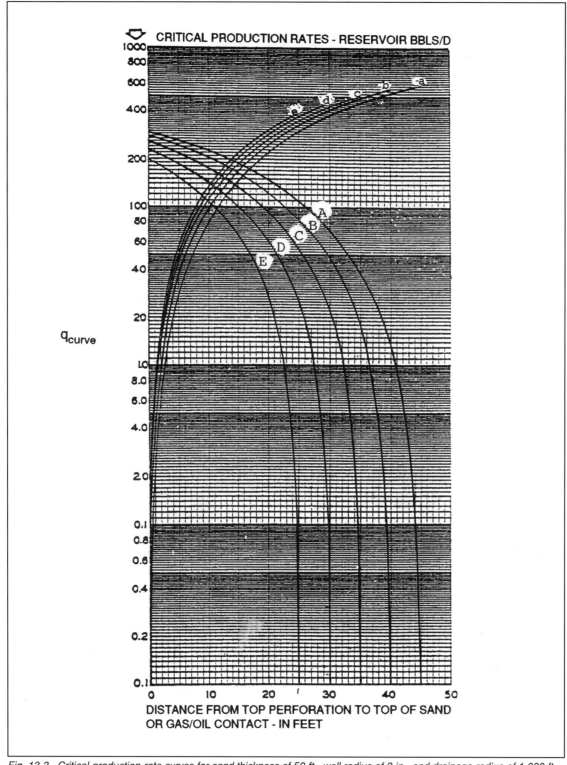

Fig. 13-3. *Critical-production-rate curves for sand thickness of 50 ft., well radius of 3 in., and drainage radius of 1,000 ft. Water coning curves: A, 5 ft. perforated interval; B, 10 ft.; C, 15 ft.; D, 20 ft.; and E, 25 ft. Gas coning curves: a, 5 ft. perforated intervals; b, 10 ft.; c, 15 ft.; d, 20 ft.; and e, 25 ft. (From Chaney et al.[2]). Permission to publish by Oil & Gas Journal.*

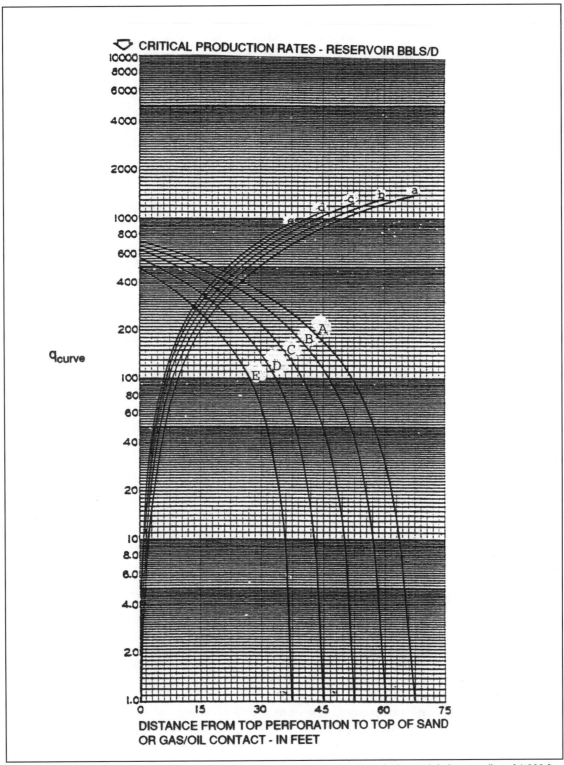

Fig. 13-4. Critical-production-rate curves for sand thickness of 75 ft., well radius of 3 in., and drainage radius of 1,000 ft. Water coning curves: A, 7.5 ft. perforated interval; B, 15 ft.; C, 22.5 ft.; D, 30 ft.; and E, 37.5 ft. Gas coning curves: a, 7.5 ft. perforated interval; b, 15 ft.; c, 22.5 ft.; d, 30 ft.; and e, 37.5 ft. (From Chaney et al.[2]). Permission to publish by Oil & Gas Journal.

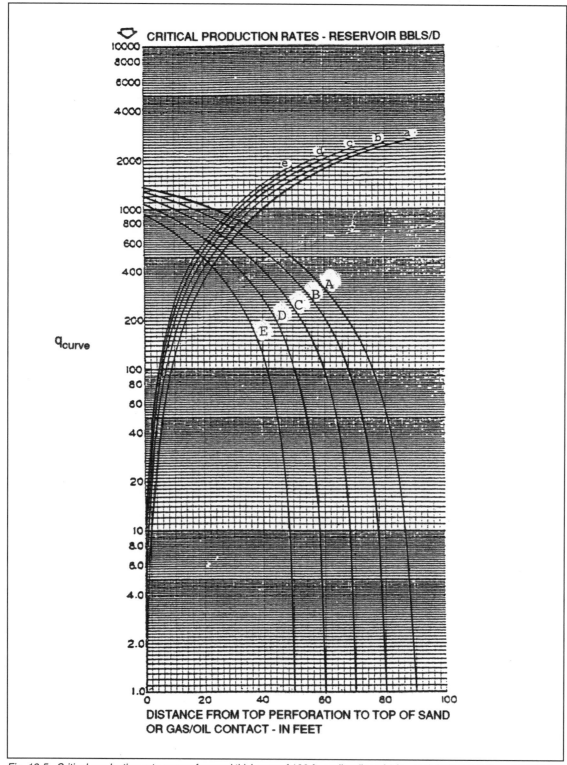

q_{curve}

Fig. 13-5. *Critical-production-rate curves for sand thickness of 100 ft., well radius of 3 in., and drainage radius of 1,000 ft. water coning curves: A, 10 ft. perforated interval; B, 20 ft.; C, 30 ft.; D, 40 ft.; and E, 50 ft. Gas coning curves: a, 10 ft. perforated interval; b, 20 ft.; c, 30 ft.; d, 40 ft.; and e, 50 ft. (From Chaney et al.[2]). Permission to publish by Oil & Gas Journal.*

Problem 13-3:

Water coning into a gas well can be considered with the Chaney *et al.* method.

Basic Data:

Permeability	100.	md
Perforated interval (at the top)	5.0	ft
Pay thickness	30	ft
Gas viscosity	0.018	cp
ρ_w - ρ_g	0.91	gm/cc
Gas formation volume factor	0.0011	B/scf

Calculation of q_{curve} (with Equation 13-4):

$$q_{curve} = (\,0.1313\,)\,(\,30^2 - 5^2\,) + (\,34\,)\left(\frac{57 - 30}{44}\right)^2$$

$$-\,(\,250\,)\left(\frac{5}{30} - 0.30\right)\left(\frac{30 - 10}{90}\right)^2 - 40 = 89.3 \; RB\,/\,D$$

$$q_c = \frac{(\,0.00333\,)\,(\,100\,)\,(\,0.91\,)\,(\,89.3\,)}{(\,0.0011\,)\,(\,0.018\,)} = 1{,}366{,}696 \; scf\,/\,D$$

$$= 1.36 \; MMscf\,/\,D$$

This problem suggests that this well can produce at a rate of about 1.36 MMscf/D without water production. As long as the gas/water contact does not move, the critical rate will remain reasonably constant. Unfortunately, under conditions of a moving gas/water contact, the effective pay thickness decreases. Therefore, the maximum rate without coning also decreases.

Problem 13-4:

The procedure for calculating the critical oil rate to preclude coning gas will be illustrated in this example.

Basic Data:

Permeability	50	md
Pay thickness (oil zone)	50	ft
Perforated interval	20	ft
Distance below gas/oil contact to top perforation	20	ft
Oil viscosity	0.8	cp
Oil formation volume factor	1.25	B/STB
Oil-Gas density difference	0.7	gm/cc

For a 50 ft oil zone, Figure 13-3 should be used. As the perforated interval is 20 ft, curve d (of the gas curves) is read. Entering this plot at a horizontal axis value of 20 (unperforated length at the top of the zone), q_{curve} = 285 RB/D.

$$q_c = \frac{(0.00333)(50)(0.7)(285)}{(0.8)(1.25)} = 33.2 \; STB/D$$

If the oil rate is less than about 33 STB/D and the gas/oil contact does not move downward, gas coning should not occur.

SOBOCINSKI AND CORNELIUS[3]

Although producing a well at or below the critical rate precludes coning, many feel that it also rules out profit. Sobocinski and Cornelius studied the water coning problem to determine the time to breakthrough (of water) when producing a well at a rate greater than the critical value. Their experimental laboratory studies were conducted with a pie-shaped, sand-packed plexiglass model illustrated in Figure 13-6. The model simulated flow near a well with the knife edge of the "pie" being the wellbore. Along the "wellbore" were located a series of valves that allowed the simulation of different perforation lengths.

Starting from an initial stable position of the water/oil contact within the model, water and oil were injected at the outside perimeter of the model to replace production exiting the system through the perforations of the wellbore. As the water and oil were colored, the position of the water/oil contact with time was easily monitored. The Sobocinski and Cornelius[3] results are summarized in Figure 13-7 using dimensionless variables defined as:

Dimensionless cone height (Z):

$$Z = \frac{(0.00307)(\rho_w - \rho_o) \; k_h \; h \; h_c}{\mu_o \; B_o \; q_o} \qquad (13\text{-}5)$$

Dimensionless time:

$$t_d = \frac{(0.00137)(\rho_w - \rho_o) k_h (1 + M^\alpha) t}{\mu_o \phi h F_k}$$

(13-6)

where:

ρ_w = formation water density, gm/cc,

ρ_o = reservoir oil density, gm/cc

k_h = horizontal formation permeability, md,

F_k = ratio of horizontal to vertical permeability,

h = oil zone thickness, ft,

h_c = cone height at breakthrough (distance from the initial oil/water contact to bottom perforation), ft,

μ_o = reservoir oil viscosity, cp,

B_o = oil formation volume factor, RB/STB,

q_o = oil rate, STB/D,

t = time, days,

ϕ = porosity, fraction,

α = exponent, such that

For $M < 1$, $\alpha = 0.5$

For $M \geq 1$, $\alpha = 0.6$

where:

M = mobility ratio

Note that the equation relating actual time and dimensionless time was derived through analytical considerations. However, when applying this relationship to their experimental results, Sobocinski and Cornelius found that the exponent, α, was needed.

Recall that M = mobility ratio = $(k_w / \mu_w) / (k_o / \mu_o)$

where:

k_w = effective permeability to water at the average water saturation in the water- invaded zone, md,

μ_w = water viscosity, cp,

k_o = effective oil permeability in the uninvaded portion of the oil zone, md,

μ_o = reservoir oil viscosity, cp.

To find the time to breakthrough, the "breakthrough" curve (top one) of Figure 13-7 is used. First, a dimensionless cone height is calculated with Equation 13-5. Then using the breakthrough curve of Figure 13-7 and the value of Z, the dimensionless time to breakthrough is determined. Equation 13-6 may be rearranged to yield actual time to breakthrough. The procedure will be illustrated in the next Example Problem.

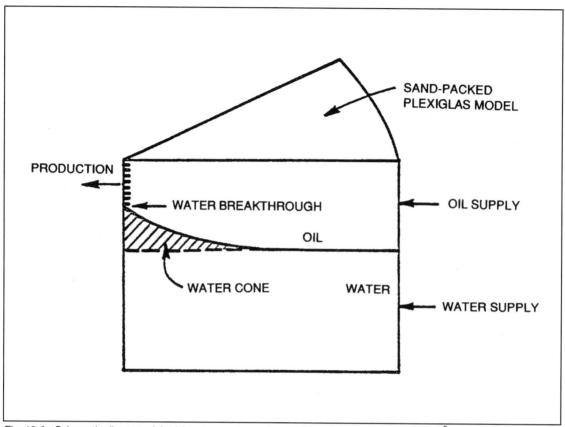

Fig. 13-6. Schematic diagram of the laboratory water coning model of Sobocinski and Cornelius[3]. Permission to publish by JPT.

In Figure 13-7 there are three different types of curves: breakthrough, departure, and basic buildup. To determine time to breakthrough, only the breakthrough curve is needed. The other two curves are not normally used; but they do allow a water cone to be studied as it builds. After the dimensionless cone height has been calculated, this allows a point on the breakthrough curve to be determined. From this point, trace a departure curve (parallel to those shown) back to the basic buildup curve. Then, to consider the first part of the water cone buildup history, the basic buildup curve is used from zero dimensionless time until reaching the traced departure curve. From this point, the departure curve is used until reaching the breakthrough curve. Sobocinski and Cornelius found that water cones normally build quite slowly until reaching a point close to the wellbore when the speed of the cone buildup accelerates. Then the cone continues to build faster and faster until breakthrough occurs.

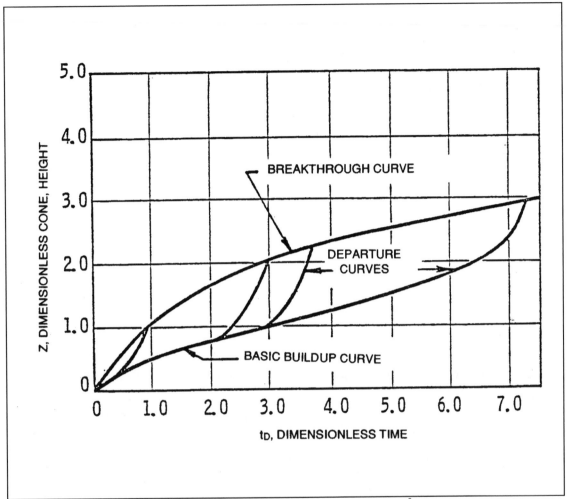

Fig. 13-7. Dimensionless cone height vs. dimensionless time (Sobocinski and Cornelius[3]). Permission to publish by JPT.

Problem 13-5:

Determine the producing time until water breakthrough occurs due to coning for the following well.

Basic Data:

Horizontal permeability	50	md
Vertical permeability	5	md
Oil producing rate	50	STB/D
Pay thickness	40	ft
Perforated interval (at top of pay)	10	ft
Reservoir oil viscosity	1.0	cp
Oil formation volume factor	1.2	B/STB
Water-oil density difference	0.2	gm/cc
Porosity	20	%
Mobility ratio	1.5	

Because $M > 1.0$ $\alpha = 0.6$

Permeability ratio: $F_k = k_h / k_v = 50 / 5 = 10$

Cone height: $h_c = 40 - 10 = 30$ ft

Dimensionless cone height, Z (Equation 13-5):

$$Z = \frac{(0.00307)(0.2)(50)(40)(30)}{(1.0)(1.2)(50)} = 0.61$$

From Figure 13-7, $t_d = 0.5$

Rearranging Equation 13-6:

$$t = \frac{\mu_o \, \phi \, h \, F_k \, t_d}{(0.00137)(\rho_w - \rho_o)(k_h)(1 + M^\alpha)}$$

$$= \frac{(1.0)(0.2)(40)(10)(0.5)}{(0.00137)(0.2)(50)[1 + (1.5)^{0.6}]} = 1283 \; days = 3.5 yr$$

The breakthrough curve (see figure 13-7) of Sobocinski and Cornelius has a hyperbolic shape. Kuo and DesBrisay[4] developed an equation that closely fits the breakthrough curve quite well:

$$(t_d)_{bt} = \frac{(Z)(16 + 7Z - 3Z^2)}{(4)(7 - 2Z)} \tag{13-7}$$

An inspection of the denominator of Equation 13-7 will show that at a value of $Z = 3.5$, the value of $(t_d)_{bt}$ becomes infinite. According to Tracy[9], this result can be used to calculate the critical rate with Equation 13-5. The value of Z that corresponds to an infinite cone breakthrough time also corresponds to the critical rate. Hence, rearranging Equation 13-5:

$$q_o = \frac{(0.00307)(\rho_w - \rho_o) \, k_h \, h \, h_c}{Z \, \mu_o \, B_o}$$

and substituting in $Z = 3.5$:

$$q_c = \frac{(0.000877)(\rho_w - \rho_o) \, k_h \, h \, h_c}{\mu_o \, B_o} \tag{13-8}$$

Problem 13-6:

As an illustration to the use of Equation 13-8, reconsider Example Problem 13-1:

$$q_c = \frac{(0.000877)(0.25)(100)(50)(40)}{(1.0)(1.2)} = 36.5 \; STB / D$$

This calculation suggests that the well with conditions given in Problem 13-1 could have a critical rate as high as 36.5 STB/D. Recall that Example 13-1 was a Meyer and Garder[1] critical rate calculation (9.6 STB/D). Example 13-2 used the same data with the method of Chaney *et al.*[2] (19.4 STB/D). Tracy[9] suggests that the most likely value for critical rate lies somewhere between the critical rate of Chaney *et al.*[2] and that calculated using Equation 13-8. So, for the data of Example 13-1, the critical rate is likely to be between 19.4 and 36.5 STB/D.

BOURNAZEL AND JEANSON[5]

After the publication of Sobocinski and Cornelius[3] (S & C), Bournazel and Jeanson[5] (B & J) critically evaluated the correlation of S & C. B & J found the basis for the method to be sound and adopted the same dimensionless variables: dimensionless cone height (Z) and dimensionless time (t_d). However, B & J found that the actual breakthrough time measured in their laboratory and field experiments was less than that predicted with the S & C correlation. They made two changes in the S & C method: (1) an equation was developed to replace the dimensionless cone height versus dimensionless time breakthrough curve; and (2) in the equation relating real time and dimensionless time, the experimentally determined exponent, α, was modified to: $\alpha = 0.7$ for all M in the range $0.14 \leq M \leq 7.3$.

The Bournazel and Jeanson[5] equations are:

Dimensionless Cone Height (from S & C):

$$Z = \frac{(0.00307)(\rho_w - \rho_o) k_h h h_c}{\mu_o B_o q_o} \tag{13-5}$$

Dimensionless Time to Breakthrough (developed by B & J):

$$(t_d)_{bt} = \frac{Z}{3.0 - 0.7 Z} \tag{13-9}$$

Calculation of Actual Breakthrough Time (developed by S & C; edited by B & J):

$$t = \frac{\mu_o \phi h F_k (t_d)_{bt}}{(0.00137)(\rho_w - \rho_o)(k_h)(1 + M^\alpha)}$$

where: $\alpha = 0.7$

By comparing Equation 13-9 and 13-7, the breakthrough time predicted by B & J is less than that predicted by S & C. For values of Z between 0.1 and 3.0, B & J water breakthrough times are only 40 to 50% of those by S & C.

Problem 13-7:

The data given in Problem 13-5 will be used to solve for time to breakthrough using the method of B & J.

Dimensionless Cone Height:

$$Z = \frac{(0.00307)(0.2)(50)(40)(30)}{(1.0)(1.2)(50)} = 0.61$$

Dimensionless TIme to Breakthrough:

$$(t_d)_{bt} = \frac{Z}{3 - (0.7)(Z)} = \frac{0.61}{3 - (0.7)(0.61)} = 0.237$$

Conversion to Real Time:

$$t = \frac{(1.0)(0.2)(40)(10)(0.237)}{(0.00137)(0.2)(50)[1 + (1.5)^{0.7}]} = 594 \; days = 1.6 \; yr$$

By inspection of Equation 13-9, it can be seen that at $Z = 4.28$, the denominator becomes zero, and the value of $(t_d)_{bt}$ increases without limit. Thus, if the time to breakthrough is infinite, $Z = 4.28$ can be used to calculate the critical rate.

Rearranging Equation 13-5:

$$q_o = \frac{(0.00307)(k_h)(\rho_w - \rho_o)(h)(h_c)}{(Z)(\mu_o)(B_o)}$$

And substituting in $Z = 4.28$:

$$q_c = \frac{0.000717 \; k_h \; (\rho_w - \rho_o) \; h \; h_c}{\mu_o \; B_o} \tag{13-10}$$

Problem 13-8:

The data of Problem 13-1 will be treated with Equation 13-10 to calculate the B & J critical rate:

$$q_c = \frac{(0.000717)(100)(0.25)(50)(40)}{(1.0)(1.2)} = 29.9 \; STB/D$$

The critical rates calculated by the Meyer and Garder, Chaney *et al.*, and the S & C methods for this problem were summarized in Example 13-6. Tracy[9] suggests that the correlation of Bournazel and Jeanson (time to breakthrough and critical rate) matches a higher percentage of field data than other methods.

Although insufficiently verified, a new method by Hoyland *et al.*[11] is particularly interesting, especially for highly permeable reservoirs like those found in the North Sea. Other notable critical rate procedures are described in Muskat and Wyckoff,[12] Chierci, *et al.*,[13] and Wheatley.[14]

KUO AND DESBRISAY[4]

Recently, Kuo and Desbrisay[4] reviewed much of the previously published literature on water coning. This work added additional conclusions and equations to those of the original authors. Using a numerical coning model, they developed a correlation for the prediction of water-cut performance in a bottomwater drive reservoir.

Kuo and DesBrisay published an equation that accurately reproduces the graphical water coning breakthrough time results of Sobocinski and Cornelius[3]. This equation was given earlier as Equation 13-7. For critical rate, Kuo and DesBrisay[4] used the calculation of Schols.[7]

The Schols equation is:

$$q_c = (Factor \; A)(Factor \; B)(Factor \; C) \; STB/D \qquad (13-11)$$

where:

$$Factor \; A = \frac{(\rho_w - \rho_o)(k)(h^2 - D^2)}{(2049)(\mu_o)(B_o)}$$

$$Factor \; B = 0.432 + (3.1416)/[\ln(r_e/r_w)]$$

$$Factor \; C = (h/r_e)^{0.14}$$

The units and use of Equation 13-11 will be illustrated with an example problem.

Problem 13-9:

Using the data of Problem 13-1:

$$Factor\ A = \frac{(0.25)(100)(50^2 - 10^2)}{(2049)(1.0)(1.2)} = 24.402$$

$$Factor\ B = 0.432 + \frac{3.1416}{\ln(745/0.25)} = 0.825$$

$$Factor\ C = (50/745)^{0.14} = 0.685$$

$$q_c = (24.402)(0.825)(0.685) = 13.79\ STB/D$$

AFTER BREAKTHROUGH PERFORMANCE — BOTTOMWATER DRIVE RESERVOIRS

Using a numerical coning model, Kuo and DesBrisay developed a simple method to predict water-cut performance for a well in a bottomwater drive reservoir. To calculate the time to water breakthrough, they used the method of Bournazel and Jeanson[5], as described earlier in this chapter. Based on the coning model results, Kuo and DesBrisay allowed water production to begin at one-half of the B & J breakthrough time.

Two dimensionless variables were defined:

$$t_d = t\ /\ t_{bt} \tag{13-12}$$

$$(WC)_d = WC\ /\ (WC)_{limit} \tag{13-13}$$

where:

t_d = dimensionless time,
t = actual time, days,
t_{bt} = Bournazel & Jeanson breakthrough time, days
$(WC)_d$ = dimensionless water cut,
WC = actual water cut, fraction,

and

$$(WC)_{limit} = \frac{Mh_w}{Mh_w + h_o} \tag{13-14}$$

where:

M = mobility ratio,
h_w = current water zone thickness, feet,
h_o = current oil zone thickness, feet.

To evaluate $(WC)_{limit}$, additional assumptions were made. Vertical flow of the water at constant pressure was assumed such that any well production causes equivalent water movement. Also, the method assumes constant cross-sectional area. With these assumptions, the following material balance equations can be written:

$$h_o = H_o \left[1 - \left(\frac{N_p}{N} \right) \frac{(1 - S_{wc})}{(1 - S_{wc} - S_{or})} \right] \qquad (13\text{-}15)$$

$$h_w = H_w + H_o \left[\left(\frac{N_p}{N} \right) \frac{(1 - S_{wc})}{(1 - S_{wc} - S_{or})} \right] \qquad (13\text{-}16)$$

where:

H_o = original oil zone thickness (original o/w contact to the top of the oil zone), feet
H_w = original water zone thickness, feet,
S_{wc} = connate water saturation, fraction,
S_{or} = residual oil saturation, fraction,
N_p = cumulative oil production, STB,
N = original oil in place, STB

Of course, Equation 13-13 relates actual water cut to dimensionless water cut. Dimensionless water cut and time are related as:

$$(WC)_d = 0.0 \qquad\qquad \text{for } t_d < 0.5$$

$$(WC)_d = 0.94 \ \log \ t_d + 0.29 \qquad \text{for } 0.5 \le t_d \le 5.7$$

$$(WC)_d = 1.0 \qquad\qquad \text{for } t_d > 5.7 \qquad (13\text{-}17)$$

Kuo and DesBrisay[4] used a trial and error procedure to relate oil production and water cut to time. Their original paper contains a program listing for the HP-41C hand-held computer.

OIL/WATER CONTACT STABILITY

Slider[8] has presented a method for analyzing the tilt of an oil/water contact as a result of production from upstructure wells. Figure 13-8 illustrates the problem. If the upstructure production rate is too high, then the water/oil contact may become unstable with the water underrunning the oil and making its way upstructure along the base of the formation. Premature water breakthrough will then occur with resulting oil productivity loss and reduction in recovery efficiency. For single stratum reservoirs, this can only happen if the mobility ratio is greater than 1.0. Of course, this indicates that water moves with greater ease than does oil. Because water is usually more dense than oil, gravity forces tend to keep water below oil. However, if the upstructure withdrawal rate is high enough with an unfavorable mobility ratio ($M > 1$), then the water/oil contact can become unstable.

If the water/oil contact is relatively plane, there are some observations that can be made about its stability as it moves upstructure. In Figure 13-8, the formation dip angle is α. β is the angle shown between the formation top and the water/oil contact. A "stable" water/oil contact is one that assumes a fixed or constant angle with the horizontal as it moves upstructure. Therefore, β is also fixed. Further, β must be greater than zero, or the water is underrunning the oil. It is interesting that a stable water/oil contact does not have to be perfectly horizontal; it can tilt somewhat.

An unstable water/oil contact is one that moves upstructure with an increasing angle with the horizontal. Of course, this indicates that β is decreasing with time.

Through arguments similar to those presented by Slider[8], it can be shown that an equation relating angle β to angle α, the gravity forces, and the mobility ratio is:

$$Tan\ \beta = \frac{[G - (M - 1)]\ (Tan\ \alpha)}{G} \tag{13-18}$$

where:

$$M = mobility\ ratio = (k_w/\mu_w)/(k_o/\mu_o)$$
$$= (k_{rw}\mu_o)/(k_{ro}\mu_w)$$

and

$$G = \frac{(0.488)(\gamma_w - \gamma_o)\ k\ A\ k_{rw}\ \sin\ \alpha}{q_t\ \mu_w}$$

where:

γ_w = the specific gravity of the formation water,
γ_o = the specific gravity of the reservoir oil,
k = formation permeability, darcies,
A = formation cross-sectional area $(ft)^2$,
k_{rw} = relative permeability to water in the invaded zone behind the water/oil contact, fraction,
k_{ro} = relative permeability to oil ahead of the water/oil contact, fraction,
α = the formation dip angle with respect to the horizontal,
q_t = Updip withdrawal rate, reservoir barrels/day,
μ_w = formation water viscosity, cp, and
μ_o = reservoir oil viscosity, cp.

Observations:

(1) If $M = 1.0$, then the oil/water contact remains perfectly horizontal, no matter what the withdrawal rate is upstructure.

(2) If $M < 1.0$, then there is no chance for instability. In fact, angle β tends to become larger than α.

(3) If $G > (M - 1)$, the water/oil contact is stable.

(4) If $G < (M - 1)$, instability of the water/oil contact occurs.

Instability cannot occur unless M > 1. Assuming that M > 1, then the point of instability just occurs when G = (M - 1).

Using this condition and substituting in the definitions for G and M, the "critical" rate can be determined as:

$$(q_t)_{crit} = \frac{0.488 \ (\gamma_w - \gamma_o) \ k \ A \ \sin \alpha}{(\mu_o / k_{ro}) - (\mu_w / k_{rw})} \quad RB / D \tag{13-19}$$

where the terms and units were defined under Equation 13-18. To ensure a stable water/oil contact, an updip withdrawal rate less than that calculated with Equation 13-19 would have to be maintained.

Problem 13-10:

Determine the critical withdrawal rate (updip) such that the water/oil contact just becomes unstable for the following reservoir conditions.

Basic Data:

Formation permeability	0.5 darcies
Cross-sectional area perpendicular to flow	50,000 sq. ft.
Water-oil specific gravity difference	0.2
Formation dip angle	10 degrees
Oil viscosity	2.0 cp
Water viscosity	0.4 cp
Oil relative permeability (uninvaded zone)	1.0
Water relative permeability (invaded zone)	0.25
Oil formation volume factor	1.35 B/STB

Using Equation 13-19:

$$(q_t)_{crit} = \frac{(0.488)(0.2)(0.5)(50{,}000)(0.174)}{(2.0/1.0) - (0.4/0.25)} = 1061.4 \ B/D$$

Assuming that the updip wells are making clean oil, then the surface rate can be calculated as:

$$(q_t)_{surface} = (q_t)_{crit} / B_o = 1061.4 / 1.35 = 786.2 \ STB / D$$

Hence, as long as the total updip clean oil rate is 786 STB/D or less, the water/oil contact should remain stable.

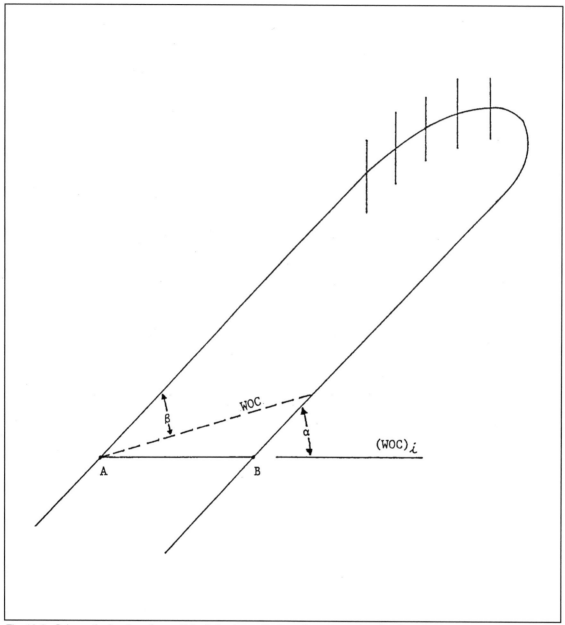

Fig. 13-8. Schematic drawing of an anticlinal structure with downdip water and updip oil production.

REFERENCES

1. Meyer, H. I. and Garder, A. O.: "Mechanics of Two Immiscible Fluids in Porous Media," *Journal of Applied Physics* (November 1954) **25**, No. 11, 1400.

2. Chaney, P. E., Noble, M. D., Henson, W. L and Rice, T. D.: "How to Perforate Your Well To Prevent Oil and Gas Coning," *Oil & Gas Journal* (May 7, 1956) **54**, 108.

3. Sobocinski, D. P. and Cornelius, A. J.: "A Correlation for Predicting Water Coning Time," *JPT* (May 1965) **234**, 594-600.

4. Kuo, M. C. T. and DesBrisay, C. L.: "A Simplified Method for Water Coning Predictions," paper 12067 presented at the 1983 SPE Annual Technical Conference and Exhibition, San Francisco (October).

5. Bournazel, C. and Jeanson, B.: "A Fast Method for Predicting Water Coning Times" paper 3628 presented at the 1971 SPE Annual Technical Conference and Exhibition, New Orleans.

6. Muskat, Morris: "Flow of Homogeneous Fluids Through Porous Media," J. W. Edwards, Inc. (1946) 263-286.

7. Schols, R. S.: "An Empirical Formula for the Critical Oil Rate," *Erdoel Erdgas, Z.* (January 1972) **88**, No. 1, 6-11.

8. Slider, H. C.: *Practical Petroleum Reservoir Engineering Methods*, Petroleum Publishing Co., (1976) 309-312.

9. Tracy, G. W.: Private Communication, 1985.

10. Howard, G. C. and Fast, C. R.: "Squeeze Cementing Operations," *Trans.*, AIME (1950) **189**, 53-64.

11. Hoyland, L. A., Papatzacos, P., and Skjaeveland, S. M.: "Critical Rate for Water Coning: Correlation and Analytical Solution," *SPE Reservoir Engineering* (November 1989) 495-502.

12. Muskat, M. and Wyckoff, R. D.: "An Approximate Theory of Water Coning in Oil Production," *Trans.*, AIME (1935) **114**, 144-161.

13. Chierci, G. L., Ciucci, G. M. and Pizzi, G.: "A Systematic Study of Gas and Water Coning by Potentiometric Models," *JPT* (August 1964) 923-929.

14. Wheatley, M. J.: "An Approximate Theory of Oil/Water Coning," paper 14210 presented at the 1985 SPE Annual Technical Conference and Exhibition, Las Vegas (September).

14

GAS-CAP DRIVE

INTRODUCTION

There are three different oil reservoir drive processes associated with an expanding gas cap. They are:

    1. Gas-cap Drive

    2. Segregation Drive

    3. Gravity Drainage

1. Gas-cap Drive

This recovery process is normally associated with oil reservoirs that are discovered with an initial gas cap present. The formation vertical permeability is usually low, generally less than 50 millidarcies. As the pressure within the oil zone is reduced due to fluid withdrawals, the initial gas cap tends to expand. This gas-cap expansion results in a larger recovery from the oil reservoir than would have been expected by solution gas drive. However, this increase in recovery is normally only a few percent of the original oil in place (OOIP). Typical recovery values associated with this mechanism range from 20 to 50 percent of the OOIP. Low recovery results if much of the gas cap is produced (depleted). Conservation considerations usually require the shutting-in of high gas/oil ratio (GOR) wells. Gas injection into the gas cap often enhances recovery from this type of reservoir, especially if the gas cap is small compared to the oil zone.

2. Segregation Drive

As pressure declines within the oil zone due to production, liberated gas migrates to the top of the structure, while oil drains downstructure. With a strict gas-cap drive, free gas liberated within the oil zone moves laterally to the nearest producing wellbore with no significant vertical movement. Segregation-drive reservoirs do not always have initial gas caps. The upward moving gas either joins the initial gas cap or forms a secondary gas cap. The segregation process requires the existence of vertical permeability, usually at least 50 md.

Segregation drive does not become active immediately. First, sufficient gas relative permeability must be developed. Of course, at the start of production, there is no free gas within the oil zone.

Then, as a result of production, pressure declines within the oil zone, and a free gas saturation develops as gas is liberated from the oil. This free gas forms in discrete globules in a random fashion. Until these free gas pockets coalesce into continuous paths through the oil zone, there is no permeability to gas flow, either vertically or laterally to the wells. The minimum gas saturation required for the free gas phase to attain a nonzero permeability is called the "equilibrium" or "critical" gas saturation, and is about four to six percent.

Where the segregation-drive mechanism is significant, oil recovery can substantially exceed solution-gas-drive recovery. The recovery range is usually between 30 and 60 percent of the original oil in place. This mechanism can be effective in relatively flat (thick) reservoirs, particularly when the vertical permeability is nearly equal to the horizontal permeability.

Tracy[1] suggests that segregation drive can occur even in low permeability formations if the reservoir withdrawal rate is low. There are numerous examples of old reservoirs where oil productivity, although low, was sustained for years at a relatively constant level. Typically, these reservoirs begin with a solution-gas-drive recovery mechanism. The wells in a dissolved-gas-drive reservoir often decline with a constant percentage decline for much of the reservoir's life. However, when the reservoir is old and the productivity is at a low level (less than five percent of the peak productivity), it is not unusual for the decline curve to become flat. In this instance, oil is draining to the lower levels in the reservoir where the remaining producing wells are completed. The draining oil maintains the oil relative permeability at a high value. Although the free gas saturation is high, most of this gas has segregated to higher structural positions in the reservoir. The presence of this gas reduces the pressure decline; and the well productivities are often sustained at a low constant level for many years.

3. Gravity Drainage

This recovery process is quite similar to segregation drive except that "horizontal" formation permeability (parallel to formation dip) is used instead of the vertical permeability. Because lateral permeability is usually higher than vertical permeability, resistance from the rock to separation of the gas and oil is typically less than in a segregation drive.

Gravity drainage is normally associated with reservoirs having a significant amount of structure. The dip angle is typically at least 15 degrees. Under such conditions the oil can drain to the bottom of the structure with the gas migrating to the top of the structure. Hence, the oil saturation within the oil zone is maintained at a high level even though the pressure is declining.

The recovery of oil from gravity drainage reservoirs is quite high: typically between 40 and 80 percent of the original oil in place. It is often desirable to inject gas into the gas cap or, in the absence of a gas cap, at the top of the structure. This tends to maintain reservoir pressure which will help sustain oil productivity of the down-structure wells. Gas injection also will often enhance the recovery from these types of reservoirs.

FACTORS WHICH ENHANCE GAS-CAP DRIVE

The reservoir characteristics which cause gas-cap expansion to recover more oil are:

1. Low oil viscosity,

2. High API oil gravity,

3. High formation permeability,

4. Large structural relief, and

5. Large density difference between the oil and gas.

Notice that each of these factors has the effect of causing the oil to be able to move downdip with greater ease. Thus, they are instrumental in effecting a higher oil rate draining down-structure. Basically, anything that enhances oil drainage will cause the ultimate recovery to be increased.

PRODUCTION CHARACTERISTICS WHICH INDICATE SEGREGATION OR GRAVITY DRAINAGE

The following are some of the important production characteristics which indicate that significant segregation or gravity drainage is occurring.

1. Gas/oil ratio (GOR) variations with structure,

2. Better apparent gas/oil relative permeability performance, and

3. A tendency for pressure maintenance.

1. Gas/Oil Ratio Variations with Structure

In fields where segregation or gravity drainage is occurring and the reservoir pressure is declining, the gas/oil ratio of individual wells is likely to show variations with structure. The highest gas/oil ratios should coincide with the wells located near the top of the structure; while the lowest GOR wells should be near the base of the structure. In such reservoirs the only place that very high GOR's should occur is at the top structural positions.

2. Better Apparent Gas/Oil Relative Permeability Performance

Figure 14-1 shows the contrast between a theoretical solution gas drive gas/oil relative permeability relationship (k_g/k_o vs S_g) and one from a field where significant gravity segregation is occurring. The use of field data to generate such relationships was discussed in the "Rock Properties" chapter.

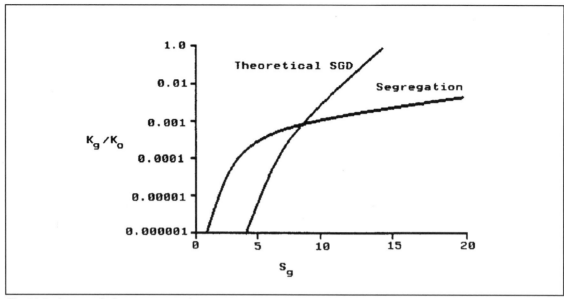

Fig. 14-1. Apparent k_g/k_o vs gas saturation.

Recalling the pertinent equations:

$$S_g = (1 - S_{wc}) \{ 1 - [(N - N_p) \, B_o \, / \, (N B_{oi})] \}$$ (14-1)

$$k_s / k_o = (GOR - R_s) (\mu_g / \mu_o) (B_g / B_o)$$ (14-2)

where:

S_g = gas saturation, fraction,
S_{wc} = connate water saturation (assumed irreducible), fraction,
N = original oil in place (at the bubblepoint pressure), STB,
N_p = cumulative oil production from the bubblepoint, STB,
B_o = oil-formation volume factor (at current pressure), RB/STB,
B_{oi} = oil-formation volume factor (at bubblepoint pressure), RB/STB,
k_g/k_o = relative permeability ratio at S_g, ratio,
R_s = solution gas/oil ratio, scf/STB,
GOR = producing gas/oil ratio, scf/STB,
μ_g = gas viscosity, cp,
μ_o = oil viscosity, cp, and
B_g = gas-formation volume factor (at current pressure), RB/scf

As discussed in the "Rock Properties" chapter, the theoretical k_g/k_o curve begins at the equilibrium gas saturation (because there is no gas permeability at all until this saturation is reached) and has a fairly steep slope. On the other hand, data taken from an actual reservoir will normally yield a curve that begins quite close to $S_g = 0$. The use of Equations 14-1 and 14-2 to generate a k_g/k_o relationship necessitates the assumption of a flat pressure surface across the reservoir; whereas, actually there is usually a significant pressure depression in the vicinity of the producing wells. This is due to the radial-flow pressure drawdown. Thus, free gas is first seen at the producing wells at calculated overall reservoir gas saturations less than the equilibrium value. If segregation or gravity drainage becomes significant, then the apparent k_g/k_o vs S_g relationship tends to flatten as shown in Figure 14-1.

Why does the curve flatten and why is the k_g/k_o vs S_g curve called an "apparent" relationship? If significant segregation of the free gas and oil is occurring, then the calculated gas saturation is not really what exists in the oil zone. Equation 14-1 assumes a constant-volume oil zone and no segregation. If segregation is occurring, then the gas is moving to the top of the structure. Hence, the actual oil-zone gas saturation is smaller than that calculated by Equation 14-1. In extreme cases, the downdip well producing GOR's tend to stabilize and even decrease with time.

3. Pressure Maintenance

It should be apparent that in a segregation-drive or gravity-drainage reservoir, for much of the producing life, the average gas/oil ratio is significantly lower than would exist if the gravity forces were not present. Lower GOR's leave more of the free gas in the reservoir. Thus, there is a tendency for pressure maintenance associated with reservoirs having gravity-controlled recovery processes. However, for a specific reservoir, it can be difficult to evaluate how much the pressure is being maintained.

GAS-CAP EXPANSION — A FRONTAL DRIVE PROCESS

Whether the reservoir is a straight gas-cap drive or one involving segregation of gas and oil, the expanding gas cap causes a downward movement of the gas/oil contact which is a frontal-drive process (immiscible displacement). To analyze the recovery from this type of mechanism, the fractional flow equation is employed. The use of this equation to study frontal-drive processes was discussed in the "Immiscible Displacement" chapter.

Assuming that only oil and free gas are flowing, at any point in the system, the fractional flow of gas is equal to the fraction of the total volumetric flow rate that is free gas flowing at that point; i.e., $f_g = q_g / (q_g + q_o)$. When the general fractional flow equation is written specifically to describe the displacement of oil by an expanding gas cap, then Equation 14-3 below results. The capillary pressure term, which is impossible to evaluate (beforehand), has been neglected. However, the solution technique will have the effect of replacing this term.

$$f_g = \frac{1 - \dfrac{0.488 \, k \, A \, \sin \alpha \, (\gamma_o - \gamma_g) \, k_{ro}}{q_{gt} \, \mu_o}}{1 + (k_o / k_g) \, (\mu_g / \mu_o)} \tag{14-3}$$

where:

k = absolute formation permeability, darcies
A = cross-sectional area of gas/oil contact, ft^2
α = flow angle (from the horizontal)**, degrees,
γ_o = specific gravity of oil (reservoir conditions), fraction,
γ_g = specific gravity of gas (reservoir conditions) referred to fresh water, fraction,
k_{ro} = relative permeability to oil, fraction,

q_{gt} = gas-cap expansion rate, reservoir bbl/D,

k_o = effective permeability to oil, darcies,

k_g = effective permeability to gas, darcies,

μ_o = oil viscosity, cp,

μ_g = gas viscosity, cp, and

f_g = fractional flow of gas in a gas-displacing-oil immiscible displacement process, fraction.

** Note: Equation 14-3 has been written for an expanding gas cap, so the flow will always be directed downward. Therefore, the equation has been adapted for positive angles in the downward direction.

CALCULATION OF RESERVOIR SPECIFIC GRAVITIES OF GAS AND OIL

Gas:

From the "Fluid Properties" chapter,

$$\rho_g = \frac{p \, M_g}{z \, R \, T} \quad lb/ft^3 \tag{3-8}$$

where:

ρ_g = gas density, lb/ft^3,

p = absolute system pressure, psia,

M_g = free gas molecular weight, lb/mole,

R = universal gas constant which equals 10.73 for these units,

T = absolute system temperature, degrees Rankine,

z = gas deviation factor, fraction.

So, the density of the gas at surface conditions is:

$$\rho_{g_{surf}} = \frac{(14.7)(M_g)}{(10.73)(520)} = 0.002635 \, M_g \quad lb/SCF$$

If B_g is equal to the reservoir gas formation volume factor in barrels/scf, then, the density of reservoir gas is:

$$\rho_{g_{res}} = \frac{0.002635 \, M_g}{B_g} \quad lb/bbl$$

The weight of one barrel of water is 350.4 lb. So, the specific gravity of the reservoir free gas relative to water is:

$$\gamma_g = \left(\frac{0.002635 M_g}{B_g} \right) / 350.4 = (7.52)(10^{-6}) \, M_g / B_g \tag{14-4}$$

Recall from the "Fluid Properties" chapter that $M_g = (\gamma_{g_{air}})(28.97)$ and:

$$B_g = 0.005034\, T\, z\, /\, p \quad bbl/SCF \tag{3-17}$$

Oil:

To calculate the reservoir oil specific gravity, most of the time:

$$\gamma_o = \left(\frac{141.5}{131.5 + {}^\circ API}\right)/B_o = \left(\gamma_{o_{surf}}\right)/B_o \tag{14-5}$$

is used. However, if the reservoir oil contains a significant amount of solution gas, then a more accurate relationship is:

$$\gamma_o = \frac{\left(\dfrac{141.5}{131.5 + {}^\circ API}\right) + \left(\dfrac{R_s M_g}{133{,}000}\right)}{B_o} \tag{14-6}$$

which considers the weight of the solution gas. B_o = oil-formation volume factor in barrels/STB, and R_s = solution GOR, scf/STB.

CALCULATION OF REMAINING OIL SATURATION IN GAS-CAP INVADED ZONE

By analyzing the fractional flow relationship of the expanding gas cap displacing oil downward, the average gas and oil saturations in the gas-cap invaded zone can be established. Because the fractional flow equation assumes the fluids involved to be incompressible, any time that the reservoir pressure changes substantially (say 100 psi or more), than the fractional flow relationship should be reinvestigated.

Consider Equation 14-3. Note that at a specific pressure in a particular reservoir, everything on the right-hand side is constant, except the fluid permeabilities which are assumed to be strict functions of saturation. Therefore, values of f_g versus S_g can be computed. After these numbers are in hand, the f_g vs S_g relationship may be graphed using standard (Cartesian) coordinate paper. An example of this plot is shown in Figure 14-2A.

For non-horizontal gas/oil displacements, it is likely that at low gas saturations, the calculated f_g values will be negative as indicated by the dotted curve in Figure 14-2A. This happens because a term involving the derivative of capillary pressure has been neglected from the numerator of Equation 14-3. As mentioned earlier, the capillary pressure term is impossible to evaluate until after the solution to the fractional flow equation has been generated. Therefore, the negative f_g's calculated at low gas saturations are not real.

GRAPHICAL SOLUTION

Assuming for the moment that the gas saturation within the oil zone is zero, then after the computed f_g vs S_g curve has been plotted, a line is drawn from the origin (0,0) such that it just touches (tangent to) the computed f_g vs S_g curve as indicated in Figure 14-2A. Then, the tangent line is extrapolated, as shown, to the point where it intersects the line $f_g = 1.0$. The intersection point is equal to S_{gavg}, the average gas saturation within the gas-cap invaded zone. References for the proof of the above procedure are given in the "Immiscible Displacement" chapter.

In the previous paragraph, the tangent construction was described for the case of an oil-zone gas saturation of zero. For the case of a small nonzero gas saturation within the oil zone, then the tangent is drawn from the point ($S_g = S_{gi}$, $f_g = 0$), not from the origin. S_{gi} is the gas saturation existing within the oil zone at the start of the displacement. This is illustrated in Figure 14-2B.

If the beginning gas saturation within the oil zone, S_{gi}, is larger than that S_g where f_g first becomes positive, then the tangent line is drawn from the computed curve at S_{gi} as shown in Figure 14-2C.

The authors have not seen any proof for the validity of the tangent construction procedure with a beginning gas saturation in the oil zone (Figures 14-2B and 14-2C).

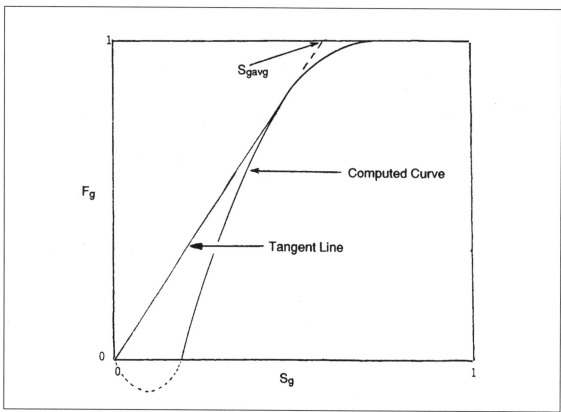

Fig. 14-2A. F_g vs. S_g and tangent solution.

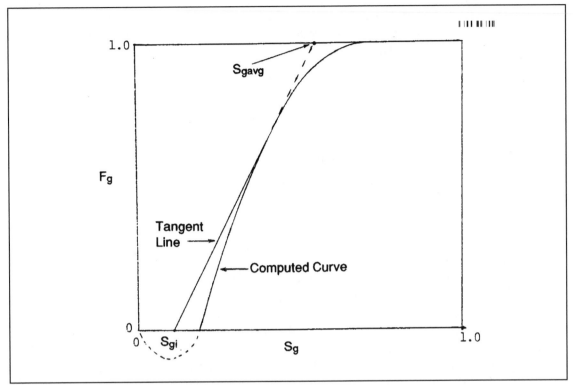

Fig. 14-2B. Gas displacing oil graphical solution.

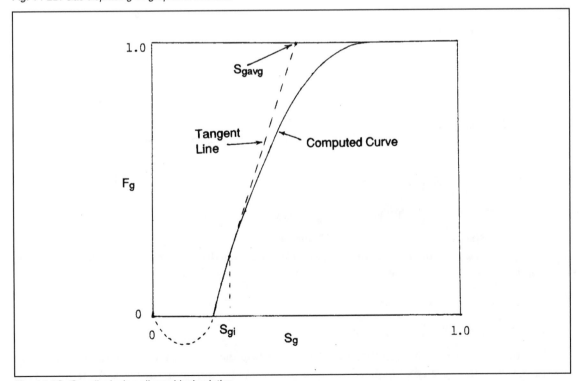

Fig. 14-2C. Gas displacing oil graphical solution.

However, beginning the tangent from S_{gi} seems to be a common practice through the oil industry. Notice that the effect of a beginning gas saturation in the oil zone is to leave less gas and, therefore, more oil in the invaded zone. This seems to be the nature of immiscible displacements, whether gas or water is used to push the oil. The lower the beginning displacing-phase saturation is in the oil zone, then the more efficient is the displacement.

AUTOMATED SOLUTION

For the case of no significant free gas within the oil zone, the average gas saturation within the gas-cap invaded zone may be found by evaluating:

$$S_{g_{avg}} = \min \ [\ S_g / f_g \ ] \tag{14-7}$$

ensuring that no negative f_g's are investigated.

To demonstrate the validity of this relationship, consider Figure 14-2D. Here is a typical f_g vs S_g relationship with three straight lines drawn from the origin and intersecting the f_g vs S_g curve. Consider the lower straight line which intersects the fractional flow curve at point (1) which is actually (S_{g1}, f_{g1}). This straight line is extrapolated up to where it intersects the line $f_g = 1.0$. Note that this intersection point is at a distance of X1 from the f_g axis. Two similar right triangles are contained within this construction. The larger one has vertices (S_g, f_g) of (0, 0), (X1, 1.0), and (0, 1.0). The smaller similar triangle is: (0, 0), (S_{g1}, f_{g1}) and $(0, f_{g1})$. From the laws of plane geometry, considering similar triangles, the ratio of two sides of one triangle is equal to the ratio of the corresponding sides of the other triangle. Hence,

$$X1 \ / \ 1 = S_{g1} / f_{g1}$$

So,

$$X1 = S_{g1} / f_{g1}$$

Now, the middle straight line is considered, which intersects the f_g vs S_g curve at point (2), which is really (S_{g2}, f_{g2}). Through similar reasoning as was used with the lower straight line, it can be shown that:

$$X2 = S_{g2} / f_{g2}$$

Notice that point (2) was higher on the f_g vs S_g curve than was point (1), but X2 < X1.

The upper straight line on Figure 14-2D intersects the f_g vs S_g curve at point (3) or $(S_{g3}, fg3)$ and is extrapolated up to the $f_g = 1.0$ line where it intersects at a distance of X3 from the f_g axis. Note that point (3) is the tangent point to the fractional flow curve; therefore,

$$X3 = S_{g_{avg}}$$

Further, notice that X3 < X2 < X1

If a fourth line were drawn from the origin intersecting the f_g vs S_g curve at a point higher than point (3), notice that when extrapolated to $f_g = 1.0$, the intersection point (distance from the f_g axis) would be greater than X3.

Hence,

$$S_{g_{avg}} = \min \ [\ S_g / f_g\ ] \tag{14-7}$$

This relationship is used as a trial and error process. To begin, some low gas saturation is chosen and the corresponding f_g is calculated. Assuming that the calculated f_g is positive, then $[S_g/f_g]$ is calculated and stored. Next, a larger S_g is selected and f_g calculated. Then, again compute $[S_g/f_g]$ and compare with the last $[S_g/f_g]$. If the current $[S_g/f_g]$ is smaller than the last one, then choose another S_g value somewhat larger than the last and calculate another $[S_g/f_g]$.

This process is continued as long as the $[S_g/f_g]$ values continue to decrease. Eventually, an S_g will be investigated where $[S_g/f_g]$ is larger than the last one. At this point, go back and find the minimum $[S_g/f_g]$ which is equal to S_{gavg}. To begin with, the difference between successive S_g's might be 10%; however, the final gas-saturation investigation increment should be no more than 1%. This automated procedure for finding S_{gavg} is illustrated in the example problem of the Combination Drive chapter.

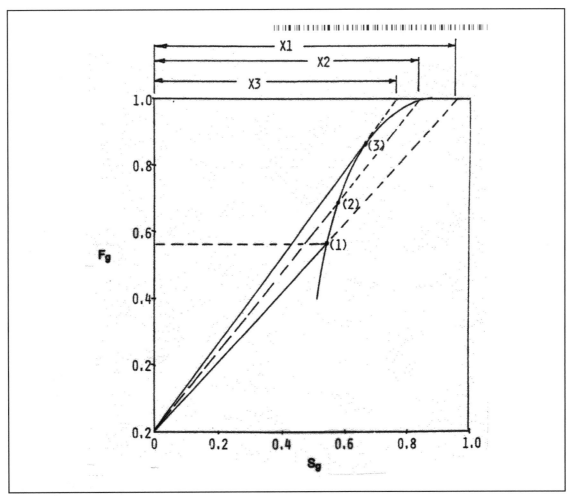

Fig. 14-2D. *F_g vs S_g graphical representation.*

REMAINING OIL SATURATION

After S_{gavg} has been determined, then the remaining oil saturation within the gas-cap-invaded zone can be calculated:

$$S_{org} = 1 - S_{wc} - S_{g_{avg}}$$

(14-8)

where:

$S_{g_{avg}}$ = the average gas saturation between the current gas/oil contact position and the original gas/oil contact, fraction

S_{wc} = connate water saturation, fraction

S_{org} = oil saturation left behind the advancing gas/oil contact, fraction

FACTORS AFFECTING OIL RECOVERY BY GAS-CAP EXPANSION

For convenience, the fractional flow equation describing the immiscible displacement of oil by gas-cap expansion will be repeated here:

$$f_g = \frac{1 - \dfrac{0.488 \; k \; A \; \sin \alpha \; (\gamma_o - \gamma_g) \; k_{ro}}{q_{gt} \; \mu_o}}{1 + (k_o / k_g)(\mu_g / \mu_o)}$$

(14-3)

Inspection of this equation will reveal that factors such as the difference in reservoir specific gravity, dip angle, etc. have a direct effect on the values of f_g. Since the values of f_g control the relative position of the f_g vs S_g curve, these factors influencing f_g also have a distinct effect on the average gas saturation within the invaded zone, and therefore on oil recovery. Obviously, as average gas saturation increases, so does the recovery.

Considering the factors that are not functions of saturation, anything that causes f_g to be smaller will also have the effect of increasing oil recovery. For instance, the coefficient of the k_{ro} term in the numerator of Equation 14-3 is:

$$Coeff_g = \frac{0.488 \; k \; A \; \sin \alpha \; (\gamma_o - \gamma_g)}{q_{gt} \; \mu_o}$$

(14-9)

Consider one variable in the numerator of the coefficient: α, and one variable in the denominator: q_{gt}. Figure 14-3 shows the relationship f_g vs S_g for two different dip angles, α_1, and α_2. Notice that the larger dip angle yields a larger average gas saturation. The greater the angle of dip is, the greater oil recovery becomes.

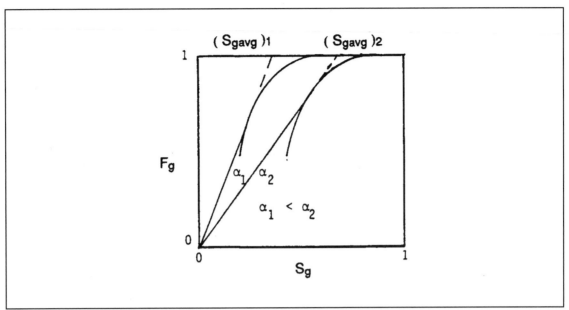

Fig. 14-3. Effect of angle of dip on average gas saturation, gas-cap movement.

Figure 14-4 illustrates that a slower rate of gas-cap expansion yields a higher recovery. The factor, q_{gt}, is the only external factor that can be influenced by well rates. Actually, q_{gt} can be modified either by external gas injection into the gas cap or by altering the producing well withdrawal rates.

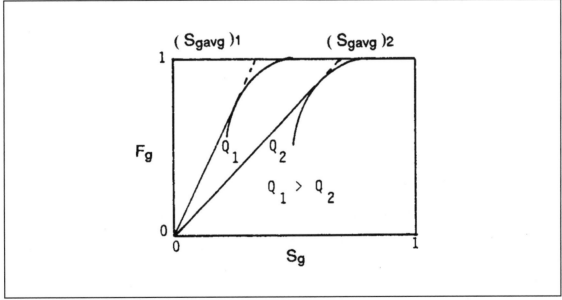

Fig. 14-4. Effect of rate of movement upon gas-cap recovery.

To generalize these results, anything that makes $Coeff_g$ larger also helps to increase oil recovery. Thus, it helps if the terms in the numerator of the k_{ro} coefficient (k, A, α, and $\Delta\gamma$) are large and the terms in the denominator (q_{gt}, μ_o) are small.

Problem 14-1:

Calculation of Gas-Cap Expansion Rate

The purpose of this problem is to illustrate the determination of the gas-cap expansion rate, which is also the total volumetric flow rate to be used in frontal drive computations. The gas-cap expansion rate is merely the average expansion volume of the original gas cap divided by the time interval.

Basic Data:

Gas cap area	2000 ac
Pay thickness (gas cap)	50 ft
Porosity	0.15 (fraction)
Connate water saturation	0.25 (fraction)
Reservoir temperature	160 °F.

Time-Pressure Data:

Time, Years.	Pressure, psia	Gas Dev. Factor, z
0	2500	0.752
1	2480	0.751

B_g calculations:

$$B_g = \frac{0.005034 \; T \; z}{p} = \frac{(0.005034)(620)z}{p} = 3.121 \; z/p \quad B/scf$$

Pressure	z-factor	B_g, Bbls / scf
2500	0.752	0.0009388
2480	0.751	0.0009451

Initial gas in place in the gas cap:

$$G = \frac{7758 \; A \; h \; \phi \, (1 - S_{wc})}{B_{gi}} = \frac{(7758)(2000)(50)(0.15)(1 - 0.25)}{(0.0009388)}$$

$$= 92.967 \; billion \; scf$$

Gas-cap expansion during the first year of production:

Expansion, Bbls = G $(B_g - B_{gi})$ = (92.967)(10^9)(0.0009451 - 0.0009388)

= 585,692 reservoir bbls.

Gas-cap expansion rate:

q_{gt} = Expansion/(Time interval) = 585,692 / 365

= 1605 reservoir bbls/day

Problem 14-2:

Calculate the Average Saturation Behind A Moving Gas-Oil Contact

An oil reservoir has been producing for a number of years. The gas cap is expanding due to pressure decline within the oil zone. It is desired to use the fractional flow equation to determine the approximate average gas saturation within the expanding gas-cap region.

Basic Data:

Formation permeability	200	md.
Formation porosity	22	%
Connate water saturation	25	%
Cross-sectional GOC area	2,178,000	sq ft
Net gas-cap expansion rate	10,000	res. B/D
Formation angle of dip	20	deg.
Specific gravity of surface oil	0.8	(fraction)
Reservoir pressure	2,000	psia
Reservoir temperature	140	°F.
Gas deviation factor	0.9	(fraction)
Oil-formation volume factor	1.35	Bbl/STB
Oil viscosity (reservoir)	1.5	cp
Gas viscosity (reservoir)	0.02	cp
Molecular weight of gas	21	lb/mole

Relative Permeability Data:

Gas Sat. Percent	k_{rg} Frac.	k_{ro} Frac.	k_g/k_o Ratio
5	0.0006	0.7588	0.0008
10	0.0044	0.5642	0.0078
15	0.0144	0.4096	0.0352
20	0.0329	0.2892	0.1138
25	0.0617	0.1975	0.3124
30	0.1024	0.1296	0.7901
35	0.1558	0.0809	1.9258
40	0.2225	0.0474	4.6941
45	0.3024	0.0256	11.8125
50	0.3951	0.0123	32.1220
55	0.4995	0.0051	97.9412
60	0.6144	0.0016	384.0000

A graph of these relative permeability data is shown in Figure 14-5. Using the fractional flow equation for gas driving oil, the average gas saturation within the gas-cap-invaded portion of the oil zone can be determined.

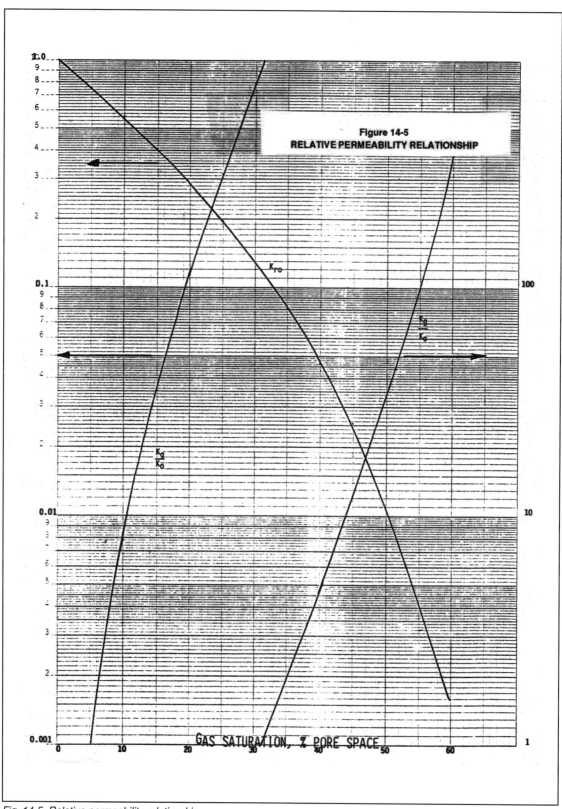

Fig. 14-5. Relative permeability relationship.

$$f_g = \frac{1 - (Coeff_g) \, k_{ro}}{1 + (\mu_g / \mu_o) \, (k_o / k_g)}$$

(14-10)

where:

$$Coeff_g = \frac{0.488 \, k \, A \, \sin \alpha \, (\gamma_o - \gamma_g)}{q_{gt} \, \mu_o}$$

Calculation Steps:

1. Determination of reservoir gas specific gravity compared to fresh water:

$$B_g = \frac{0.005034 \, T \, z}{P} = \frac{(0.005034)(140 + 460)(0.9)}{2000} = 0.001359 \; Bbl / scf$$

$$\gamma_g = (7.52)(10^{-6}) \, M_g / B_g = \frac{(7.52)(10^{-6})(21)}{0.001359} = 0.1162$$

2. Calculate $Coeff_g$:

$$(\gamma_o)_{res} = \gamma_o = \frac{(\gamma_o)_{surf}}{B_o} = 0.8 / 1.35 = 0.5926$$

$$Coeff_g = \frac{(0.488)(0.2)(2,178,000)(0.342)(0.5926 - 0.1162)}{(10,000)(1.5)} = 2.309$$

3. Determine the ratio μ_g / μ_o:

$$\mu_g / \mu_o = 0.02 / 1.5 = 0.0133$$

This completes the evaluation of the non-saturation-dependent parameters in the fractional flow equation, so:

$$f_g = \frac{1 - (Coeff_g) \, k_{ro}}{1 + (\mu_g / \mu_o) / (k_g / k_o)} = \frac{1 - (2.309) \, k_{ro}}{1 + (0.0133) / (k_g / k_o)}$$

4. Assume a series of gas saturation values, such as:

15%, 20%, 25%, . . . ,55%, 60%, etc.

5. Determine the values for k_{ro} and k_g/k_o for each of the saturation values.

6. Multiply each value of k_{ro} by $Coeff_g$.

7. For each of the saturation values, evaluate the numerator of f_g:

numerator = [1.0 - (Coeff$_g$)(k$_{ro}$)].

8. For each saturation, divide the viscosity ratio by the relative permeability ratio:
$(\mu_g/\mu_o)/(k_g/k_o) = 0.0133)/(k_g/k_o)$

9. Evaluate the denominator of f_g for each saturation.

   denominator = [1.0 + (0.0133)/(k_g/k_o)]

10. Divide the results of step 7 by the results of step 9 to yield f_g.

11. Use either the graphical or automated procedure to solve for S_{gavg}. Then,

   $S_{org} = 1 - S_{gavg} - S_{wc}$

(1) Gas Sat Percent	(2) k_{ro} Frac.	(3) k_g/k_o Ratio	(4) 2.309*(k_{ro})	(5) 1.0 - (4)	(6) 0.0133/(k_g/k_o)	(7) 1.0 + (6)	(8) f_g = (5)/(7) Fraction
15	0.4096	0.0352	0.9458	0.0542	0.3778	1.3778	0.0394
20	0.2892	0.1138	0.6678	0.3322	0.1169	1.1169	0.2975
25	0.1975	0.3124	0.4560	0.5440	0.0426	1.0426	0.5218
30	0.1296	0.7901	0.2992	0.7008	0.0168	1.0168	0.6892
35	0.0809	1.9258	0.1868	0.8132	0.0069	1.0069	0.8076
40	0.0474	4.6941	0.1094	0.8906	0.0028	1.0028	0.8880
45	0.0256	11.8125	0.0591	0.9409	0.0011	1.0011	0.9398
50	0.0123	32.1220	0.0284	0.9716	0.0004	1.0004	0.9712
55	0.0051	97.9412	0.0118	0.9882	0.0001	1.0001	0.9881

The graphical solution is used (Figure 14-6) by plotting column (8) against column (1). Then, the Welge tangent is drawn from the origin, which assumes no significant free gas within the oil zone, and extrapolated to the point where it intersects the horizontal line $f_g = 1.0$. The average gas saturation within the invaded zone is determined to be 42.6%. Finally, $S_{org} = 1.0 - S_{wc} - S_{gavg} = 1.0 - 0.25 - 0.426 = 0.324$. So, the remaining oil saturation within the gas-cap-invaded zone is 32.4%. The rather low S_{gavg} is partly due to the rather flat displacement; i.e., the small angle of dip.

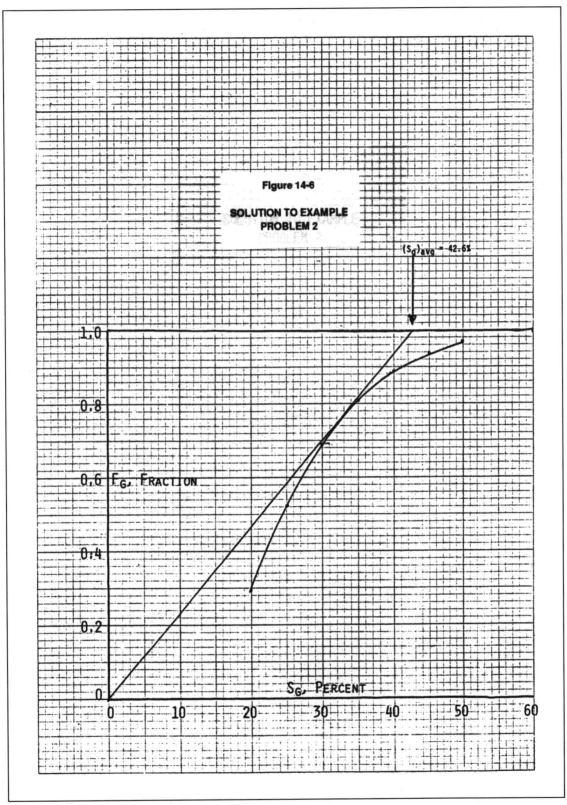

Fig. 14-6. Solution to example problem 2

CRITERION FOR STABILITY IN THE GAS-OIL CONTACT MOVEMENT

By inspection of Equation 14-3, the fractional flow equation, it should be apparent that the rate of movement of the gas/oil contact, as measured by q_{gt}, can be so fast that the gravity term (involving the specific gravity difference) becomes negligible. Under these conditions, the oil recovery from the gas contacted portion of the oil zone is lowered until it is approximately the same as that from a horizontal displacement.

Rapid movement of the gas/oil contact also can cause the gas to override the oil (due to the unfavorable mobility ratio) and finger down-structure along the top of the oil zone. This leads to premature breakthrough of gas at the producing wells with resultant poor recovery efficiency and oil productivity loss. It is beneficial for the gas/oil contact to move downward as a stable front with a reasonably constant velocity. A stable gas-oil contact is not necessarily perfectly horizontal; it can tilt some, but at least the gas does not override and move down-structure, bypassing much of the oil.

The diagram below presents a steeply dipping oil zone with an expanding gas cap. Notice the slight tilt of the gas/oil contact. Because the density of the gas is less than that of the oil, gravity forces act to keep the gas on top of the oil. Due principally to the low viscosity of gas, free gas usually has a higher mobility than oil ($k_g/\mu_g > k_o/\mu_o$) and therefore gas moves more easily in the reservoir. For low gas/oil contact velocities, the gas stays on top of the oil; but if the gas-cap-expansion rate becomes too high, then the viscous force will exceed the gravity force, and the gas will override the oil.

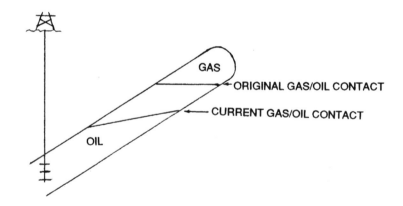

The criterion for stable gas/oil contact movement is:

$$G > M - 1 \qquad (14\text{-}11)$$

where:

$$G = \frac{0.488\ k\ A\ \sin\alpha\ (\gamma_o - \gamma_g)\ k_{rg}}{q_g\ \mu_g} \qquad (14\text{-}12)$$

and

$$M = \left(\frac{k_g}{\mu_g}\right) / \left(\frac{k_o}{\mu_o}\right) = \left(\frac{k_{rg}}{\mu_g}\right) / \left(\frac{k_{ro}}{\mu_o}\right) = \frac{k_{rg}\ \mu_o}{k_{ro}\ \mu_g} \qquad (14\text{-}13)$$

where:

      $k, A, \alpha, \gamma_o, \gamma_g, \mu_o$, and μ_g were defined under Equation 14-3;

      G = the gravity term, dimensionless,

      M = the mobility ratio, dimensionless,

      q_g = gas-cap flow rate, RB/D,

      k_{rg} = free gas relative permeability in the gas-cap-invaded zone, fraction,

      k_{ro} = oil relative permeability in the uninvaded oil zone, fraction.

To calculate the maximum gas-cap-expansion rate that maintains a stable gas/oil contact, the inequality in relationship 14-11 is replaced with an equal sign: $G = M - 1$. Then, the definitions presented in Equations 14-12 and 14-13 are substituted into the relationship, and the following results:

$$(q_g)_{\max} = \frac{0.488 \; k \; A \; \sin \; \alpha \; (\gamma_o - \gamma_g)}{(\mu_o / k_{ro}) - (\mu_g / k_{rg})} \quad Bbl \, / \, Day \qquad (14\text{-}14)$$

Problem 14-3:

Maximum Rate of Gas Flow to Maintain a Stable Gas/Oil Contact

Basic Data:

Formation permeability	0.1	darcies
Gas-oil contact area	4,356,000	sq ft
Angle of dip	30	degrees
Specific gravity of surface oil	0.85	(fraction)
Oil formation volume factor	1.30	bbl/STB
Reservoir pressure	2500	psia
Reservoir temperature	180	°F
Gas deviation factor	0.88	(fraction)
Gas molecular weight	21	lbs/mole
Oil viscosity	0.8	cp
Gas viscosity	0.015	cp
Free gas relative permeability (invaded zone)	0.35	(fraction)
Oil relative permeability (uninvaded oil zone)	0.90	(fraction)

$$(q_g)_{\max} = \frac{(0.488)(0.1)(4,356,000)(0.5)[(0.85 \, / \, 1.30) - ((7.52)(10^{-6})(21) \, / \, B_g)]}{(0.8 / 0.9) - (0.015 / 0.35)}$$

$$= \frac{106,286.4 \, (0.65385 - (1.5792)(10^{-4}) / B_g)}{0.84603}$$

$$B_g = \frac{0.005034 \, T \, z}{p} = \frac{(0.005034)(180 + 460)(0.88)}{2500} = 0.001134 \; Bbl \, / \, scf$$

$$(q_g)_{\max} = \frac{(106,286.4)(0.65385 - (1.5792)(10^{-4}) / 0.001134)}{0.84603} = 64,647 \; Bbl \, / \, Day$$

Equivalent surface rate:

$$[(q_g)_{max}]_{surface} = (q_g)_{max} / B_g = 64,647 / 0.001134 = 57,008\ MSCF/D$$

Equation 14-14 can also be used to calculate the maximum stable reservoir displacement rate when displacing oil by gas injection. To obtain the corresponding surface injection rate, simply divide by B_g.

CRITERIA FOR THE SEGREGATION OF GAS AND OIL

As the pressure declines within the oil zone of a solution-gas-drive reservoir, gas evolves from solution in the oil and forms a free-gas phase. Initially this free gas exists in discrete, unconnected gas globules. These free-gas sites are randomly distributed through the oil zone. Free-gas flow is not possible until sufficient gas has come out of solution that the gas globules interconnect and form a continuous gas phase. At this "equilibrium" gas saturation, flow of free gas begins.

Once free gas begins to flow, the path is usually to the nearest wellbore. Free gas situated within an oil zone will be subject to capillary, gravitational, and viscous forces. Assuming that the capillary force is small compared to the others, then the gravity force tends to cause the gas to move upward and the oil to drain downward. The viscous force tends to cause the gas to move laterally along with the oil to the nearest wellbore.

Early in the reservoir life, the free gas effective permeability is quite low due to the small gas saturation. Therefore, the only significant free gas moving is that near the wells (where the gas saturation and viscous force are both higher than away from the wells). And, this free gas moves with the oil to the nearest well. Away from the wells, where the pressure gradient is small, the oil phase moves slowly. And due to the low gas permeability, the free gas is moving even more slowly, if at all.

As the pressure declines, a greater amount of free gas develops. The resulting increased flow of gas to the wells is evidenced by the growing GOR's seen at the surface. As the gas permeability increases, the volume of the reservoir contributing significant amounts of flowing free gas also increases around each well. Away from the wells, assuming that the free gas saturation is greater than the equilibrium saturation, the gas mobility is also increasing.

When the gas mobility (away from the wells) has increased to the point that it is equal to the oil mobility, then it is possible for segregation of the gas from the oil to become a significant mechanism in the reservoir. If this happens, then the gas moves upward towards the top of the structure, while the oil drains downward. With segregation, the GOR's from the oil-zone producers tend to stabilize.

For segregation to occur, there must be continuous pay from the top of the structure to the bottom. Pressure forces within the reservoir dictate that the volumetric rate of free gas flow upward must be equal to the oil drainage rate downward. Therefore:

$$q_g(up) = q_o(down) \tag{14-15}$$

Considering Darcy's law for each of the two phases and simplifying for segregation to start:

$$k_g / \mu_g = k_o / \mu_o \qquad (14\text{-}16)$$

Equation (14-16) can be rearranged as:

$$k_g / k_o = \mu_g / \mu_o \qquad (14\text{-}17)$$

The implication is that when the free gas saturation has grown to the point that the gas/oil relative permeability ratio equals the gas/oil viscosity ratio, then segregation can begin. This result is interesting in that the right-hand side of 14-17, the viscosity ratio, is strictly a function of reservoir pressure; while, the left-hand side, the permeability ratio, is a function of saturation.

With Equation 14-17, it is possible to consider how gas/oil segregation is related to reservoir pressure. With pressure decline below the bubblepoint pressure, gas saturation increases with cumulative production and pressure reduction. A criterion for segregation is that the relative permeability ratio must equal the viscosity ratio. And since the viscosity ratio, μ_g / μ_o, normally decreases with decreasing pressure, this indicates that the amount of gas saturation required for segregation to begin also decreases with decreasing reservoir pressure.

The solution-gas-drive saturation Equation 14-1 may be used to develop a saturation vs. reservoir pressure relationship assuming no segregation. Then, Equation 14-17 may be used to generate a relationship of gas saturation needed for segregation to start vs. pressure. When plotted, the intersection of these two S_g vs. pressure curves will yield the reservoir pressure at which it is possible for segregation to begin. This procedure is illustrated in the following example problem.

Problem 14-4:

Determination of Pressure and Gas Saturation Necessary for Gas/Oil Segregation

Basic Data:

Original oil in place	10 million STB
Connate water saturation	25 percent
Bubblepoint pressure	2200 psia

Pressure - Production Data:

Pressure psia	B_o B/STB	μ_o, cp	μ_g, cp	Oil Prod MMSTB	N_p/N Fraction
2200	1.300	1.200	0.0266	0.0	0.0
2125	1.289	1.208	0.0258	0.129	0.0129
2025	1.275	1.218	0.0245	0.484	0.0484
1950	1.264	1.226	0.0239	0.812	0.0812
1850	1.250	1.250	0.0228	1.264	0.1264
1750	1.236	1.265	0.0217	1.604	0.1604

The first step is to assume no segregation and calculate the solution gas drive saturation versus reservoir pressure relationship. This will be done by rearranging Equation 14-1:

$$S_g = (1.0 - S_{wc}) \{ 1.0 - [1.0 - (N_p / N)] (B_o / B_{oi}) \} \qquad (14\text{-}18)$$

No Segregation — S_g vs Pressure

Pressure psia	N_p/N	B_o/B_{oi}	$[1 - (N_p/N)](B_o/B_{oi})$	S_g Fraction
2125	0.0129	0.9915	0.9787	0.016
2025	0.0484	0.9808	0.9333	0.050
1950	0.0812	0.9723	0.8933	0.080
1850	0.1264	0.9615	0.8400	0.120
1750	0.1604	0.9508	0.7983	0.151

Equation 14-17 permits determination of the saturation needed for segregation of the free gas and oil at each reservoir pressure. The procedure is simple and straightforward. The gas and oil viscosities are determined at the pressure being investigated; μ_g/μ_o is calculated, and then set equal to the gas/oil relative-permeability ratio. Then, the corresponding gas saturation is determined from the relative permeability relationship (Figure 14-5).

$$k_g / k_o = \mu_g / \mu_o \qquad (14\text{-}17)$$

Segregation — S_g vs Pressure

Pressure psia	μ_o, cp	μ_g, cp	μ_g / μ_o	k_g / k_o	S_g, fraction
2125	1.208	0.0258	0.0213	0.0213	0.130
2025	1.218	0.0245	0.0201	0.0201	0.128
1950	1.226	0.0239	0.0195	0.0195	0.127
1850	1.250	0.0228	0.0182	0.0182	0.125
1750	1.265	0.0217	0.0171	0.0171	0.123

These two pressure vs. S_g relationships are plotted in Figure 14-7. It can be seen that the two curves intersect at 12.45% gas saturation and 1837 psia. The significance of this intersection point is that the producing performance of the reservoir would follow a normal depletion behavior (no segregation) down to a reservoir pressure of about 1840 psia. At pressures below this level, there may be gas and oil segregation.

Notice that the pressure vs S_g relationship for segregation indicates a decreasing gas saturation with decreasing pressure. However, after segregation begins, it is unlikely that the gas saturation will decrease significantly. Yet to be discussed is a method for the calculation of average gas saturation over a well's drainage area undergoing segregation. For the purposes of modeling the performance of a segregation-drive (or gravity-drainage) reservoir, the gas saturation is often held constant after the onset of segregation.

THE DISTANCE FROM A WELL WHERE SEGREGATION IS LIKELY TO OCCUR

As previously discussed, significant segregation of the free gas from the oil is not possible until the gas saturation increases to the point where the gas mobility is equal to the oil mobility. Another criterion is that the gravity force tending to cause segregation must be greater than the viscous force acting to cause flow to the nearest well. To investigate these forces, Darcy's law for non-horizontal radial flow is written for the oil phase and for the free-gas phase:

$$q_o = -7.08 \, \frac{k \, k_{ro} \, h \, r}{\mu_o \, B_o} \left[\frac{\partial p}{\partial r} - 0.433 \, \gamma_o \, \sin \, \alpha \right] STB/D$$

$$q_g = -7.08 \, \frac{k \, k_{rg} \, h \, r}{\mu_g \, B_g} \left[\frac{\partial p}{\partial r} - 0.433 \, \gamma_g \, \sin \, \alpha \right] STB/D$$

Considering the oil equation, the pressure gradient, $\partial p / \partial r$, is a measure of the viscous force, while the gravity gradient term, $0.433 \, \gamma_o \, \sin \, \alpha$, is a measure of the force of gravity acting on the oil. Now, the assumption is made that only the viscous force is contributing to flow to the well, while the gravity force (difference between the oil and gas gravity gradients) is acting to cause segregation. Hence, the viscous force can represented as:

$$Viscous \; Force = \frac{q_o \, \mu_o \, B_o}{7.08 \, k \, h \, k_{ro}} \; (1/r)$$

(14-19)

Notice that the viscous force decreases with distance from the well. The gravity force is equal to:

$$Gravity \; Force = 0.433 \sin \, \alpha \; (\gamma_o - \gamma_g)$$

(14-20)

Although the oil and gas specific gravities are related to reservoir pressure, the gravity force is relatively constant over the drainage area of the well. However, the viscous force is decreasing with distance from the well; so, there is likely to be some radial position where the viscous force has decreased to the level of the gravity force. This distance can be determined by equating 14-19 and 14-20 and solving for radius (and setting equal to "R").

$$r = \frac{q_o \, \mu_o \, B_o}{3.07 \, k \, h \, k_{ro} \, \sin \alpha \, (\gamma_o - \gamma_g)} = R$$

(14-21)

Thus, for all radial positions greater than R, if the other criteria have been met, segregation is expected. So, when segregation begins, it will occur out away from the well; whereas, liberated free gas close to the well (distances less than R), will flow along with the oil to the wellbore. Notice that R is directly related to formation permeability and q_o. Therefore, different wells within the same reservoir will likely have different amounts of segregation (different R's) over their drainage areas.

Equations 14-20 and 14-21 contain the term sin α. Both of these equations assume that the segregation of the oil and gas is occurring via the "horizontal" (parallel to the dip angle) formation

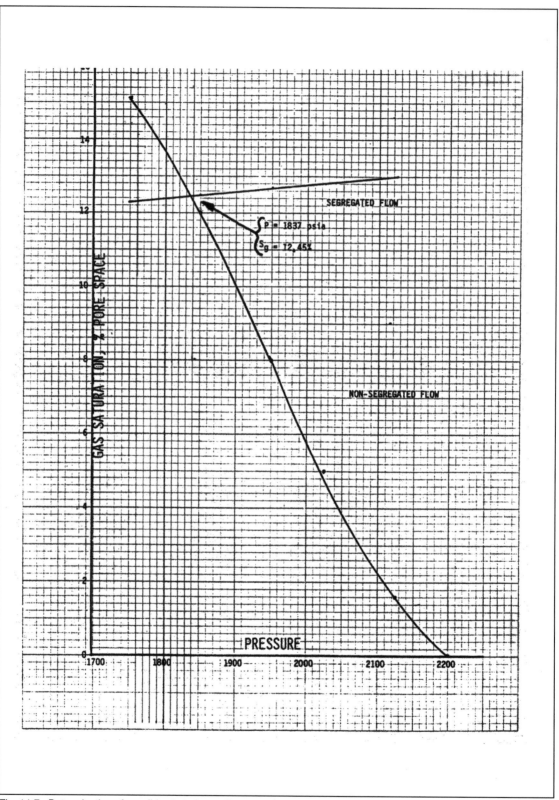

Fig. 14-7. Determination of possible start of gas-oil segregation.

permeability. If there is significant vertical permeability ($k_v \geq 50$ *md*), than the segregation can occur using the vertical formation permeability. In this case, the sin α in Equations 14-20 and 14-21 becomes one, and may be deleted from these relationships.

The units and use of Equation 14-21 will be illustrated with an example.

Problem 14-5:

Determination of the Distance from a Well where Segregation is Likely to Occur

This example is a continuation of example 14-4 which found that segregation could occur below reservoir pressures of about 1840 psia and at a gas saturation of 12.45%. It is assumed that even though the current pressure is 1750 psia, the gas saturation is still approximately 12.45%.

Basic Data:

Formation horizontal permeability	0.1	darcies
Pay thickness	20	ft
Angle of dip	30	degrees
Oil rate	100	STBO/D
Oil viscosity	1.265	cp
Reservoir pressure	1750	psia
Oil formation volume factor	1.236	bbl/STB
Oil API gravity	45.37	degrees
Reservoir temperature	150	°F
Gas deviation factor	0.9	(fraction)
Field reported gas gravity (compared to air)	0.76	(fraction)
Relative permeability to oil (k_{ro} @ S_g = 12.45 %)	0.47	(fraction)

Calculation of reservoir oil specific gravity:

$$Tank\text{–}oil \ specific \ gravity = \frac{141.5}{131.5+45.37} = 0.8000$$

$$Reservoir \ specific \ gravity = \gamma_{o_{surf}} / B_o = 0.8 \ / \ 1.236 = 0.6472$$

Calculation of reservoir-gas specific gravity (compared to water):

$$B_g = 0.005034 \, T z \, / p = \frac{(\,0.005034\,)\,(\,150+460\,)\,(\,0.9\,)}{1750} = 0.0015792 \ bbl \, / \, scf$$

$$Gas \ molecular \ weight = M_g = (\,28.97\,)\,(\gamma_g\,)_{air} = (\,28.97\,)\,(\,0.76\,) = 22.0$$

$$\gamma_g = (\,7.52\,)\,(\,10^{-6}\,)\,(\,M_g\,)\,/\,B_g = \frac{(\,7.52\,)\,(\,10^{-6}\,)\,(\,22.0\,)}{0.0015792} = 0.1048$$

Calculation of radial distance from well for gas-oil segregation:

$$R > \frac{q_o\, \mu_o\, B_o}{3.07\; k\; h\; k_{ro}\; \sin \alpha\; (\gamma_o - \gamma_g)}$$

$$R > \frac{(100)(1.265)(1.236)}{(3.07)(0.1)(20)(0.47)(0.5)(0.6472 - 0.1048)}$$

$$R > 199.8\; ft$$

AVERAGE GAS SATURATION OVER WELL'S DRAINAGE AREA

Even if a well's drainage area is undergoing segregation, there will still be a volume of reservoir close to the well, where the viscous force is high, in which the free gas and oil do not segregate. Therefore, to calculate average gas saturation, these two different reservoir volumes (segregation and non-segregation) should be taken into account. Assuming a circular drainage area, a volumetrically-weighted average yields:

$$(S_g)_{avg} = (S_g)_{seg} + [(S_g)_{noseg} - (S_g)_{seg}]\,(R^2)/(r_e^2) \qquad (14\text{-}22)$$

Problem 14-6:

Determination of Average Gas Saturation

This problem is a continuation of examples 14-4 and 14-5.

Basic Data:

   Well drainage area = 40 acres

   Reservoir pressure = 1750 psia

To calculate r_e:

$$r_e = \sqrt{\frac{(40)(43{,}560)}{\pi}} = 744.7\; ft$$

From example 5, R was calculated to be 199.8 ft

From example 4, at 1750 psia:

$(S_g)_{seg} = 0.123$ and

$(S_g)_{noseg} = 0.151$

Therefore, using Equation 14-22:
$(S_g)_{avg} = 0.123 + (0.151 - 0.123)(199.8^2)/(744.7^2)$
$\qquad = 0.125$

The average gas saturation in the drainage area of wells subject to segregation will change slowly as pressure declines. Such performance will tend to cause the recovery of oil to be enhanced.

Recap:

Significant segregation of free gas from the oil requires that three criteria be met:

(1) There must be sufficient formation permeability. A minimum of about 50 md usually is needed.

(2) The free gas saturation has to build up to the point where the mobility of the gas is equal to the oil mobility; i.e., $k_g/\mu_g = k_o/\mu_o$, or $k_g/k_o = \mu_g/\mu_o$.

(3) Over part of the drainage area, the gravity forces tending to cause segregation must exceed the viscous force acting to cause flow to the well.

REFERENCES

1. Tracy, G. W.: Chapter 14, "Gas-Cap Drive", *Applied Reservoir Engineering*, OGCI public course manual, Tulsa (1981).

15 COMBINATION DRIVE RESERVOIRS

INTRODUCTION

Few oil reservoirs produce by only one drive mechanism. There are usually at least two, and sometimes more, different forms of reservoir energy causing the expulsion of fluids from the wells.

For instance, consider a hydrocarbon reservoir in contact with an extremely active aquifer such that the principal drive mechanism is water drive. As soon as pressure declines, fluid and rock expansion occurs within the reservoir. This represents another mechanism that helps to drive fluids to wells. If the reservoir has an original gas cap, a decrease in pressure with production causes the gas cap to expand, driving oil downward. Thus, the gas cap would also be acting as a source of reservoir energy.

It is the objective in this chapter to use the material balance equation to isolate and quantify the effects of different drive mechanisms. A set of criteria (drive indices) are defined to evaluate the relative amount of each recovery mechanism. The drive indices can be calculated at different points in reservoir history to determine how the active forms of energy are changing with time. The immiscible displacements within the invaded zones (gas-cap expansion and water influx) will be considered with Buckley-Leverett techniques. Thus, knowing the relative amount of energy from each drive mechanism, together with the associated recovery efficiencies in each of the three parts of the oil zone (gas-cap-invaded, water-invaded, and undisturbed), it is possible to investigate how recovery efficiency is varying with time. This includes the ability to extrapolate to ultimate recovery efficiency. Finally, the benefits of injecting either gas or water can be investigated.

These concepts will be illustrated with an example problem, which feature an oil reservoir with an aquifer and an initial gas cap.

DEFINITION OF DRIVE INDICES

I. Oil Reservoirs

The material balance equation for an oil reservoir can be written as follows:

$$N_p B_o = N [(R_{si} - R_s) B_g - (B_{oi} - B_o)] - N_p (R_{poz} - R_s) B_g +$$

$$G (B_g - B_{gi}) - G_{pc} B_g + Q_{gin} B_g + (W_e - W_p B_w) + W_{in} \tag{15-1}$$

where:

Q_{gin} = cumulative gas injection into the gas cap, scf,
W_{in} = cumulative water injection (that has entered the original hydrocarbon pore volume), reservoir barrels,
N_p = cumulative oil production, STB,
N = the original oil in place, STB,
G = the original, free gas in place in the gas cap, scf,
G_{pc} = gas cumulative production from the gas cap, scf,
W_e = cumulative water influx, reservoir barrels,
W_p = cumulative water production, STB,
B_o = oil-formation volume factor, bbl/STB,
B_g = gas-formation volume factor, bbl/scf,
B_w = water-formation volume factor (often assumed to be 1.0), bbl/STB,
R_s = solution gas/oil ratio, scf/STB,
R_{poz} = cumulative produced gas/oil ratio from the oil zone, scf/STB, and
i = subscript indicating initial conditions (R_{si}, B_{oi}, B_{gi}).

Comments:

1. Equation 15-1 does not take into account the effects of rock and interstitial water expansion. This could be included if desired. Usually, when free gas exists in the reservoir, the large gas compressibility will cause the effect of the rock and water compressibilities to be negligible.

2. Notice that the term R_{poz} considers cumulative production volumes from the oil zone only. No gas production from the gas cap is included here.

$$R_{poz} = G_{poz} / N_p$$

where:

G_{poz} = cumulative gas production from the oil zone only, scf. This includes solution gas and liberated free gas. Because gas produced from the gas cap (G_{pc}) affects recovery differently than gas produced from the oil zone, they must be considered separately.

3. Unlike the equations given in the "Oil Reservoir Drive Mechanisms" chapter, G is used in Equation 15-1 instead of the gas cap factor, m. Comparison of Equation 15-1 with other material balance equations in this book is a simple task through the definition of the gas-cap factor:

$$m = (G\,B_{gi}) / (N\,B_{oi}) \qquad (15\text{-}2)$$

Recall that this is the ratio of the hydrocarbon pore volume of the gas cap to the hydrocarbon pore volume of the oil zone at the original conditions. The gas-cap factor is usually determined on the basis of results from well logs, drillstem tests, and mapping (volumetric calculations).

Considering Equation 15-1, if all of the oil-zone withdrawal terms are placed on the right-hand side and all of the other terms put on the left-hand side; then by dividing both sides of the equation by the right-hand side, the following relation results:

$$SDI + GDI + WDI = 1.0 \qquad (15\text{-}3)$$

where:

SDI = solution-gas-drive index,
GDI = gas-cap-drive index, and
WDI = water-drive index

These drive indices are defined as:

$$SDI = \frac{N\,[\,(R_{si} - R_s)\,B_g - (B_{oi} - B_o)\,]}{N_p B_o + N_p (R_{poz} - R_s)\,B_g} \qquad (15\text{-}4)$$

$$GDI = \frac{G\,(B_g - B_{gi}) - G_{pc}\,B_g + Q_{gin}\,B_g}{N_p B_o + N_p (R_{poz} - R_s)\,B_g} \qquad (15\text{-}5)$$

$$WDI = \frac{W_e - W_p\,B_w + W_{in}}{N_p B_o + N_p (R_{poz} - R_s)\,B_g} \qquad (15\text{-}6)$$

Notice that Equation 15-3 stipulates that at a particular point in time, the sum of the drive indices must be equal to 1.0. However, these indices will usually vary with time or cumulative production.

Notice that the denominator for each index is the same: total cumulative oil-zone production on a reservoir volume basis. This is the factor used to normalize the energy (expansion) associated with each of the three drive mechanisms.

The solution-gas-drive index has for its numerator the expansion of the oil (including liberated free gas) from the original pressure down to the current pressure. When production begins from a new reservoir, the pressure declines. So, every reservoir operates, in the beginning, predominantly by expansion drive. This is illustrated through the drive indices. Even for a reservoir which develops into a strong water drive, during the early stages, the SDI will be larger than the WDI.

The numerator of the water-drive index is the net "expansion" (into the original oil volume) of encroached and injected water. This is simply the cumulative volume of net water influx that has occurred across the original oil/water contact. Whether much water drive occurs depends upon the proximity of water (bottom or edge), the volume of water, and the permeability-area product available to the water.

The numerator of the gas-cap-drive index is the net expansion (including injection) of the original gas cap. The magnitude of this index is largely influenced by the size of the original gas cap and by the amount of gas-cap production.

Therefore, a drive index may be considered as the fraction of total oil zone withdrawals due to a particular drive mechanism.

II. Gas Reservoirs

For a gas reservoir, the material balance equation can be written as:

$$G_p B_g = G (B_g - B_{gi}) + (W_e - W_p B_w) \tag{15-7}$$

where:

G_p = cumulative gas production, scf,
B_g = gas formation volume factor, bbl/scf,
G = original gas in place, scf,
W_e = cumulative water influx, reservoir barrels,
W_p = cumulative water production, STB,
B_w = water-formation volume factor (often taken to be 1.0), bbl/STB, and
i = subscript indicating initial conditions.

Equation 15-7 can be rearranged into:

$$GDI + WDI = 1.0 \tag{15-8}$$

where:

GDI = the gas (expansion) drive index
WDI = the water drive index

The indices are defined as:

$$GDI = \frac{G (B_g - B_{gi})}{G_p B_g} \tag{15-9}$$

$$WDI = \frac{W_e - W_p B_w}{G_p B_g} \tag{15-10}$$

Similar to oil reservoirs, each index is the net expansion associated with that particular drive mechanism divided by cumulative gas production expressed as a reservoir volume.

EVALUATION OF DRIVE INDICES — OIL RESERVOIRS

To determine the magnitude of the drive indices, it is necessary to carefully analyze past reservoir performance data. This includes past reservoir static pressure data and cumulative production data. Also needed are evaluations of the original oil in place (N) and the gas cap factor (m). The original oil in place can be determined by either volumetric (pore volume) or material balance calculations.

Equations 15-1, 15-4, 15-5 and 15-6 include two terms: R_{poz} and G_{pc} which may be difficult to evaluate. R_{poz} is the cumulative produced gas/oil ratio from the oil zone; while G_{pc} is the cumulative gas production from the gas cap. To evaluate these factors, it is necessary to separate the surface gas production into two classes: (1) gas from the oil zone (solution and liberated), and (2) gas from the gas cap. This is a difficult, arbitrary, and only approximate procedure. However, it is necessary to determine ultimate recovery. The method to be described has yielded results that are generally acceptable for most engineering calculations.

Figure 15-1 shows a typical gas/oil ratio map. This is a map showing well locations with each well's recent producing GOR recorded above the well location and the daily oil production rate below the location. Similar gas/oil ratio maps are made for different times in the past to span the reservoir history.

At this point, it is necessary to divide the wells into two classes: (1) those which have consistently produced with gas/oil ratios considered to be low, and (2) the remainder of the wells. Wells in the second group are assumed to be producing some gas-cap gas. It is not an easy task to make this division between the two classes of wells. In making this choice it is sometimes helpful to have some idea where the gas cap is in relation to the completion interval of each well. For the example shown in Figure 15-1, a vertical dashed line has been drawn separating the low GOR wells (on the right from the others (left). So, the wells on the right are assumed to be producing gas only from the oil zone; while those to the left of the dashed line are thought to be producing some gas-cap gas.

Next, for each map, an average producing gas/oil ratio from the oil zone is calculated. Considering the low GOR wells only, for each well, determine the product of the daily oil rate and the producing gas/oil ratio. Then, sum these products, and divide by the total of the oil rates from these wells. As shown in Figure 15-1, this process yields the average GOR from the oil zone, $(GOR)_{oz}$.

With a value for $(GOR)_{oz}$ available, the rate of gas production from the gas cap can be calculated. First the oil zone gas production rate is determined by multiplying $(GOR)_{oz}$ by the total reservoir oil rate (all wells). Then subtracting the oil zone gas production rate from the total reservoir gas production rate yields the gas production rate from the gas cap. These calculations are illustrated in Figure 15-1.

Thus far, we have discussed how to determine the gas cap production rate and the average producing GOR from the oil zone at a particular point in time. Gas-cap production rate and $(GOR)_{oz}$ need to be determined at each of the points in reservoir history at which reservoir static pressure data is available. Then, graphs are prepared as shown in Figure 15-2. Considering the first plot, the gas cap production rate is plotted versus time. This graph may be used to determine G_{pc}, the cumulative gas cap gas production, by calculating the area (integrating) under the curve to the time of interest.

The second graph in Figure 15-2 is prepared by plotting $(GOR)_{oz}$ versus cumulative oil production from the entire reservoir. The area under the curve to a particular N_p of interest, yields cumulative gas production from the oil zone. By dividing the oil-zone cumulative gas production by the cumulative oil production, R_{poz} is obtained.

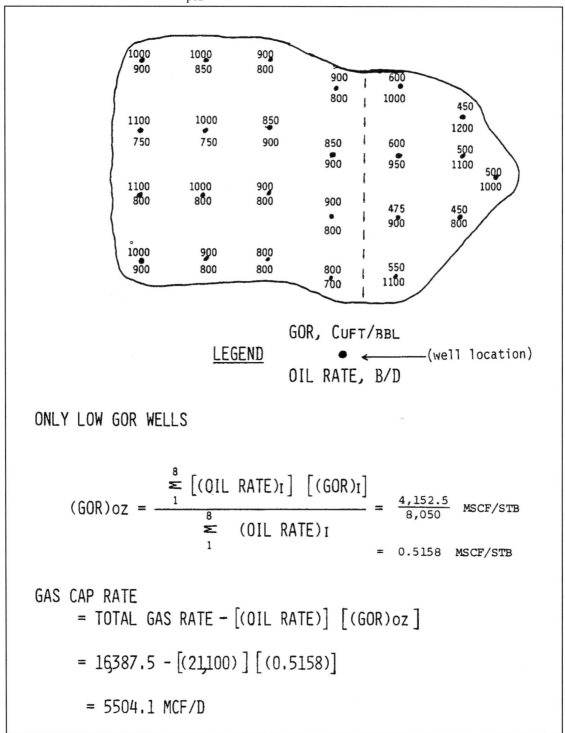

Fig. 15-1. Allocation of gas production between the oil zone and gas cap.

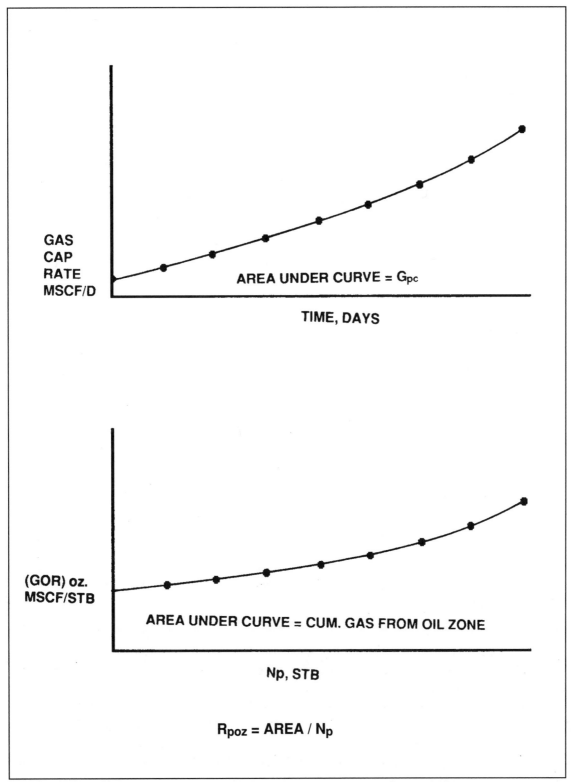

Fig. 15-2. Evaluation of gas-cap production, G_{pc} and cumulative gas/oil ratio from the oil zone, R_{poz}.

W_e, or cumulative water encroachment across the original oil/water contact, is needed at each point in reservoir history to calculate the water-drive index. The original oil/water contact is often determined from a study of openhole logs and initial well test data. Note that initial water production can be due to filtrate from water-base drilling fluids. Merely producing water following drilling is not sufficient evidence that the producing interval is near the original oil/ water contact. Comparison of the chlorine content of the produced water with that of drilling fluid filtrate can sometimes be diagnostic in this situation.

W_e (at each point in the past) can be calculated by rearranging Equation 15-1. This assumes that good values of N and m are available.

DEFINITION OF RECOVERY FACTOR

Recovery factor, or fractional recovery efficiency, is a fraction representing the portion of the original oil in place that is recovered. Mathematically, recovery factor is computed by dividing the change in oil content in the reservoir by the original oil content. Considering one barrel of reservoir pore volume:

$$Oil\ content,\ STB = S_o\ /\ B_o =$$

$$= \frac{\left(\dfrac{Oil\ Content,\ Reservoir\ Barrels}{Pore\ Volume,\ Reservoir\ Barrels} \right)}{\left(\dfrac{Oil\ Content,\ Reservoir\ Barrels}{Oil\ Content,\ Stock\ Tank\ Barrels} \right)}$$

$$= \frac{Oil\ Content,\ Stock\ Tank\ Barrels}{Pore\ Volume,\ Reservoir\ Barrels}$$

Therefore,

$$Recovery\ Factor = RF =$$

$$= \frac{(S_o\ /\ B_o\ )_i - (S_o\ /\ B_o)}{(S_o\ /\ B_o\ )_i} \tag{15-11}$$

The major difficulty in evaluating recovery factor is the determination of the oil saturation, S_o.

EFFECT OF DRIVE INDICES ON OIL RECOVERY

Recovery from oil reservoirs producing mostly by solution-gas drive is low: typically from 2 to 25% of the original oil in place. For those reservoirs subject to an effective gas cap expansion, recovery can be considerably higher: normally from 20 to 60% of the OOIP. However, if gravity drainage becomes effective, then recovery can exceed 60%. With water drive, recovery normally ranges from 30 to 60% of the OOIP. Thus, each of the drive mechanisms is associated with a different range of recovery efficiencies.

The amount of influence of a particular drive mechanism on overall recovery is directly related to the corresponding drive index. Therefore, to effect the greatest recovery, it is desirable to increase either the WDI or the GDI. To do this, the production of excess water or gas should be minimized.

The production of water eliminates some water which can be used to displace oil. However, there is usually an even larger negative effect on the economics of operation: increased lifting, handling, treating, and disposal costs.

It is likely that of all the parameters affecting recovery, excess gas production has the greatest negative impact. This is true whether the principal recovery mechanism is solution gas, gas cap, or water drive. At times it is necessary to inject some of the produced gas into the gas cap (or at the top of the structure) to maximize recovery. This is especially important with a very active aquifer. In this case, gas-cap production without reinjection can cause oil to be pushed into the gas cap, which creates a residual oil saturation where one formerly did not exist.

EFFECTIVE RECOVERY FACTOR

After production begins and some pressure drop occurs in a combination drive reservoir, then the original oil zone can be considered as three different parts: the gas-cap-invaded (expansion) zone, the water-invaded zone, and the undisturbed oil zone. Since the recovery efficiency in each of these three zones is different, a method of averaging is needed to arrive at a total reservoir recovery efficiency.

Recall that the drive indices at a particular point in time must sum to 1.0. Further, each drive index represents the fraction of total oil-zone withdrawals due to that particular drive mechanism. For example, a solution-gas-drive index value of 0.021 would indicate that 2.1% of the total oil zone withdrawals were due to the solution-gas-drive fluid-expulsion mechanism.

Since the drive indices are based on actual reservoir withdrawals, they serve as excellent weighting factors for the recovery factors for each of the mechanisms that are effective in the reservoir. Thus, the effective recovery factor is defined as:

$$(RF)_{eff} = (SDI)(RF_{oz}) + (GDI)(RF_{gc}) + (WDI)(RF_{wd}) \qquad (15\text{-}12)$$

where:

RF_{oz} = recovery factor in the undisturbed oil zone,
RF_{gc} = recovery factor in the gas-cap-invaded zone,
RF_{wd} = recovery factor in the water-invaded zone

Notice that each of six parameters on the right is likely to be changing with time. Thus, the drive indices and recovery factors must all be evaluated at the same point in time. The effective recovery factor also usually changes with time.

DETERMINATION OF THE OIL SATURATION IN THE GAS-CAP-EXPANSION ZONE

With pressure depletion, the expanding gas cap causes a displacement of oil downward. Because the gas and oil are not miscible, some oil is left behind in the expansion zone. To determine the efficiency of the displacement, this remaining oil saturation must be determined. This will be done using the fractional flow equation in exactly the same manner described in the "Gas-Cap Drive" chapter. The fractional flow Equation 14-3 for oil being displaced by an expanding gas cap is repeated here for convenience.

$$f_g = \cfrac{1 - \cfrac{0.488 \ k \ A \ \sin \alpha \ (\gamma_o - \gamma_g) \ k_{ro}}{q_{gt} \ \mu_o}}{1 + (k_o / k_g)(\mu_g / \mu_o)} \qquad (15\text{-}13)$$

The terms in Equation 15-13 were defined under Equation 14-3. Notice that at a pressure of interest, each of the variables on the right is a constant except for k_{ro} and k_o/k_g which are functions of saturation. Thus, an f_g versus S_g relationship can be calculated and then graphed (Figure 14-2 of the "Gas-Cap Drive" chapter).

It should be noted that in Equation 15-13 the term involving capillary pressure has been left out because it was too difficult to evaluate. Therefore, to replace the neglected capillary pressure term, the Welge tangent line is drawn. This tangent begins from the gas-saturation value in the undisturbed oil zone (usually zero is used), and then drawn such that it touches the computed curve at only one point. The Welge tangent is extrapolated up to the point of intersection with the $f_g = 1.0$ horizontal line. This intersection on the gas-saturation axis is the average gas saturation within the gas-cap-expansion zone. S_{gavg} can also be determined with an automated solution ("Gas-Cap Drive" chapter). Then, the remaining oil saturation within the gas cap invaded zone is:

$$S_{org} = 1 - S_{wc} - S_{gavg} \qquad (15\text{-}14)$$

The gas-cap-expansion rate, q_{gt}, which appears in Equation 15-13 is equal to the gas expansion rate across the gas/oil contact. This may be calculated by dividing the incremental change in the gas-cap-expansion volume by the incremental time. A simple example calculation (Problem 14-1) for q_{gt} was given in the "Gas-Cap Drive" chapter. A more general equation allowing for gas-cap production and injection into the gas cap is:

$$q_{gt} = \frac{\Delta [G (B_g - B_{gi}) - G_{pc} B_g]}{\Delta t} + q_{gi} B_g \qquad (15\text{-}15)$$

where:

q_{gt} = gas cap expansion rate, reservoir barrels / day
q_{gi} = gas cap injection rate, scf / day
Δt = time increment, days

Every time that the static pressure in the reservoir changes by 100 psi, some of the terms in Equation 15-13 change enough that a new calculation should be made for average gas saturation and remaining oil saturation.

DETERMINATION OF THE OIL SATURATION IN THE WATER INVADED ZONE

As water encroaches across the original oil/water contact, oil is immiscibly displaced upward by the water. As in the gas/oil displacement case, some oil is left behind in the water-invaded zone due to capillary forces and the oil/water viscosity contrast. To calculate recovery efficiency in the water-invaded zone, the remaining oil saturation is needed, and the fractional flow equation is used. Considering water displacing oil, the fractional flow equation (neglecting the term involving capillary pressure) is:

$$f_w = \frac{1 - \dfrac{0.488 \; k \; A \; \sin \alpha \, (\gamma_w - \gamma_o) \; k_{ro}}{q_{wt} \, \mu_o}}{1 + (k_o / k_w)(\mu_w / \mu_o)} \qquad (15\text{-}16)$$

where:

f_w = fraction of total flow rate that is water,
k = absolute formation permeability, darcies,
A = cross-sectional area of water/oil contact, ft^2,
α = dip angle (from the horizontal), degrees,
γ_o = specific gravity of oil (reservoir conditions), fraction,
γ_w = specific gravity of reservoir water, fraction,
k_{ro} = relative permeability to oil, fraction,
q_{wt} = net water influx rate, reservoir barrels/day,
k_o = effective permeability to oil, darcies,
k_w = effective permeability to water, darcies,
μ_o = oil viscosity, cp, and
μ_w = water viscosity, cp.

The theory behind Equation 15-16 assumes incompressible, steady state flow. Therefore, to compensate for this restriction, a system pressure is stipulated. At a particular pressure (and at reservoir temperature), the specific gravities of reservoir oil and water can be approximated as:

$$\gamma_o = (\gamma_{o\,tank}) / B_o \qquad (15\text{-}17)$$

$$\gamma_w = (\gamma_{w\,surf}) / B_w \qquad (15\text{-}18)$$

where:

$\gamma_{o\,tank}$ = specific gravity of stock tank oil, fraction,
B_o = oil formation volume factor, bbl/STB,
$\gamma_{w\,surf}$ = specific gravity of produced water, fraction,
B_w = water-formation volume factor, bbl/STB

The net water influx rate used in Equation 15-16 is determined by dividing the change in the net water influx volume by the time increment. If water is being injected, this must also be considered. The net influx equation is:

$$q_{wt} = \frac{\Delta(W_e - W_p B_w)}{\Delta t} + q_{wi}$$

or

$$q_{wt} = \frac{\Delta W_e}{\Delta t} + q_{wi} - q_{wp} B_w \qquad (15\text{-}19)$$

where:

q_{wt} = net water influx rate, reservoir barrels/day,
ΔW_e = change in water influx volume during a given time increment, bbl,
Δt = time increment, days
q_{wi} = water injection rate, reservoir barrels/day,
q_{wp} = water production rate, surface barrels/day,
B_w = water formation volume factor, bbl/STB,
W_p = cumulative water production, STB

Because the fractional flow equation considers the fluids to be incompressible, each time that the reservoir pressure changes by 100 psi, the fractional flow relationship should be reinvestigated.

At a particular pressure, values on the right hand side of Equation 15-16 are constant (for a given reservoir) except the fluid permeabilities, which are dependent only on saturation. Therefore, if a value of S_w is stipulated, then the fluid permeabilities can be determined, and the corresponding f_w calculated. Hence, at a given pressure, with relative (or effective) permeability data in hand, an f_w versus S_w relationship can be calculated. If this relationship is then graphed, it should resemble that shown in Figure 15-3. Notice that the computed curve begins at the irreducible water saturation and ends at: $S_w = 1 - S_{or}$.

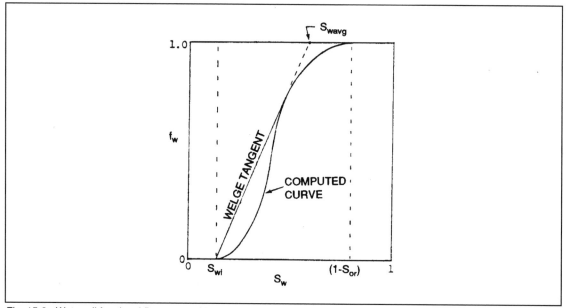

Fig. 15-3. Water-oil fractional flow curve with Welge tangent and S_{wavg}.

Recall that the computed curve, f_w versus S_w, in Figure 15-3 neglected capillary pressure. To replace the capillary pressure term, a tangent line is drawn from the beginning water saturation in the oil zone (normally the irreducible water saturation) such that it just touches the computed curve at one point. This line is called the Welge tangent line, and is illustrated in Figure 15-3. To determine the average water saturation in the water-invaded zone, the Welge tangent is extrapolated to the point of intersection with the $f_w = 1.0$ line. As shown in Figure 15-3, this point is equal to the average water saturation within the water-invaded zone, S_{wavg}.

It is also possible to determine the average water saturation through an automated procedure (suitable for computer use) by using the following equation:

$$S_{wavg} = S_{wi} + Min \ [(S_w - S_{wi}) \ / f_w]$$ (15-20)

The proof of Equation 15-20 is similar to the proof given in the "Gas-Cap Drive" chapter for the "automated solution" Equation 14-7 to determine the average gas saturation within the gas-cap-invaded zone.

To use Equation 15-20, some low water saturation, S_{wl}, which is greater than the irreducible value, is chosen. Then, using Equation 15-16, the fractional flow, f_{wl}, at S_{wl} is calculated. Then evaluate the function:

$$x = (S_w - S_{wi})/f_w + S_{wi}$$

At this point, a slightly larger water saturation, S_{w2}, is chosen, and f_{w2} is calculated. Now, calculate x_2. If the first S_w investigated was low enough, then x_2 should be smaller than x_1. In this manner, investigate larger and larger S_w's. With each new S_w, calculate f_w, and then x. Continue this process as long as the x's are decreasing. As soon as the x's start to increase, go back and determine the minimum x (Equation 15-20). The minimum x is equal to the average water saturation within the water invaded zone.

After S_{wavg} is determined, then the remaining oil saturation within the water invaded zone can be calculated:

$$S_{orw} = 1 - S_{wavg}$$ (15-21)

DETERMINATION OF OIL SATURATION WITHIN THE UNINVADED OIL ZONE

Unfortunately, the oil saturation within the undisturbed oil zone must be determined indirectly. Consider Figure 15-4, which represents a simplified cross-sectional diagram of a combination drive reservoir. At the initial conditions, within the hydrocarbon portion of the reservoir system, there was only the initial gas cap and the original oil zone. However, after producing the reservoir long enough to obtain some pressure depletion, the original oil zone is comprised of three different parts: the gas-cap-expansion zone, the uninvaded-oil zone, and the water-invaded zone. GCE is defined to be the pore volume of the gas-cap-expansion zone, and WCM (water/oil contact movement) is defined to be the pore volume of the water invaded zone.

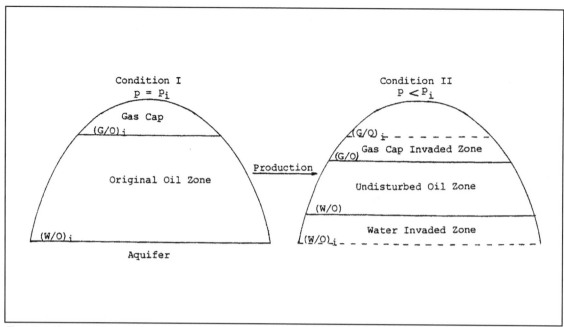

Fig. 15-4. Combination-drive reservoir with initial conditions and conditions after some pressure depletion.

The pore volume of the gas-cap-invaded zone may be approximated as:

$$GCE = \frac{G\,(B_g - B_{gi}) - G_{pc}\,B_g + Q_{gin}\,B_g}{S_{gavg}} \qquad (15\text{-}22)$$

where:

Q_{gin} = cumulative gas injection into the gas cap, scf
GCE = the pore volume within the gas cap invaded zone, bbl,
S_{gavg} = average gas saturation within the gas cap invaded zone, frac, and
B_g = gas formation volume factor, bbl/scf.

The pore volume of the water-invaded zone is:

$$WCM = \frac{W_e - W_p\,B_w + W_{in}}{S_{wavg} - S_{wc}} \qquad (15\text{-}23)$$

where:

WCM = the pore volume within the water-invaded zone, bbl,
S_{wavg} = the average water saturation of the water-invaded zone, fraction
W_{in} = cumulative water injection (that has entered the water invaded zone), reservoir barrels.

After the values of GCE and WCM are determined, then the oil saturation within the uninvaded-oil zone can be computed:

$$S_o = \frac{(N - N_p)B_o - (GCE)(S_{org}) - (WCM)(S_{orw})}{\dfrac{NB_{oi}}{1 - S_{wc}} - (GCE + WCM)} \qquad (15\text{-}24)$$

where:

S_{org} = remaining oil saturation in gas-cap-invaded zone, fraction, and
S_{orw} = remaining oil saturation in water-invaded zone, fraction.

It is instructive to consider the different terms in Equation 15-24. In the denominator, the term: $(NB_{oi})/(1\text{-}S_{wc})$ is equal to the total original oil-zone pore volume. Hence, after subtracting the pore volume of the two invaded zones, the pore volume of the undisturbed oil zone remains.

In the numerator, the term: $(N - N_p)B_o$ represents the reservoir volume of the total remaining oil in the reservoir. Unfortunately, this oil is spread over the two invaded zones and the undisturbed zone. Therefore, the volume of oil within the two invaded zones $(GCE)(S_{org})$ and $(WCM)(S_{orw})$ must be subtracted off to yield the remaining oil in the uninvaded zone.

Problem:

A reservoir has been producing for 11 years. There was an initial gas cap and underlying water. Sufficient permeability exists to support a water drive; however, inspection of the pressure data indicates that only a partial water drive is possible.

Basic Data:

Initial pressure	2200	psia
Gas cap factor (m)	0.20	fraction
Average dip angle	30	degrees
Reservoir temperature	140	°F
Average oil zone productive area	524	ac
Area of initial gas/oil contact	15,217,000	sq ft
Area of initial water/oil contact	30,434,000	sq ft
Formation permeability	100	md
Formation porosity	0.20	fraction
Average oil pay thickness	100.0	ft
Connate water saturation	0.20	fraction

Table 15-1 presents the static pressure and production date for the 11 years of operations. Table 15-2A provides fluid properties data, and Table 15-2B includes additional rock properties data.

Table 15-1
Pressure - Production Data.

| TIME | PRESSURE | CUM. OIL PROD. | CUM GAS PROD. | | CUM. WATER PROD. |
| | | | Oil Zone | Gas Cap | |
Years	Psia	MMSTB	BSCF	BSCF	MMSTB
0.00	2200	0.0000	0.0000	0.0000	0.0000
0.50	2100	1.1617	0.7126	0.0410	0.0040
1.35	2000	2.4577	1.6839	0.0910	0.0500
2.60	1900	4.3391	2.7674	0.2110	0.0850
4.20	1800	5.6093	4.5599	0.4400	0.1500
6.00	1700	7.1542	6.4258	0.7800	0.3500
8.25	1600	8.8336	8.8159	1.1500	0.5000
11.00	1500	10.8233	11.6022	1.6000	0.7000

Table 15-2A
Fluid Properties.

PRESSURE psia	B_o BBL/STB	R_s MSCF/STB	Z FRACTION	μ_o CP	μ_g CP
2200	1.3010	0.5750	0.8350	0.70	0.015
2100	1.2850	0.5430	0.8270	0.71	0.014
2000	1.2710	0.5110	0.8200	0.72	0.013
1900	1.2570	0.4790	0.8170	0.73	0.013
1800	1.2430	0.4460	0.8160	0.74	0.012
1700	1.2290	0.4140	0.8180	0.75	0.011
1600	1.2150	0.3820	0.8200	0.76	0.011
1500	1.2000	0.3500	0.8240	0.77	0.010

Gas Molecular Weight	22.0	Lbs/Mole
Water Viscosity	0.40	Cps
Tank-Oil Specific Gravity	0.80	Fraction
Water-Formation Volume Factor (B_w)	1.00	bbl/STB
Produced-Water Specific Gravity	1.074	Fraction

Table 15-2B
Rock Properties.

Water - Oil (Imbibition)				Gas - Oil (Drainage)		
S_w	k_{ro}	k_{rw}		S_g	k_g/k_o	k_{ro}
0.21	0.9300	0.00012		0.01	0.0001	0.98000
0.25	0.72000	0.00120		0.02	0.0008	0.97000
0.30	0.43000	0.00710		0.05	0.0045	0.95000
0.35	0.22000	0.01830		0.08	0.0120	0.92000
0.40	0.10000	0.03200		0.10	0.0200	0.90000
0.45	0.03600	0.05100		0.15	0.0600	0.82000
0.50	0.00800	0.07400		0.20	0.1500	0.66000
0.55	0.00092	0.10800		0.25	0.4000	0.50000
0.58	0.00020	0.13300		0.30	1.0000	0.36000
0.60	0.00000	0.15000		0.35	1.6000	0.23800
				0.40	2.8000	0.18000
				0.45	5.5000	0.05200
				0.50	12.3000	0.03100
				0.55	31.0000	0.01180
				0.60	100.0000	0.00310
				0.65	400.0000	0.00064
				0.70	1600.0000	0.00010
				0.75	6400.0000	0.00002

CALCULATION OF ORIGINAL OIL IN PLACE: PORE VOLUME METHOD

$$N = 7758 \; \frac{\phi \; h \; A \; (1 - S_{wc})}{B_{oi}}$$

where:

N = original oil in place, STB,
ϕ = porosity, fraction,
h = net pay thickness, ft,
S_{wc} = connate water saturation, fraction, and
B_{oi} = initial oil-formation volume factor bbl/STB

Substituting the appropriate data:

$$N = \frac{(7758)(0.2)(100)(524)(1.0 - 0.2)}{1.301} = 50,000,000 \; STB$$

CALCULATION OF GAS-FORMATION VOLUME FACTOR

Equation 3-17 presented in the "Fluid Properties" chapter to calculate gas formation volume factor in bbl/SCF is:

$$B_g = 0.005034 \, [\, T \, z \, / \, p \,] \quad bbl \, / \, SCF \tag{3-17}$$

For this problem, it is convenient to use a gas formation volume factor in bbl/MSCF. Hence:

$$B_g = 5.034 \, [\, T \, z \, / \, p \,] \; bbl \, / \, MSCF = (\, 5.034 \,)(\, 140 + 460 \,)(\, z \, / \, p \,) = 3020.4 \, (\, z \, / \, p \,)$$

PRESSURE, psia	Z, FRACTION	B_g, BBL/MCF
2200	0.835	1.1464
2100	0.827	1.1895
2000	0.820	1.2384
1900	0.817	1.2988
1800	0.816	1.3692
1700	0.818	1.4533
1600	0.820	1.5480
1500	0.824	1.6592

CALCULATION OF ORIGINAL GAS-CAP VOLUME

$$G = \frac{m \, N \, B_{oi}}{B_{gi}} = \frac{(\, 0.2 \,)(\, 50,000,000 \,)(\, 1.301 \,)}{1.146} = 11,353,000 \; MSCF$$

CALCULATION OF WATER ENCROACHMENT USING MATERIAL BALANCE

Because water production has increased during the life of the reservoir, it is likely that water has encroached across the original oil/water contact from the associated aquifer. Assuming that good values of N and m are known, it is possible to rearrange Equation 15-1 and calculate water encroachment at each of the history points. Because there has been no injection:

$$W_e = N_p \, [\, (\, R_{poz} - R_s \,) B_g + B_o \,] + W_p B_w$$

$$- \, N \, [\, (\, R_{si} - R_s \,) B_g - (\, B_{oi} - B_o \,) \,] - [\, G \, (\, B_g - B_{gi} \,) - G_{pc} B_g \,] \tag{15-25}$$

Notice that Equation 15-25 can be stated in words as:

Water influx

> *= (cumulative withdrawals from the oil zone, including water)*

> *- (original oil-zone expansion)*

> *- (net expansion of the gas cap)*

These calculations of water influx volumes are shown in Table 15-3. The computations to column 13 are to determine the cumulative oil-zone withdrawals (including water):

$$N_p [(R_{poz} - R_s) B_g + B_o] + W_p B_w$$

The original oil-zone expansion:

$$N [(R_{si} - R_s) B_g - (B_{oi} - B_o)]$$

is computed in columns 14 through 17. Columns 18 through 22 are used to calculate the net gas-cap expansion:

$$[G (B_g - B_{gi}) - G_{pc} B_g]$$

Water influx volumes are shown in column 24.

As some columns in Table 15-3 are needed in later sections of this problem, they will be mentioned here. Column 11 is the reservoir withdrawals of oil and associated gas from the oil zone. Column 12 is cumulative reservoir water production; however, since $B_w = 1.0$, this is also cumulative surface water production. Column 17 is the expansion of the original oil in place and the evolved solution gas. Column 19 is the gross expansion of the original gas cap (no reduction yet for gas cap production). Column 21 is the reservoir volume of the cumulative withdrawals from the gas cap. Column 22 is the net gas cap expansion. Notice that many of the columns in Table 15-3 have units of MMRB (millions of reservoir barrels).

Table 15-3
Water Influx Calculations.

(1) TIME YEARS	(2) PRESSURE PSIA	(3) B_o DIM	(4) R_s MCF/B	(5) R_{poz} MCF/B	(6) (5) - (4) MCF/B	(7) B_g B/MCF	(8) (6)*(7) DIM
0.00	2200	1.301	0.575	0.5750	0.0000	1.1464	0.0000
0.50	2100	1.285	0.543	0.6134	0.0704	1.1895	0.0838
1.35	2000	1.271	0.511	0.6852	0.1742	1.2384	0.2157
2.60	1900	1.257	0.479	0.6378	0.1588	1.2988	0.2062
4.20	1800	1.243	0.446	0.8129	0.3669	1.3692	0.5024
6.00	1700	1.229	0.414	0.8982	0.4842	1.4533	0.7037
8.25	1600	1.215	0.382	0.9980	0.6160	1.5480	0.9535
11.00	1500	1.200	0.350	1.0720	0.7220	1.6592	1.1979

(9) (3)+(8) DIM	(10) N_p MMB	(11) (9)*(10) MMRB	(12) W_p*B_w MMRB	(13) (11)+(12) MMRB	(14) $B_{oi} - B_o$ DIM	(15) $(R_{si} - R_s)B_g$ DIM	(16) (15)-(14) DIM
0.0000	0.0000	0.0000	0.0000	0.0000	0.000	0.0000	0.000
1.3688	1.1617	1.5901	0.0040	1.5941	0.016	0.0381	0.022
1.4867	2.4577	3.6538	0.0500	3.7038	0.030	0.0793	0.049
1.4632	4.3391	6.3491	0.0850	6.4341	0.044	0.1247	0.081
1.7454	5.6093	9.7905	0.1500	9.9405	0.058	0.1766	0.119
1.9327	7.1542	13.8268	0.3500	14.1768	0.072	0.2340	0.162
2.1685	8.8336	19.1560	0.5000	19.6560	0.086	0.2988	0.213
2.3979	10.8233	25.9531	0.7000	26.6531	0.101	0.3733	0.272

(17) N*(16) MMRB	(18) $B_g - B_{gi}$ B/MCF	(19) G*(18) MMRB	(20) G_{pc} BCF	(21) (20)*(7) MMRB	(22) (19)-(21) MMRB	(23) (17)+(22) MMRB	(24) W_e MMRB
0.0000	0.0000	0.0000	0.000	0.0000	0.0000	0.0000	0.0000
1.1031	0.0431	0.4891	0.041	0.0488	0.4404	1.5435	0.0506
2.4628	0.0920	1.0443	0.091	0.1127	0.9316	3.3944	0.3094
4.0341	0.1524	1.7301	0.211	0.2740	1.4561	5.4902	0.9439
5.9316	0.2229	2.5302	0.440	0.6025	1.9278	7.8594	2.0811
8.0994	0.3070	3.4850	0.780	1.1336	2.3514	10.4508	3.7260
10.6378	0.4016	4.5591	1.150	1.7801	2.7789	13.4167	6.2393
13.6161	0.5128	5.8221	1.600	2.6547	3.1674	16.7835	9.8696

$$W_e = N_p [(R_{poz} - R_s) B_g + B_o] + W_p B_w - N [(R_{si} - R_s) B_g - (B_{oi} - B_o)] - [G (B_g - B_{gi}) - G_{pc} B_g]$$

MATERIAL BALANCE DETERMINATION OF ORIGINAL OIL IN PLACE

The Table 15-3 calculations of water influx were based on a volumetric oil in place calculation. The purpose of Table 15-4 is twofold: (1) material balance determination of original oil in place, and (2) calibration of the aquifer model. The procedure for using material balance in conjunction with an aquifer model to determine oil in place and the aquifer model constants was discussed in the "Water Drive" chapter.

Neglecting the injection terms, Equation 15-1 can be solved for N as:

$$N = \frac{N_p\,[\,B_o + B_g\,(R_{poz} - R_s)\,] + G_{pc}\,B_g - W_e + W_p\,B_w}{B_g\,(R_{si} - R_s) - (B_{oi} - B_o) + m\,\dfrac{B_{oi}}{B_{gi}}\,(B_g - B_{gi})} \qquad (15\text{-}26)$$

Then, if W_e is assumed to be zero in 15-26, instead of N, the relationship becomes the equation for apparent oil in place:

$$N_a = \frac{N_p\,[\,B_o + B_g\,(R_{poz} - R_s)\,] + G_{pc}\,B_g + W_p\,B_w}{B_g\,(R_{si} - R_s) - (B_{oi} - B_o) + m\,\dfrac{B_{oi}}{B_{gi}}\,(B_g - B_{gi})} \qquad (15\text{-}27)$$

Therefore:

$$N_a = N + (W_e / D) \qquad (15\text{-}28)$$

where:
$$D = B_g\,(R_{si} - R_s) - (B_{oi} - B_o) + m\,(B_{oi}/B_{gi})\,(B_g - B_{gi})$$

In this problem, the Schilthuis steady-state aquifer model was chosen:

$$W_e = C_s\,\Sigma\Delta\bar{p}\,\Delta t \qquad (15\text{-}29)$$

Combining Equations 15-29 and 15-28 results in:

$$N_a = N + C_s\left[\frac{\Sigma\,\Delta\bar{p}\,\Delta t}{D}\right] \qquad (15\text{-}30)$$

Table 15-4
Determination of OOIP (N) and Aquifer Constant C$_s$

(1) Pressure Psia	(2) Oil Zone Cum. Prod. MMRB	(3) $G_{pc}B_g$ MMRB	(4) W_pB_w MMRB	(5) (2)+(3)+(4) Num. of N$_a$ MMRB	(6) D, Denom of N$_a$ BBL/STB	(7) (5)/(6) N$_a$ MMSTB
2100	1.5901	0.0488	0.0040	1.64285	0.03184	51.59431
2000	3.6538	0.1127	0.0500	3.81647	0.07013	54.41712
1900	6.3491	0.2740	0.0850	6.70811	0.11527	58.19393
1800	9.7905	0.6025	0.1500	10.54295	0.16922	62.30366
1700	13.8268	1.1336	0.3500	15.31045	0.23166	66.08948
1600	19.1560	1.7801	0.5000	21.43611	0.30390	70.53599
1500	25.9531	2.6547	0.7000	29.30780	0.38872	75.39553

$$Oil\ Zone\ Cum.\ Prod. = N_p\ [\ B_o\ +\ (\ R_{poz}\ -\ R_s\ )\ B_g\ ]$$

(8) Time Years	(9) Time Interval Days	(10) Avg. Press. Drop Psi	(11) $\Delta\bar{p}\,\Delta t$ Psi-Days	(12) $\Sigma\,\Delta\bar{p}\,\Delta t$ Psi-Days	(13) $\dfrac{\Sigma\,\Delta\bar{p}\,\Delta t}{D}$ x 10^{-6}	(14) W_e MMRB (Model)	(15) W_e MMRB (Mt bal.)
0.50	182.5	50	9125.0	9125.0	0.28657	0.0508	0.0506
1.35	310.2	150	46537.5	55662.5	0.79366	0.3098	0.3094
2.60	456.3	250	114062.5	169725.0	1.47239	0.9445	0.9439
4.20	584.0	350	204400.0	374125.0	2.31090	2.0820	2.0811
6.00	657.0	450	295650.0	669775.0	2.89117	3.7273	3.7260
8.25	821.3	550	451687.5	1121463.0	3.69020	6.2409	6.2393
11.00	1003.8	650	652437.5	1773900.0	4.56343	9.8718	9.8696

$$N_a = \left\{\, N_p\,[\,B_o + B_g\,(R_{poz} - R_s)\,] + G_{pc}\,B_g + W_p\,B_w \right\} / D$$

$$D = B_g\,(R_{si} - R_s) - (B_{oi} - B_o) + m\,\frac{B_{oi}}{B_{gi}}\,(B_g - B_{gi})$$

$$N_a = N + W_e\,/\,D$$

$$W_e = C_s\ \Sigma\,\Delta\bar{p}\,\Delta t \qquad (Schilthuis)$$

$$N_a = N + C_s\left[\frac{\Sigma\,\Delta\bar{p}\,\Delta t}{D}\right]$$

Thus, if Na is plotted versus $(\,\Sigma\,\Delta\bar{p}\,\Delta t\,)\,/\,D$ (Figure 15-5), and a reasonably straight line results, the assumptions of the Schilthuis model (see "Water Drive" chapter) conform closely enough to the reservoir-aquifer system being considered. In this case, the y-intercept can be determined and is equal to N. Also, the slope of the straight line is equal to the Schilthuis aquifer constant, C_s. Of course, another aquifer model such as that by van Everdingen and Hurst could also be used, but the Schilthuis model was chosen due to its simplicity to illustrate the combination-drive analysis procedure.

Equations 15-27 through 15-30 will be solved at each of the points in history using Table 15-4. In Table 15-4, notice that column 2 contains the cumulative oil-zone withdrawals (not including the water). Column 3 represents cumulative gas-cap production, and column 4 is cumulative water production. Column 5 is the sum of columns 2, 3, and 4 and is the numerator of the apparent oil in place Equation 15-27. Column 6 is the denominator of Equation 15-27. Column 7 is column 5 divided by column 6, and is the apparent oil in place at each of the points in history. Notice that the apparent oil in place is growing with time, which is an indication of water encroachment.

The second part of Table 15-4, columns 8 to 13, is devoted to computing $\Sigma \, \Delta \bar{p} \, \Delta t \, / \, D$. First, the time interval in days is calculated between each of the static reservoir pressures being considered. These incremental times, or Δt's, are in column 9. Column 10 contains the average pressure drop (between the original pressure and the current average static pressure) for the time interval being considered. As explained in the "Water Drive" chapter, these $\Delta \bar{p}$'s are computed as:

$$\Delta \bar{p}_1 = p_i - \frac{(p_i + p_1)}{2} = \frac{p_i - p_1}{2}$$

$$\Delta \bar{p}_2 = p_i - \frac{(p_1 + p_2)}{2}$$

$$\Delta \bar{p}_j = p_i - \frac{(p_{j-1} + p_j)}{2} \quad for \ j \geq 2$$

Next, $\Delta \bar{p} \, \Delta t$, column 11, is calculated by multiplying column 9 by column 10. Column 12 is obtained by adding the current value in column 11 to all of the previous values in column 11. Column 13 results by dividing column 12 by column 6. At this point, N_a is plotted versus $(\Sigma \, \Delta \bar{p} \, \Delta t) \, / \, D$; or column 7 is plotted against column 13. This graph is shown in Figure 15-5.

Notice that the points are reasonably collinear. Thus, the plot can be extrapolated back to the y-intercept to obtain N, which is equal to 50 MMSTB. We are fortunate because the material balance calculated N agrees with that calculated through volumetric means. Consequently, Table 15-3, which depended on the value of N, does not have to be reworked. Also, notice that in Figure 15-5, the slope of the line is equal to C_s (productivity index of the aquifer). Here, $C_s = 5.565$ bbl/D/psi.

Now that C_s is in hand, the Schilthuis Equation 15-29 can be used to calculate water encroachment at each of the history points. Hence, C_s is multiplied by each value in column 12, and the resulting W_e is placed in column 14. For comparison purposes, the material balance calculated W_e from Table 15-3 has been placed in column 15.

Table 15-5 is used to calculate the drive indices with Equation 15-4, 15-5, and 15-6 at each of the history points. Columns 2, 3, and 4 represent the numerators of the SDI, GDI, and WDI, respectively. Oil and gas cumulative withdrawals from the oil zone are contained in column 5, which is the denominator of the drive index equations. Then column 6 is the SDI, column 7 is the GDI, and column 8 is the WDI.

Notice that the SDI is the highest, especially at the start of production. However with the passing

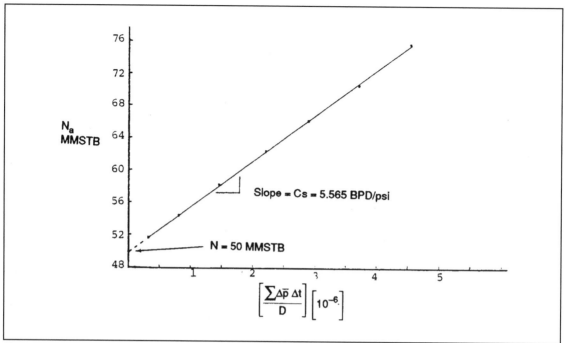

Fig. 15-5. Combination drive problem. Material balance - Schilthuis aquifer model determination of N and C_s.

of time, the WDI increased dramatically. It should be evident that water drive is an ever-increasing source of energy in this reservoir. The value of the GDI depends largely on the gas-cap factor (m), the amount of gas-cap production, and gas-cap injection.

Tables 15-6 and 15-7 are devoted to the calculation of the average gas saturation in the gas-cap-invaded zone. The fractional flow Equation 15-13 will be used in this analysis. First Equation 15-13 is written as:

$$f_g = \frac{1.0 - (Coeff_g)\, k_{ro}}{1.0 + \left(\dfrac{\mu_g}{\mu_o}\right)\left(\dfrac{k_o}{k_g}\right)} \tag{15-31}$$

where:

$$Coeff_g = \frac{0.488\ k\ A\ Sin\ \alpha\ (\gamma_o - \gamma_g)}{q_{gt}\ \mu_o} \tag{15-32}$$

The $Coeff_g$ is evaluated in Table 15-6 at each of the static pressures being considered in history. The gas-cap-expansion rate is computed in columns 2 through 5 using Equation 15-15. The area of the current gas/ oil contact, column 6, is evaluated based on gas-cap-invaded zone pore volume (GCE). GCE is evaluated in Table 15-10. This calculation (gas/oil area) is a simple linear interpolation between the original gas/oil contact area and the original water/oil contact area based on the GCE and the original oil zone pore volume. The rest of the columns in Table 15-6 finish the calculation of $Coeff_g$. Note that column 11 is the gas specific gravity (referred to water) obtained using:

$$\gamma_g = 0.00752\ (M_g)/B_g = (0.00752)(22)/B_g \qquad (15\text{-}33)$$

where:

γ_g = gas specific gravity at reservoir conditions referred to water,
B_g = gas-formation volume factor, bbl/MSCF.

Table 15-5
Calculations of the Reservoir Drive Indices

(1) Pressure PSIA	(2) SDI Numerator MMRB	(3) GDI Numerator MMRB	(4) WDI Numerator MMRB	(5) Oil Zone Cum. Prod. MMRB	(6) SDI (2)/(5)	(7) GDI (3)/(5)	(8) WDI (4)/(5)
2100	1.1031	0.4404	0.0466	1.5901	0.69376	0.27694	0.02930
2000	2.4628	0.9316	0.2594	3.6538	0.67403	0.25497	0.07100
1900	4.0341	1.4561	0.8589	6.3491	0.63539	0.22934	0.13528
1800	5.9316	1.9278	1.9311	9.7905	0.60586	0.19690	0.19724
1700	8.0994	2.3514	3.3760	13.8268	0.58578	0.17006	0.24417
1600	10.6378	2.7789	5.7393	19.1560	0.55532	0.14507	0.29961
1500	13.6161	3.1674	9.1696	25.9531	0.52464	0.12204	0.35331

$$SDI = \frac{N\,[\,B_o - B_{oi} + B_g\,(R_{si} - R_s)\,]}{Cum.\ Oil\ Zone\ Withdrawals}$$

$$GDI = \frac{G\,(B_g - B_{gi}) - G_{pc}\,B_g + Q_{gin}\,B_g}{Cum.\ Oil\ Zone\ Withdrawals}$$

$$WDI = \frac{W_e - W_p\,B_w + W_{in}}{Cum.\ Oil\ Zone\ Withdrawals}$$

Cum. Oil Zone
Withdrawals $= N_p\,[\,B_o + (R_{poz} - R_s)\,B_g\,]$

Table 15-6
Calculation of "Coeff$_g$" in f$_g$ Equation 15-31

(1)	(2)	(3)	(4)	(5)	(6)
	Net Gas Cap Expans.	Incr. Net Gas Cap Expans.,	Incremental Time	Gas Cap Expansion	Gas/Oil Area
Pressure psia	MMRB	BBLS	Days	Rate, BPD	Sq. Ft.
2100	0.4404	440356.9	182.5	2412.9	15,217,000
2000	0.9316	491257.6	310.2	1583.4	15,336,080
1900	1.4561	524462.3	456.3	1149.5	15,464,620
1800	1.9278	471686.5	584.0	807.7	15,598,810
1700	2.3514	423617.6	657.0	644.8	15,715,220
1600	2.7789	427563.7	821.3	520.6	15,818,850
1500	3.1674	388454.9	1003.8	387.0	15,922,150

Net Gas Cap Expansion = $G(B_g - B_{gi}) - G_{pc}B_g$ (no injection)

Gas Cap Expansion Rate = $\Delta [(B_g - B_{gi}) - G_{pc}B_g] / \Delta t$ = (3) / (4)

(7) (6) / (5) A/q_{gt} Ft^2/B/D	(8) μ_o C_p	(9) (7)/(8) $A/(q_{gt}\,\mu_o)$	(10) $\gamma_o =$ γ_{otank}/B_o	(11) $\gamma_g = \left[\dfrac{0.00752\,M_g}{B_g} \right]$	(12) (10) - (11) γ_o - γ_g	(13) Coeff$_g$
6306.48	0.71	8882.4	0.6266	0.1387	0.4838	104.8647
9685.39	0.72	13451.9	0.6294	0.1332	0.4962	162.8614
13453.27	0.73	18429.1	0.6364	0.1270	0.5094	229.0591
19313.05	0.74	26098.7	0.6436	0.1205	0.5231	333.1146
24373.16	0.75	32497.5	0.6509	0.1135	0.5374	426.1296
30384.33	0.76	39979.4	0.6584	0.1066	0.5518	538.3220
41142.13	0.77	53431.3	0.6667	0.0994	0.5672	739.5007

$$\gamma_g = \frac{0.00752\,M_g}{B_g} = \frac{(0.00752)(22)}{B_g} = 0.165 / B_g$$

$$\gamma_o = \gamma_{o\,tank} / B_o = 0.8 / B_o$$

$$f_g = \frac{1.0 - (Coeff_g)(k_{ro})}{1.0 + \left(\dfrac{\mu_g}{\mu_o}\right)\left(\dfrac{k_o}{k_g}\right)}$$

$$Coeff_g = \frac{0.488 \; k \, A \, \sin\alpha \; (\gamma_o - \gamma_g)}{q_{gt}\,\mu_o}$$

$$Coeff_g = \frac{(0.488)(0.1)(A)(0.5)(0.8/B_o - 0.165/B_g)}{q_{gt}\,\mu_o}$$

$$Coeff_g = \frac{(0.0244)(A)(0.8/B_o - 0.165/B_g)}{q_{gt}\,\mu_o}$$

$$Coeff_g = [0.0244][column(9)][column(12)]$$

Table 15-7
Calculation of F_g vs. S_g

(1) S_g	(2) k_{ro}	(3) k_g/k_o	(4) CFG*k_{ro}	(5) 1-(4)	(6) $\{\mu_g/[\mu_o * Col(3)]\}$	(7) f_g	(8) S_g/f_g
Pressure = 2100;		Coeff$_g$ = CFG = 104.86470;			μ_g = 0.014;	μ_o = 0.71	
0.64	0.0009	303.142	0.0920	0.9080	0.00007	0.9080	0.7048
0.65	0.0006	399.998	0.0671	0.9329	0.00005	0.9329	0.6968
0.66	0.0004	527.800	0.0463	0.9537	0.00004	0.9537	0.6920 *
0.67	0.0003	696.436	0.0319	0.9681	0.00003	0.9681	0.6921
0.68	0.0002	918.955	0.0220	0.9780	0.00002	0.9780	0.6953
Pressure = 2000;		Coeff$_g$ = CFG = 162.86140;			μ_g = 0.013;	μ_o = 0.72	
0.66	0.0004	527.800	0.0719	0.9281	0.00003	0.9281	0.7111
0.67	0.0003	696.436	0.0496	0.9504	0.00003	0.9504	0.7050
0.68	0.0002	918.954	0.0342	0.9658	0.00002	0.9658	0.7041 *
0.69	0.0001	1212.565	0.0236	0.9764	0.00001	0.9764	0.7067
0.70	0.0001	1599.992	0.0163	0.9837	0.00001	0.9837	0.7116
Pressure = 1900;		Coeff$_g$ = CFG = 229.05910;			μ_g = 0.013;	μ_o = 0.73	
0.66	0.0004	527.800	0.1011	0.8989	0.00003	0.8989	0.7342
0.67	0.0003	696.436	0.0698	0.9302	0.00003	0.9303	0.7202
0.68	0.0002	918.954	0.0481	0.9519	0.00002	0.9519	0.7144
0.69	0.0001	1212.565	0.0332	0.9668	0.00001	0.9668	0.7137 *
0.70	0.0001	1599.992	0.0229	0.9771	0.00001	0.9771	0.7164
Pressure = 1800;		Coeff$_g$ = CFG = 333.11460;			μ_g = 0.012;	μ_o = 0.74	
0.68	0.0002	918.954	0.0700	0.9300	0.00002	0.9300	0.7312
0.69	0.0001	1212.565	0.0483	0.9517	0.00001	0.9517	0.7250
0.70	0.0001	1599.989	0.0333	0.9667	0.00001	0.9667	0.7241 *
0.71	0.0001	2111.198	0.0231	0.9769	0.00001	0.9769	0.7268
0.72	0.0000	2785.745	0.0160	0.9840	0.00001	0.9840	0.7317
Pressure = 1700;		Coeff$_g$ = CFG = 426.12960;			μ_g = 0.011;	μ_o = 0.75	
0.68	0.0002	918.954	0.0895	0.9105	0.00002	0.9105	0.7468
0.69	0.0001	1212.565	0.0618	0.9382	0.00001	0.9383	0.7354
0.70	0.0001	1599.989	0.0426	0.9574	0.00001	0.9574	0.7311 *
0.71	0.0001	2111.198	0.0295	0.9705	0.00001	0.9705	0.7316
0.72	0.0000	2785.745	0.0205	0.9795	0.00001	0.9795	0.7350
Pressure = 1600;		Coeff$_g$ = CFG = 538.32200;			μ_g = 0.011;	μ_o = 0.76	
0.70	0.0001	1599.989	0.0538	0.9462	0.00001	0.9462	0.7398
0.71	0.0001	2111.198	0.0373	0.9627	0.00001	0.9627	0.7375 *
0.72	0.0000	2785.742	0.0259	0.9741	0.00001	0.9741	0.7391
0.73	0.0000	3675.807	0.0179	0.9821	0.00000	0.9821	0.7433
0.74	0.0000	4850.262	0.0124	0.9876	0.00000	0.9876	0.7493
Pressure = 1500;		Coeff$_g$ = CFG = 739.50070;			μ_g = 0.010;	μ_o = 0.77	
0.70	0.0001	1599.989	0.0740	0.9260	0.00001	0.9261	0.7559
0.71	0.0001	2111.198	0.0513	0.9487	0.00001	0.9488	0.7483
0.72	0.0000	2785.742	0.0355	0.9645	0.00000	0.9645	0.7465 *
0.73	0.0000	3675.807	0.0246	0.9754	0.00000	0.9754	0.7484
0.74	0.0000	4850.262	0.0171	0.9829	0.00000	0.9829	0.7528

$$f_g = \frac{1.0 - (Coeff_g)(k_{ro})}{1.0 + \mu_g/[(\mu_o)\left(\dfrac{k_g}{k_o}\right)]} \quad ; \qquad S_{g\ avg} = Min\ (S_g/f_g)$$

Table 15-7 makes the calculation of gas fractional flow versus saturation. There is a separate table for each of the historical pressures. For a particular pressure, such as p = 2100 psia, column 1 is the gas saturation being considered. Column 2 contains k_{ro} and column 3 is k_g/k_o. Column 4 is the $Coeff_g$ times k_{ro}. The numerator of the fractional flow Equation 15-31 is found in Column 5, which is 1 minus column 4. Column 6 begins the calculation of the denominator. Then gas fractional flow (column 7) is obtained by adding 1 to column 6 and dividing this result into column 5. So, for each of the saturations considered in column 1, the corresponding f_g is found in column 7.

To find the average gas saturation in the gas-cap-invaded zone, S_{gavg}, the "automated solution", Equation 14-7, from the "Gas Cap Drive" chapter is used:

$$S_{gavg} = min \ (S_g / f_g) \tag{15-34}$$

This relationship is investigated in column 8 by dividing column 1 by column 7. The minimum value is found (the one with an *), and this is equal to the average gas saturation in the gas-cap-invaded zone. S_{gavg} could have also been determined graphically as previously discussed.

Notice what is happening to S_{gavg} as the pressure declines in this reservoir. It is increasing. Why? The principal reason for this behavior is the decreasing rate of gas-cap expansion (Column 5 of Table 15-6). In immiscible displacements that are using gravity (such as a gas cap expanding downward), the slower the rate is, the more efficient is the displacement. If the S_{gavg} is increasing with time, this means that the remaining oil saturation in the gas-cap-invaded zone is decreasing: $S_{org} = 1 - S_{gavg} - S_{wc}$. Here, the remaining oil saturation has decreased from 0.108 to 0.054.

Tables 15-8 and 15-9 are used to calculate the average water saturation in the water-invaded zone. Once again the fractional flow equation is used. Recall that Equation 15-16 is the fractional flow equation for water displacing oil. Here we write it as:

$$f_w = \frac{1.0 \ - \ (\ Coeff_w \ ) \ k_{ro}}{1.0 \ + \ \left(\dfrac{\mu_w}{\mu_o}\right)\left(\dfrac{k_o}{k_w}\right)} \tag{15-35}$$

where:

$$Coeff_w = \frac{0.488 \ k \ A \ \sin \alpha \ (\ \gamma_w - \gamma_o \ )}{q_{wt} \ \mu_o} \tag{15-36}$$

Table 15-8 is concerned with calculating $Coeff_w$, and Table 15-9 computes the f_w versus S_w relationship at each pressure. Then, Equation 15-20 is used to determine S_{wavg}, the average water saturation within the water invaded zone.

$$S_{wavg} = S_{wi} + Min \ [(S_w - S_{wi}) / f_w] \tag{15-20}$$

Because Table 15-8 and 15-9 are analogous respectively to Tables 15-6 and 15-7, discussion of the calculation details has been omitted. Notice the column 5 of Table 15-8 is the net water influx rate. The rate of water encroachment is increasing with time, which is a typical expression of a water drive. The effect of increasing influx rate is a decreasing average water saturation in the water invaded zone. This is seen in Table 15-9. The S_{wavg} at each pressure is the value in column 8 with an * beside it. For $\bar{p}$ =2100, S_{wavg} = 0.5933. And for $\bar{p}$ = 1500, S_{wavg} = 0.5830. The remaining oil saturation in the water invaded zone is increasing with time. Recall that $S_{orw} = 1 - S_{wavg}$.

Table 15-10 computes the oil saturation in the uninvaded oil zone using Equations 15-22 through 15-24. Note that column 2 is the net gas cap expansion: $G(B_g - B_{gi}) - G_{pc}B_g$, and column 7 is the net water encroachment: $W_e - W_pB_w$. Column 19 is the oil saturation within the uninvaded zone. Notice that it is decreasing with time, which indicates free gas buildup.

Table 15-8
Calculation of "Coeff$_w$" in the F$_w$ Equation 15-35

(1) PRESSURE PSIA	(2) NET WATER INFLUX MMRB	(3) INCREMENTAL WATER INFLUX RES BBLS	(4) INCREMENTAL TIME DAYS	(5) WATER INFLUX RATE RB/DAY	(6) AREA SQ FT
2100	0.0466	46583.7	182.5	255.3	30,434,000
2000	0.2594	212816.6	310.2	686.0	30,411,830
1900	0.8589	599483.1	456.3	1313.9	30,309,900
1800	1.9311	1072185.0	584.0	1835.9	30,022,490
1700	3.3760	1444965.0	657.0	2199.3	29,508,350
1600	5.7393	2363225.0	821.3	2877.6	28,833,610
1500	9.1696	3430345.0	1003.8	3417.5	27,664,130

Net Water Influx = $W_e - W_pB_w$ (no injection)

Water Influx Rate = $q_{wt} = [\Delta (W_e - W_pB_w)] / \Delta t$

(7) AREA/RATE SQ FT/B/D	(8) μ_o cp	(9) COL(7)/COL(8) A/(qwt μ_o)	(10) 0.8/Bo	(11) 1.074 − COL(10)	(12) COEFF$_w$
119230.7	0.71	167930.5	0.6226	0.4514	1849.7440
44335.2	0.72	61576.7	0.6294	0.4446	667.9603
23068.0	0.73	31600.0	0.6364	0.4376	337.3798
16352.7	0.74	22098.3	0.6436	0.4304	232.0684
13416.9	0.75	17889.2	0.6509	0.4231	184.6664
10020.0	0.76	13184.3	0.6584	0.4156	133.6852
8094.8	0.77	10512.7	0.6667	0.4073	104.4850

$$\gamma_w = \gamma_{w\,surf} / B_w = 1.074 / 1 = 1.074$$

$$\gamma_o = \gamma_{o\,tank} / B_o = 0.8 / B_o$$

$$f_w = \frac{1.0 - (Coeff_w)(k_{ro})}{1.0 + \left(\dfrac{\mu_w}{\mu_o}\right)\left(\dfrac{k_o}{k_w}\right)}$$

$$Coeff_w = \frac{0.488 \; k \; A \; \sin\alpha \; (\gamma_w - \gamma_o)}{q_{wt}\,\mu_o}$$

$$Coeff_w = \frac{(0.488)(0.1)(A)(0.5)(1.074 - 0.8/B_o)}{q_{wt}\,\mu_o}$$

$$Coeff_w = \frac{(0.0244)(A)(1.074 - 0.8/B_o)}{q_{wt}\,\mu_o} = [0.0244][col.(9)][col.(11)]$$

Table 15-9
Calculation of F_W vs. S_W

(1)	(2)	(3)	(4)	(5)	(6)	(7)	(8) 0.2 +
S_w	k_{ro}	k_{rw}	$[CFW][k_{ro}]$	1.0 - (4)	$(\mu_w * k_{ro})/(\mu_o * k_{rw})$	f_w	$(S_w - 0.2)/f_w$
Pressure = 2100;		Coeff$_w$ = CFW = 1849.74;			$\mu_w = 0.4$;	$\mu_o = 0.71$	
0.57	0.0003	0.1241	0.6153	0.3847	0.0015	0.3842	1.1632
0.58	0.0002	0.1330	0.3700	0.6300	0.0008	0.6295	0.8036
0.59	0.0000	0.1412	0.0083	0.9917	0.0000	0.9917	0.5933*
0.60	0.0000	0.1500	0.0002	0.9998	0.0000	0.9998	0.6001
Pressure = 2000;		Coeff$_w$ = CFW = 667.96;			$\mu_w = 0.4$;	$\mu_o = 0.72$	
0.57	0.0003	0.1241	0.2222	0.7778	0.0015	0.7767	0.6764
0.58	0.0002	0.1330	0.1336	0.8664	0.0008	0.8657	0.6390
0.59	0.0000	0.1412	0.0030	0.9970	0.0000	0.9970	0.5912*
0.60	0.0000	0.1500	0.0001	0.9999	0.0000	0.9999	0.6000
Pressure = 1900;		Coeff$_w$ = CFW = 337.38;			$\mu_w = 0.4$;	$\mu_o = 0.73$	
0.57	0.0003	0.1241	0.1122	0.8878	0.0015	0.8865	0.6174
0.58	0.0002	0.1330	0.0675	0.9325	0.0008	0.9318	0.6078
0.59	0.0000	0.1412	0.0015	0.9985	0.0000	0.9985	0.5906*
0.60	0.0000	0.1500	0.0000	1.0000	0.0000	1.0000	0.6000
Pressure = 1800;		Coeff$_w$ = CFW = 232.07;			$\mu_w = 0.4$;	$\mu_o = 0.74$	
0.57	0.0003	0.1241	0.0772	0.9228	0.0014	0.9215	0.6015
0.58	0.0002	0.1330	0.0464	0.9536	0.0008	0.9528	0.5998
0.59	0.0000	0.1412	0.0010	0.9990	0.0000	0.9989	0.5904*
0.60	0.0000	0.1500	0.0000	1.0000	0.0000	1.0000	0.6000
Pressure = 1700;		Coeff$_w$ = CFW = 184.67;			$\mu_w = 0.4$;	$\mu_o = 0.75$	
0.57	0.0003	0.1241	0.0614	0.9386	0.0014	0.9372	0.5948
0.58	0.0002	0.1330	0.0369	0.9631	0.0008	0.9623	0.5947
0.59	0.0000	0.1412	0.0008	0.9992	0.0000	0.9992	0.5903*
0.60	0.0000	0.1500	0.0000	1.0000	0.0000	1.0000	0.6000
Pressure = 1600;		Coeff$_w$ = CFW = 133.69;			$\mu_w = 0.4$;	$\mu_o = 0.76$	
0.55	0.0009	0.1080	0.1230	0.8770	0.0045	0.8731	0.6009
0.56	0.0006	0.1158	0.0740	0.9260	0.0025	0.9237	0.5897
0.57	0.0003	0.1241	0.0445	0.9555	0.0014	0.9542	0.5878*
0.58	0.0002	0.1330	0.0267	0.9733	0.0008	0.9725	0.5907
0.59	0.0000	0.1412	0.0006	0.9994	0.0000	0.9994	0.5902
Pressure = 1500;		Coeff$_w$ = CFW = 104.48;			$\mu_w = 0.4$;	$\mu_o = 0.77$	
0.55	0.0009	0.1080	0.0691	0.9039	0.0044	0.8999	0.5889
0.56	0.0006	0.1158	0.0578	0.9422	0.0025	0.9399	0.5830*
0.57	0.0003	0.1241	0.0348	0.9652	0.0014	0.9639	0.5839
0.58	0.0002	0.1330	0.0209	0.9791	0.0008	0.9783	0.5884
0.59	0.0000	0.1412	0.0005	0.9995	0.0000	0.9995	0.5902

$$f_w = \frac{1.0 - (Coeff_w)(k_{ro})}{1.0 + \left(\dfrac{\mu_w}{\mu_o}\right)\left(\dfrac{k_{ro}}{k_{rw}}\right)}; \quad S_{wavg} = 0.2 + Min\left[\,(S_w - 0.2)/f_w\,\right]$$

Table 15-10
Calculations of Oil Saturation in Uninvaded Oil Zone

(1)	(2) GAS CAP EXPANSION	(3)	(4) GCE	(5) $S_{org} =$	(6) (4)*(5)	(7)
PRESSURE PSIA	MMRB	S_{gavg} FRAC	(2)/(3) MMRB	$1 - S_{gavg} - S_{wc}$ FRAC	(GCE)(S_{org}) MMRB	$W_e - W_p B_w$ MMRB
2100	0.4404	0.6920	0.6363	0.1080	0.0687	0.0466
2000	0.9316	0.7041	1.3231	0.0959	0.1269	0.2594
1900	1.4561	0.7137	2.0402	0.0863	0.1761	0.8589
1800	1.9278	0.7241	2.6622	0.0759	0.2020	1.9311
1700	2.3514	0.7311	3.2160	0.0689	0.2214	3.3760
1600	2.7789	0.7375	3.7680	0.0625	0.2355	5.7393
1500	3.1674	0.7465	4.2429	0.0535	0.2269	9.1696

(8)	(9)	(10) $S_{orw} =$	(11) (9)*(10)	(12)	(13)	(14)
S_{wavg} FRACTION	WCM MMRB	$1 - S_{wavg}$ FRACTION	(S_{orw})(WCM) MMRB	(11) + (6) MMRB	N_p MMB	$N - N_p$ MMRB
0.5933	0.1185	0.4067	0.0482	0.1169	1.1617	48.8383
0.5912	0.6631	0.4088	0.2711	0.3980	2.4577	47.5423
0.5906	2.1989	0.4094	0.9002	1.0763	4.3391	45.6609
0.5904	4.9462	0.4096	2.0259	2.2279	5.6093	44.3907
0.5903	8.6499	0.4097	3.5438	3.7652	7.1542	42.8458
0.5878	14.8009	0.4122	6.1015	6.3369	8.8336	41.1664
0.5830	23.9395	0.4170	9.9820	10.2089	10.8233	39.1767

(15)	(16)	(17) (4) + (9)	(18) PV_i - (17)	(19)
(14) * B_o MMRB	(15) - (12) MMRB	(GCE + WCM) MMRB	PV OIL ZONE MMRB	S_o = (16) / (18) FRACTION
62.7572	62.6403	0.7548	80.5577	0.7776
60.4263	60.0283	1.9863	79.3262	0.7567
57.3958	56.3194	4.2391	77.0734	0.7307
55.1776	52.9497	7.6085	73.7040	0.7184
52.6575	48.8923	11.8659	69.4466	0.7040
50.0172	43.6803	18.5689	62.7436	0.6962
47.0120	36.8031	28.1824	53.1301	0.6927

$$s_{org} = 1 - S_{gavg} - S_{wc}; \quad S_{orw} = 1 - S_{wavg}$$

$$PV_i = \frac{NB_{oi}}{1 - S_{wc}} = \frac{(50)(1.301)}{(1 - 0.2)} = 81.3125 \ MMRB$$

$$GCE = \frac{G(B_g - B_{gi}) - G_{pc} B_g + Q_{gin} B_g}{S_{gavg}}; \quad WCM = \frac{W_e - W_p B_w + W_{in}}{S_{wavg} - S_{wc}}$$

$$S_o = \frac{(N - N_p) B_o - [(GCE)(S_{org}) + (WCM)(S_{orw})]}{PV_i - (GCE + WCM)}$$

The use of Equation 15-24 to calculate the uninvaded-oil-zone oil saturation assumes negligible gas/oil segregation throughout reservoir history. For the problem here this may not be entirely true. The pressure and gas saturation where significant gas/oil segregation can start can be investigated with the techniques given in the "Gas-Cap Drive" chapter. The only difference is that for a combination drive, the solution gas drive saturation equation should not be used to calculate free-gas saturation buildup with time. Equation 15-24 should be used (as in Table 15-10) to first calculate oil saturation, S_o. Then $S_g = 1 - S_o - S_{wc}$.

For this problem, segregation of the free gas and oil in the uninvaded zone could begin after free-gas saturation exceeded 9%. Therefore, some gas segregation might have occurred during the last two time steps. This effect has not been considered. The problem was felt to be sufficiently complex.

To adequately handle free gas/oil segregation, the time steps should be made much smaller. It is likely that the needed time-step size should be 180 days or less (to avoid oscillations in calculated results).

The necessary modifications to be made to handle gas/oil segregation (or gas percolation updip) are:

(1) Perform the analysis to determine the free-gas saturation (and static pressure) at which significant segregation begins.

(2) Make future time steps from the segregation point small enough to prevent oscillations in computed results.

(3) Set the oil saturation in the uninvaded oil zone at the value required to cause gas segregation, and this S_o value normally is maintained constant for all future time steps.

(4) Allocate the excess gas saturation to the gas cap as a pseudo gas injection volume.

(5) Re-evaluate the gas-cap growth as a result of the added gas.

(6) Re-evaluate the drive indices (Table 15-5). The cumulative pseudo gas injection volume, $(Q_{gi,seg})(B_g)$, should be added to the GDI numerator and subtracted from the SDI numerator.

(7) Re-evaluate gas-cap-invaded-zone average gas saturation (Tables 15-6 and 15-7).

(8) Re-evaluate GCE and Table 15-10.

(9) These adjustments are needed for each future time step.

Table 15-11 uses Equation 15-11 to investigate the variations in recovery factor in the water invaded zone, the gas cap invaded zone, and the undisturbed oil zone. Each of these recovery factors is calculated as if the associated drive mechanism were acting alone in the reservoir.

Table 15-12 uses the results of Table 15-11 along with Equation 15-12 to compute the effective (or modified) recovery efficiency.

Recall that this calculation involves weighting each of the recovery factors by the corresponding drive index. Actually, the recovery factors computed in Table 15-11 are idealistic since they were determined assuming no outside interference from the other drive mechanisms. However, when the individual recovery factors are weighted by the drive indices, which are measures of the fraction of total withdrawals due to a particular mechanism, then a more realistic ultimate recovery factor can be determined. Notice that the effective recovery factor is a measure of the ultimate fractional oil recovery if the drive indices were to remain constant from that point forward.

Considering Table 15-11, the water-drive recovery factors are presented in Column 6, gas-cap-drive recovery factors in Column 10, and undisturbed oil-zone recovery factors in Column 14. The water drive recovery efficiencies have decreased with time due to the increasing rate of water influx; while the gas-cap-drive recovery factor has increased due to the decrease in the expansion rate of the gas cap. The undisturbed oil-zone recovery factor is relatively insignificant compared to the other two. Most of the cumulative oil production has been displaced from the two invaded zones. Table 15-10, column 17 presents the combined pore volume of the two invaded zones, while column 18 contains the pore volume of the undisturbed oil zone. Notice that the displacement processes at work in this reservoir are significant.

Table 15-11
Calculations of Individual Recovery Factors

(1)	(2)	(3)	(4)	(5) INCR (4) [0.6149 - (4)]	(6)
PRESSURE	B_o	S_{orw}	S_{orw}/B_o		(RF)WD
2200	1.3010	0.8000	0.6149	0.0000	0.0000
2100	1.2850	0.4067	0.3165	0.2984	0.4852
2000	1.2710	0.4088	0.3217	0.2933	0.4769
1900	1.2570	0.4094	0.3257	0.2892	0.4703
1800	1.2430	0.4096	0.3295	0.2854	0.4641
1700	1.2290	0.4097	0.3334	0.2816	0.4579
1600	1.2150	0.4122	0.3393	0.2756	0.4482
1500	1.2000	0.4170	0.3475	0.2674	0.4349

(7)	(8)	(9) INCR (8) [0.6149 - (8)]	(10)
S_{org}	S_{org}/B_o		(RF)GC
0.8000	0.6149	0.0000	0.0000
0.1080	0.0840	0.5309	0.8634
0.0959	0.0755	0.5395	0.8773
0.0863	0.0687	0.5462	0.8883
0.0759	0.0611	0.5539	0.9007
0.0689	0.0560	0.5589	0.9089
0.0625	0.0514	0.5635	0.9164
0.0535	0.0446	0.5703	0.9275

(11)	(12)	(13) INCR (12) [0.6149 - (12)]	(14)
S_{ooz}	S_{ooz}/B_o		(RF)OZ
0.8000	0.6149	0.0000	0.0000
0.7776	0.6051	0.0098	0.0159
0.7567	0.5954	0.0195	0.0318
0.7307	0.5813	0.0336	0.0546
0.7184	0.5780	0.0369	0.0601
0.7040	0.5728	0.0421	0.0684
0.6962	0.5730	0.0419	0.0682
0.6927	0.5772	0.0377	0.0612

$$RF = \frac{(S_o/B_o)_i - (S_o/B_o)}{(S_o/B_o)_i} = \frac{0.6149 - (S_o/B_o)}{0.6149}$$

Table 15-12
Calculations of Effective Recovery Factor

(1) PRESSURE	(2) WDI	(3) RF(WD)	(4) (2)*(3)	(5) GDI	(6) RF(GC)	(7) (5)*(6)
2100	0.0293	0.4852	0.0142	0.2769	0.8634	0.2391
2000	0.0710	0.4769	0.0339	0.2550	0.8773	0.2237
1900	0.1353	0.4703	0.0636	0.2293	0.8883	0.2037
1800	0.1972	0.4641	0.0915	0.1969	0.9007	0.1774
1700	0.2442	0.4579	0.1118	0.1701	0.9089	0.1546
1600	0.2996	0.4482	0.1343	0.1451	0.9164	0.1329
1500	0.3533	0.4349	0.1537	0.1220	0.9275	0.1132

(8) SDI	(9) RF(OZ)	(10) (8)*(9)	(11) (4) + (7) + (10) RF(EFF)
0.6938	0.0159	0.0110	0.2644
0.6740	0.0318	0.0214	0.2790
0.6354	0.0546	0.0347	0.3021
0.6059	0.0601	0.0364	0.3053
0.5858	0.0684	0.0401	0.3065
0.5553	0.0682	0.0379	0.3051
0.5246	0.0612	0.0321	0.2990

$$(RF)_{eff} = (SDI)(RF_{oz}) + (GDI)(RF_{gc}) + (WDI)(RF_{wd})$$

Column 11 of Table 15-12 presents the effective recovery factor with time. In a reservoir where recovery is mainly due to a displacement process, it has been found that the effective recovery factor usually changes with time. On the other hand, in a reservoir where recovery is due mostly to solution-gas drive, the effective recovery factor will remain fairly constant. So, there are two indicators which may be used to gauge whether oil recovery by displacement (water drive and/or gas-cap drive) is significant or not: invaded zone pore volume growth and effective recovery efficiency behavior with time.

According to Tracy[3], the following relationship, from decline curve analysis, is true for combination-drive oil reservoirs in which recovery is principally due to a displacement process:

$$Ln [(RF)_{eff}] = C_1 (PV_{unf}) + C_2 \tag{15-37}$$

where:

$(RF)_{eff}$ = effective recovery factor, fraction,
PV_{unf} = unflooded pore volume, barrels,
C_1 = equation constant, and
C_2 = second equation constant

Therefore, extrapolating to the point of zero unflooded pore volume,

$$(RF)_{ult} = e^{C_2} \tag{15-38}$$

where:

$(RF)_{ult}$ = the ultimate recovery factor for the reservoir, fraction

Equation 15-37 will be used in conjunction with the last four points in history to evaluate the constant, C_2. When possible, only the last four points are used because water drives often need some time (or Δp) to develop.

If we define:

$y = Ln\ [(RF)_{eff}]$ *and*

$x = (PV)_{unf}$

then to solve for C_2, the least squares relationship is:

$$C_2 = \frac{\Sigma x\ \Sigma xy\ -\ \Sigma y\ \Sigma x^2}{\Sigma x\ \Sigma x\ -\ 4\ \Sigma x^2} \qquad\qquad (15\text{-}39)$$

Then, using Equation 15-38, the ultimate recovery factor can be calculated. These calculations are addressed in Table 15-13. For this example problem, using the last four points, the extrapolated ultimate fractional recovery is 28.3%, or an ultimate recovery of 14.2 million stock-tank barrels.

Table 15-13
Extrapolation To Likely Ultimate Recovery

$$Ln\ [\ (RF\ )_{eff}] = C_1\ (PV_{unf}) + C_2$$

$$C_2 = (\ \Sigma x\ \Sigma xy\ -\ \Sigma y\ \Sigma x^2\ )\ /\ (\ \Sigma x\ \Sigma x\ -\ n\Sigma x^2\ )$$

where:

$y = Ln[(RF)_{eff}]$ and $x = (PV)_{unf}$
$(RF)_{eff}$ from Table 15-12, column (11); $(PV)_{unf}$ from Table 15-10, column (18).

(1)	(2)	(3) $Ln[RF_{(eff)}]$	(4) PV(UNF)	(5)	(6)
PRESSURE	RF_{eff}	Y	X	XY	X^2
1800	0.3053	-1.1865	73.704	-87.4471	5432.283
1700	0.3065	-1.1825	69.447	-82.1211	4822.886
1600	0.3051	-1.1871	62.744	-74.4845	3936.760
1500	0.2990	-1.2073	53.130	-64.1454	2822.808
SUMMATIONS =		-4.7634	259.025	-308.1981	17014.737

$$C_2 = \frac{[\ (\ 259.025\ ) \times (\ -308.198\ ) - (\ -4.7634\ ) \times (\ 17014.74\ )\ ]}{[\ (\ 259.025\ ) \times (\ 259.025\ ) - (\ 4.0\ ) \times (\ 17014.74\ )\ ]}$$

$C_2 = \text{-}1.2612$ $RF(ULT) = 0.2833$ $RECOVERY = 14.17\ MMB$

Discussion

Normally, for a "recovery-due-to-displacement" type oil reservoir, the effective recovery efficiency increases with time. The reason is simple. Early production is mostly due to expansion (SDI high and WDI low), with a recovery efficiency normally less than 25%. As time goes on, production is caused more and more by encroaching water and possibly by the expanding gas cap. Hence, the WDI is increasing with time, while the SDI decreases. Because the displacement recovery factors are usually at least 30%, the effective recovery efficiency tends to increase with time. However, in this example, something different happened. The effective recovery factors increase down to a pressure of 1700 psi but then begin to decrease slightly! In fact, the last five effective recovery efficiencies are actually all around 30%. The pore volume of the undisturbed oil zone, (Table 15-10, column 18) has continued to decrease substantially with time, so it is clear that recovery by displacement (mostly water drive) is an important drive mechanism. The fairly constant behavior of the effective recovery factor is caused by the decreasing pressure within the oil zone. The water drive is weak. With the decreasing oil zone pressure, the water influx rate is increasing, which causes the water-drive recovery efficiency to decrease. Hence, even though the WDI is increasing, its effect is nullified somewhat by the deteriorating water displacement efficiency. Further, the decreasing oil-zone pressure is causing expansion (SDI) to continue to be a significant mechanism in this reservoir. After 11 years, the SDI is still the largest drive index.

In this example, after the effective recovery factor begins to decrease, there are really not enough data points (3) to be able to extrapolate with confidence. As shown in Table 15-13, if Equation 15-37 is used with the last four points and extrapolated to ultimate recovery efficiency, then 28.3% is determined. When the effective recovery efficiency tends to stabilize, Tracy[3] suggests just averaging the last four or five values. The average of the last four values (Table 15-12) yields a recovery efficiency of 30.4%.

Gas injection would likely be quite beneficial in the example reservoir. One goal would be to help maintain the reservoir pressure which would tend to inhibit the water influx rate. Thus, the water displacement efficiency would be increased. Further, this would allow the gas cap displacement to affect more of the oil volume: a larger GCE and GDI. Of course, with the increase in gas displacement rate, the gas-displacing-oil efficiency would decrease. The combined effect of all of these mechanisms can be investigated with the concepts of this chapter.

REFERENCES

1. Cole, F. W.: *Reservoir Engineering Manual*, Gulf Publishing Company, 179 - 182.

2. Slider, H. C.: *Practical Petroleum Reservoir Engineering Methods*, Petroleum Publishing Company, Tulsa (1976) 328 - 360.

3. Tracy, G. W.: Unpublished work.

4. Pirson, S. J. *Oil Reservoir Engineering*, McGraw-Hill Book Co., New York City (1958) 694 - 710.

16 PRESSURE TRANSIENT ANALYSIS

To understand well test analysis, three basic terms must be defined: steady state, semi-steady state, and unsteady state or transient flow.

Steady-state flow exists where there is no change in density at any position within the reservoir as a function of time. Practically, this means that there is no change in pressure at any position as a function of time or cumulative production. Obviously, this situation never really exists within an actual reservoir. However, the concept of steady state is a very practical one and is the basis for Darcy's Law.

Semi-steady state (also called quasi or pseudosteady state) flow exists for conditions wherein the pressure declines linearly with time or in direct proportion to the reservoir withdrawals. Any time that the pressure declines as a linear function of time at any reservoir position, or as a direct function of the withdrawal rate, semi-steady state conditions exist.

Unsteady state or transient flow conditions exist whenever the pressure changes non-linearly with time.

INFORMATION THAT CAN BE OBTAINED FROM TRANSIENT WELL TESTS

1. Interwell flow capacity of a reservoir.

Flow capacity is the product of the permeability times pay thickness. It is directly related to the ability of a reservoir to transmit fluids. It is used in predicting the maximum rate of production from a well.

2. Static well pressure.

The static well pressure can be used as a measure of the stage of depletion of a reservoir. It is an essential feature in material balance calculations. The static well pressure is that pressure that would be measured if a well were shut-in for a long period of time without the external influence of adjacent wells.

3. Extent of well damage.

If a well has altered permeability in the near-wellbore vicinity, then a measure of the amount of change in the near well conductivity (from the virgin state) can be computed from an analysis of transient well test data. The concept of well damage or stimulation is translated into quantitative terms with the "skin effect." This is a dimensionless quantity. A positive skin effect indicates a condition of reduced permeability near the well; whereas, a negative skin usually relates to a near-wellbore region of increased permeability (relative to the interwell permeability).

4. Distance to nearest boundary.

If a fault or pinchout exists near a well, then the distance to this boundary can often be calculated through an analysis of well test data. At times multiple barriers can be seen and analyzed.

5. Fluid volume in place.

Under certain test conditions, the volume of the fluids within the drainage area of a well can be computed from an analysis of the well test data.

6. Detecting heterogeneities within the pay zone.

Such heterogeneities include man-made fractures, layered conditions, naturally fractured conditions, and lateral changes in mobility of the flowing fluids.

WHAT TYPES OF TESTS ARE USED

Although a few comments will be made concerning interference testing (multiple wells), this discussion will mainly address single well tests.

The first and the most popular test is the pressure buildup. This test involves a well which has produced for a period of time. The analysis is easier if sufficient production time has occurred for the well to reach stabilization. After the production period, the well is shut-in. The bottomhole pressures during the shut-in period are monitored and recorded. One example of a pressure buildup test is the drillstem test. This test usually involves pressures measured over a short period of time (minutes to a few hours). A standard pressure buildup test normally lasts a few tens of hours to several days.

The second type of transient test is the pressure drawdown test. This involves measuring the bottomhole pressure during a period of time while the well is produced at a constant rate. From analysis of the pressure data, reservoir information can be obtained. An example of a pressure drawdown test is the reservoir limit test. This test has as its primary objective, the determination of the pore volume within the drainage area of a well. In most instances, analysis of reservoir limit test data yields a conservative estimate of the pore volume and therefore volume of hydrocarbon in place.

The next type of well test is a pressure falloff test. This test is performed on an injection well that has been injecting into the formation for a period of time at or near a constant rate. The well is shut-in with pressure data collected from the end of the injection period through the shut-in time.

This test is analogous to a pressure buildup test on a producing well. For the falloff test, the pressure data during the shut-in, or period of falling pressures, can be analyzed to calculate essentially the same reservoir characteristics that are obtained from a pressure buildup test.

Interference testing will be described as the fourth type of test. This test usually involves two wells: an active well (which is either producing or injecting) and a shut-in pressure observation well. The primary use for this test is to prove pressure communication between the two wells. If there is pressure communication, then this indicates sufficient permeability for fluid flow to occur. Interwell kh can usually be calculated.

Another use for the data collected while interference testing is to determine the pore volume within the drainage area between the test wells. Unfortunately, if the communication between the test wells is limited, then the value of the computed pore volume may be larger than actually exists.

Although an interference test may be designed only for two wells, considerations must be made if there are other non-shut-in wells in the vicinity of the two test wells. In this situation, the production and/or injection rate of the active well must be varied in a specific way. Often the rate is varied in a cyclic manner: maximum rate for one time period (such as a day), then shut-in for a second time period; and then the cycle repeated. The pressure observation well remains shut-in, collecting pressure data. All other wells in the neighborhood of the test wells are operated at their normal rates (as close to constant as possible). These tests often last weeks or even months. So, the cost can be quite high, especially when the deferred oil production from the shut-in observation well is considered.

A fifth type of well test falls into the classification of deliverability testing. This usually involves gas wells, although some advocate that certain types of oil wells should be rate-tested in this manner. The purpose usually is not to obtain reservoir information, but to estimate a well's flow rate under stabilized conditions, and how this will vary with changing static and flowing bottomhole pressures.

In deliverability testing, there are two basic procedures that are commonly used. At times a flow-after-flow procedure is used. The well is produced at four different rates, one right after another, with each flow period being the same amount of time, e.g., 4 hours. The last rate is often maintained for an extended period of time, e.g., 72 hours.

The second kind of deliverability test is often called an "isochronal" test procedure. Once again, typically four different rates are used. However, with the isochronal test, there are shut-in periods between the flow periods. The flow periods each last the same length of time; e.g., 4 hours. And the shut-in periods are also normally of equal time lengths. The last flow period is usually allowed to continue until the well is stabilized. Deliverability tests are described in detail in the "Gas Reservoirs" chapter.

TRANSIENT FLOW THEORY

The beginning or fundamental basis of transient flow theory is the Continuity Equation. The form of this equation is dependent on the geometry of the flow system. As the flow into or out of wells is generally convergent or divergent, the theory for application to wells usually assumes radial flow.

A radial reservoir element (doughnut-shaped with internal radius "r" and external radius "r +Δr") some distance from the well is considered. The Continuity Equation compares the mass rate of flow across the internal radius with that at the external radius. If there is a difference between these two mass flow rates across the concentric rings, then with time, a net accumulation or drainage of mass within the radial element is occurring. This is stated mathematically as follows:

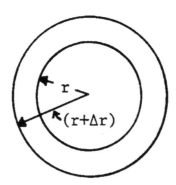

(Mass rate of flow across the inner radius) - (Mass rate of flow across the outer radius) = Mass rate of accumulation within the volume between the rings.

$$(\rho q)_r - (\rho q)_{r+\Delta r} = Volume \times (Change \ of \ density / Change \ of \ time)$$

$$= \left[(r + \Delta r)^2 - (r)^2 \right] \pi h \phi \frac{\partial \rho}{\partial t}$$

$$= \Delta r (2r + \Delta r) \pi h \phi \frac{\partial \rho}{\partial t}$$

Dividing both sides by Δr, and taking the limit as Δr shrinks to zero:

$$- \frac{\partial}{\partial r} (\rho q) = 2 \pi h \phi r \frac{\partial \rho}{\partial t} \qquad (16\text{-}1)$$

Equation 16-1 states the Continuity Equation concept in mathematical terms.

The density term (ρ) and the flow rate (q) must be expressed using a common variable. The variable traditionally used is pressure. The density term will be transformed using an equation of state. To consider liquid flow, the assumption is made that the fluid is a slightly compressible liquid (i.e., water, oil, etc.). The flow rate term (q) will be transformed using Darcy's Law.

Equation of state for a slightly compressible liquid:

Compressibility, $c = - \left[\frac{1}{V} \frac{\partial V}{\partial p} \right]_T$

If we consider a fixed mass of liquid, then the volumes may be related to density, or:

$$c = \left[\frac{1}{\partial} \frac{\partial p}{\partial p} \right]_T$$

Separating the variables and integrating:

$$\rho = \rho_{ref} \; e^{\,c\,(p - p_{ref})} \tag{16-2}$$

Derivatives of density (ρ) are needed for substitution into Equation 16-1. Therefore:

$$\frac{\partial \rho}{\partial r} = [\{\rho_{ref} \; e^{\,c\,(p - p_{ref})}] \; c \; \frac{\partial p}{\partial r} = \rho c \; \frac{\partial p}{\partial r} \tag{16-3}$$

$$\frac{\partial \rho}{\partial t} = \rho c \; \frac{\partial p}{\partial t} \tag{16-4}$$

Darcy's Law, reservoir flow rate, radial geometry:

$$q = - \frac{2\pi hk}{\mu} \; r \; \frac{\partial p}{\partial r} \tag{16-5}$$

Substituting Equations 16-2, 16-3, 16-4 and 16-5 into Equation 16-1 and simplifying:

$$\frac{\partial}{\partial r} \left(\rho r \; \frac{\partial p}{\partial r} \right) = \left(\frac{\phi \mu c}{k} \right) \left(\rho r \; \frac{\partial p}{\partial t} \right)$$

Further expanding and substituting Equation 16-3 one more time:

$$\frac{1}{r} \left[\frac{\partial}{\partial r} \left(r \; \frac{\partial p}{\partial r} \right) \right] + c \left(\frac{\partial p}{\partial r} \right)^2 = \frac{\phi \mu c}{k} \left(\frac{\partial p}{\partial t} \right)$$

For a slightly compressible liquid, "c" is quite small, so the term,

$\left[c \left(\frac{\partial p}{\partial r} \right)^2 \right]$ is negligible. Therefore:

$$\frac{\partial^2 p}{\partial r^2} + \frac{1}{r} \; \frac{\partial p}{\partial r} = \left(\frac{\phi \mu c}{k} \right) \frac{\partial p}{\partial t} \tag{16-6}$$

This is the standard form of the Continuity Equation. Some analysts refer to Equation 16-6 as the diffusivity equation.)

In order to make solution treatments of Equation 16-6 as general as possible, dimensionless variables are defined:

$$\Delta P_d = \frac{p_i - p}{\left(\dfrac{q \, \mu \, B}{2 \, \pi \, k \, h} \right)} ; \quad R_d = r \, / \, r_w ; \quad t_{dw} = \frac{kt}{\phi \, \mu \, c \, r_w^2}$$

Different types of dimensionless times used in well testing. The one shown above is based on the wellbore radius (indicated by the subscript "w"). Other dimensionless times are based on the well's drainage radius (r_e) or on the well's drainage area (A). If the type of dimensionless time is not specified (only a subscript "d" given), then this normally indicates dimensionless time with respect to the wellbore radius.

When these dimensionless variables are introduced into Equation 16-6, the Continuity Equation in terms of dimensionless variables results:

$$\frac{\partial^2 (\Delta P_d)}{\partial R_d^2} + \frac{1}{R_d} \frac{\partial (\Delta P_d)}{\partial R_d} = \frac{\partial (\Delta P_d)}{\partial t_{dw}} \tag{16-7}$$

The principal assumptions made in the development of Equation 16-7 are as follows:

1. Constant porosity

2. Constant thickness

3. Constant and isotropic permeability

4. Constant viscosity

5. Single phase flow

6. Small and constant compressibility

7. Radial flow

8. Laminar flow

9. Isothermal conditions

10. Gravity effects are negligible

11. $(c) \, (\partial p \, / \, \partial r)^2$ is negligible.

SOLUTIONS TO THE CONTINUITY EQUATION

Steady State

For steady state flow conditions, the right hand side of Equation 16-7 becomes zero because the pressure at any point within the reservoir is not changing with time. The general analytic solution is:

$$\Delta P_d = c_1 \, Ln \, (R_d) + c_2 \tag{16-8}$$

Evaluating the inner (at the wellbore) and outer (at the radius of drainage) boundary conditions, Equation 16-8 becomes:

$$\Delta P_d = \Delta P_{dw} - Ln(R_d)$$

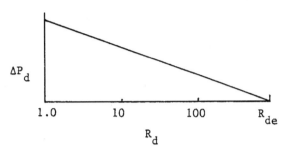

Considering the steady state equations and the graphical solution, we have the following observations:

1. Constant slope throughout the entire drainage area.

2. $(p_i - p_w)$ proportional to flow rate.

3. Reservoir flow rate is the same at each radial position.

Semi-Steady State

For semi-steady state the right-hand side of Equation 16-7 becomes a constant due to the linear change in pressure with time. The general analytic solution is then:

$$\Delta P_d = c_1 Ln(R_d) + c_2 + 0.5(R_d/R_{de})^2 \qquad (16\text{-}9)$$

Evaluating the boundary conditions, we obtain:

$$\Delta P_d = \Delta P_{dw} - [Ln(R_d) - 0.5(R_d/R_{de})^2]$$

Graphically:

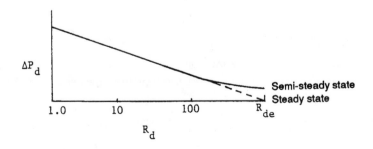

Observations:

1. The slope of the semi-steady state solution is essentially the same as the steady state solution near the wellbore.

2. The flow rate is zero at r_e. The flow rate increases with decreasing radial positions to a maximum at the wellbore. The flow rate at any point is proportional to the volume of the reservoir between the radius of interest and the outside radius, r_e; i.e., $q \propto (r_e 2 - r^2)$.

TRANSIENT FLOW SOLUTIONS

The solutions of practical value are those which assume that the well is draining from an area with an infinite drainage radius. Also, the well is assumed to be a point sink or source. So, the wellbore radius, r_w, is considered to be of negligible size. With these two additional assumptions, the "line source" solution can be obtained. This solution is an exponential integral evaluation with ∞ as one of the limits. It is tedious to work with, so normally approximations are used which are specific to a particular time range.

$$0 \le t_{dw} \le 0.01$$

$$\Delta P_{dw} = 2\sqrt{\frac{t_{dw}}{\pi}} \tag{16-10}$$

The applicable time range is normally over too quickly to be of much use in radial flow systems. However, Equation 16-10 describes transient flow in linear systems until the transition into semi-steady state is reached.

$$0.01 < t_{dw} \le 25.0$$

Due to the finite wellbore size of a real well, the actual pressure response differs from that given by the line source solution for radial positions close to the wellbore during early times. Fortunately, this range of dimensionless time will normally be over within one hour of real time during a well test. If for some reason a solution is needed during this dimensionless time range, reference 3 contains a graphical solution).

$$25.0 < t_{dw} \le 0.25 R_{de}^2$$

$$\Delta P_{dw} = -0.5 E_i \left(\frac{1}{4 t_{dw}} \right) \tag{16-11}$$

This is the exponential integral solution which is valid from the time that wellbore effects have become negligible to the beginning of the transition into semi-steady state flow. The exponential integral is defined as:

$$E_i(-x) = -\int_x^\infty \frac{e^{-u}}{u} \, du$$

The graph of the Ei(-x) function is presented in Figure 16-1.

$$25.0 < t_{dw} < 0.3\, R_{de}^2$$

$$\Delta P_{dw} = 0.5\, (\, Ln\, t_{dw} + 0.80907\,) \qquad\qquad (16\text{-}12)$$

Equation 16-12 is an excellent approximation to the exponential integral solution. Fortunately, from the time that the exponential solution begins to represent the behavior of a real well (at the well itself) in a radial system, Equation 16-12 is sufficiently accurate that it is used for 95% of all single well pressure transient analyses. Equation 16-12 is the basis for the classical Horner and Miller-Dyes-Hutchinson methods.

$$t_{dw} > 0.3\, R_{de}^2$$

$$\Delta P_{dw} = \frac{2\, t_{dw}}{R_{de}^2} + Ln\, (\, R_{de}\,) - 0.75 + 0.84\, e^{-x}$$

$$\qquad\qquad (16\text{-}13)$$

where:

$$x = 14.682\, \frac{t_{dw}}{R_{de}^2}$$

Equation 16-13 describes both late transient response (pressure is starting to be influenced by the drainage boundaries) and semi-steady state behavior.

DISCUSSION

Equation 16-10 is applicable to radial flow for such a short time that it is likely to be of little practical value for true radial systems. There are times that it does find use in very low permeability gas reservoirs. However, this solution is applicable to linear flow for the entire range of transient behavior. It will be shown later that this solution is applicable to well test data in fractured systems and in long and narrow reservoirs.

Equation 16-11 is the exponential integral solution to the Continuity Equation. For a single well test, this solution describes the pressure response at the wellbore from the time that the effects of the finite wellbore have become negligible out to the time that the drainage boundaries are felt. Because the logarithmic approximation Equation 16-12 is easier to use, Equation 16-11 finds little use in the analysis of single well tests. However, whenever the pressure data are collected at another well at a remote location from the active well (interference testing), Equation 16-11 must be used.

Equation 16-12 is such an accurate approximation to the exponential integral solution that it is applicable to almost all transient (single) well test data except at very late times.

Equation 16-13 is applicable only to late time transient well test data. This solution sometimes can be used to estimate the hydrocarbon pore volume.

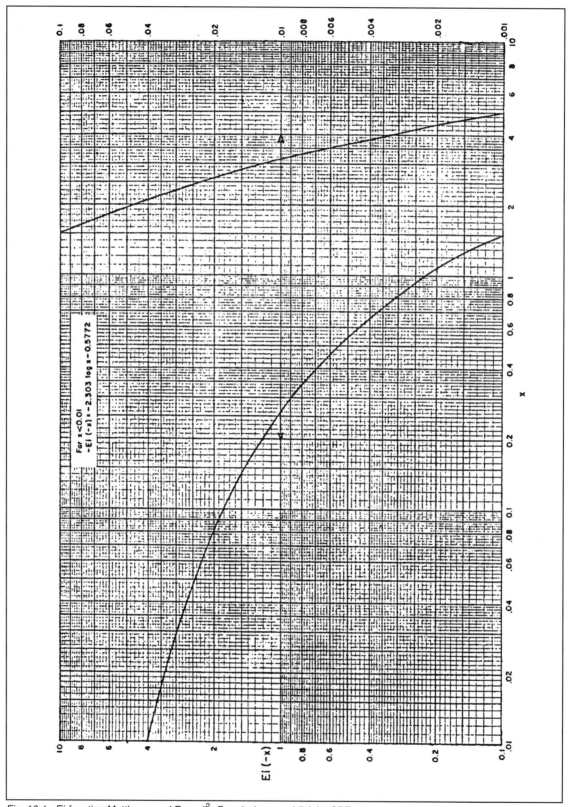

Fig. 16-1. Ei-function Matthews and Russell[2]. Permission to publish by SPE.

DRAWDOWN PRESSURE BEHAVIOR

Consider a well that has been shut-in long enough to achieve a flat pressure surface, p_i, across its entire drainage area. Now, the well is opened up on a constant producing rate. Figure 16-2 is a plot of pressure versus the log of radial distance from the well. The pressure response at different times is shown.

At any particular point in time, while the well is infinite acting, pressure varies linearly with the log of the radial distance from the well out to the radius of investigation, r_n. Notice that the radius of investigation (distance out to where pressure is undisturbed) is increasing with increasing time. The r_n associated with time t_n is related through the following dimensionless time relation:

$$t_{dn} = \frac{0.000264\, k\, t_n}{\phi\, \mu\, c_e\, r_n^2} = 0.25 \tag{16-14}$$

This equation requires permeability in md. and time in hours. With this equation, we can determine at a particular producing time what distance away from the well has been affected by the production test. This relation is valid only until the radius of investigation reaches the radius of drainage; i.e., until a dimensionless time of:

$$t_{de} = \frac{0.000264\, k\, t_e}{\phi\, \mu\, c_e\, r_e^2} = 0.25 \tag{16-15}$$

or

$$t_{dw} = 0.25\, (R_{de})^2$$

After this time, the transition between infinite acting and semi-steady state begins. It lasts until a dimensionless time of:

$$t_{de} = 0.30$$

is reached which is the beginning of semi-steady state for wells with true circular drainage areas. From this time on, the pressure declines uniformly for every point in the reservoir.

Considering Figure 16-2, assume that the well is shut-in at time t_4. Notice that the well is still infinite acting; i.e., the pressure transient due to placing the well on production has not reached the boundary at the time of shut-in. At shut-in, the pressure at the well begins to rise.

Pressure versus the log of radius plotted at the time of shut-in and at four different shut-in times (Δt_n) is shown in Figure 16-3. Notice that the pressure transient due to putting the well on production is still moving outward. There is also a portion of the reservoir that has not been subjected to pressure decline. And, the area where the pressure is increasing due to shut-in is growing with time. It should be evident that the earlier production period is affecting the shut-in pressure response. This is exactly the situation that we have with a buildup test where there has only been a short producing time, such as a drillstem test.

Consider the case where the well was produced longer before it was shut-in (at least long enough for the pressure transient due to production to reach the drainage boundary):

$$t_{de} \geq 0.25$$

This time, when the well is shut-in, the pressure response with time would look similar to that shown in Figure 16-4. In this case, the production time is not affecting the shut-in response. This situation represents a long-producing-time buildup test.

Thus, there are two different types of buildup tests: (1) the well is shut-in while it is producing in an infinite acting state (short producing time), and (2) the well is shut-in after it has produced long enough to be past the infinite acting state (long producing time). Before this can be considered mathematically, the superposition principle must be addressed.

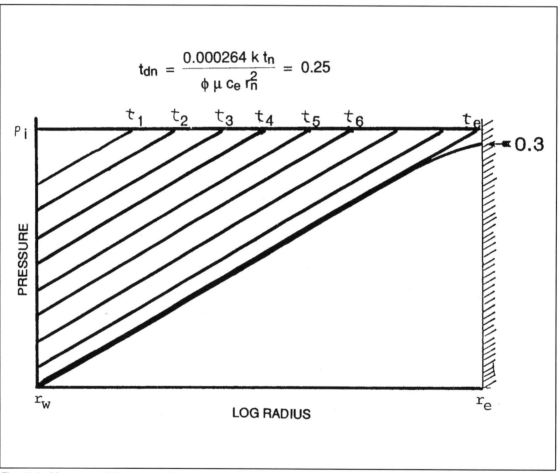

$$t_{dn} = \frac{0.000264\, k\, t_n}{\phi\, \mu\, c_e\, r_n^2} = 0.25$$

Fig. 16-2. Movement of drainage radius.

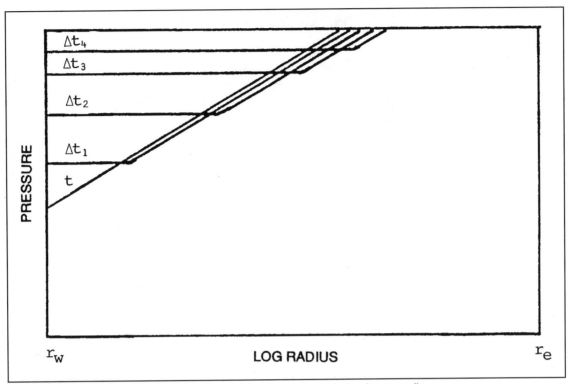

Fig. 16-3. Pressure distribution during pressure buildup operation, producing time, t-small.

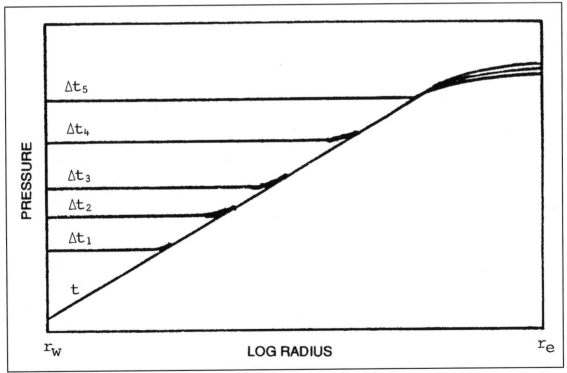

Fig. 16-4. Pressure distribution during pressure buildup operation, producing time, t-large.

PRINCIPLE OF SUPERPOSITION

This principle can be used to treat conditions of varying rate by using combinations of solutions that assume constant rate conditions. By using the principle of superposition, simpler mathematical solutions (based on constant well rate) are obtained than those that allow the rate to vary continuously. The procedure involves dividing a variable rate history into a number of time intervals. Although not strictly required, the application of the superposition principle is easier if the time intervals are each of the same length. The time interval is chosen such that a constant rate can be assumed for each time interval, which is not necessarily the same rate as that during the previous or the next time intervals. So, although the actual rate may be continuously varying with time, for use with the superposition principle, the rate history is represented as a series of constant rates.

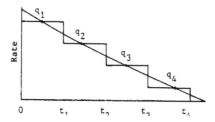

Actually, what is important in the use of the superposition principle is the rate change that exists from one time interval to the next. To start, let's consider the pressure over a well's drainage area to be everywhere at the static pressure, or $p = p_i$. Then, if we start producing the well as shown in the above diagram, the pressure response at the well can be described as follows:

$0 \leq t \leq t_1$

$$\Delta p_w = \left(\frac{\mu B}{2 \pi h k} \right) q_1 \Delta P_{dw} \left(t_{dw} \right) \qquad \text{where,} \quad \Delta p_w = p_i - p_w$$

$t_1 \leq t \leq t_2$

$$\Delta p_w = \left(\frac{\mu B}{2 \pi h k} \right) [\, q_1 \Delta P_{dw} \left(t_{dw} \right) + \left(q_2 - q_1 \right) \Delta P_{dw} \left(t_{dw} - t_{1dw} \right)]$$

$t_2 < t \leq t_3$

$$\Delta p_w = \left(\frac{\mu B}{2 \pi h k} \right) [q_1 \Delta P_{dw} \left(t_{dw} \right) + \left(q_2 - q_1 \right) \Delta P_{dw} \left(t_{dw} - t_{1dw} \right)$$

$$+ \left(q_3 - q_2 \right) \Delta P_{dw} \left(t_{dw} - t_{2dw} \right)]$$

$t_{n-1} \leq t \leq t_n$

$$\Delta p_w = \left\{ \frac{\mu B}{2 \pi h k} \right) [\, q_1 \Delta P_{dw} \left(t_{dw} \right) + \sum_{j=1}^{n-1} \left(q_{j+1} - q_j \right) \Delta P_{dw} \left(t_{dw} - t_{jdw} \right)]$$

(16-16)

Specifically, the principle of superposition will be used to develop the fundamental equation for pressure buildup analysis. It will be shown that a pressure buildup operation is a two-rate-change operation. The well begins with a flat pressure surface over the drainage area with the pressure everywhere equal to the static value. Then, a constant rate, q_o, is imposed on the well. Notice that this is also a rate change of $+q_o$. Then, after a specific producing time, the well is shut-in to observe the bottomhole pressure as it builds up with time. Mathematically, what is done is to add a second rate change (equal to $-q_o$) to the pressure solution at the time that the well is physically shut-in. Of course, the two rate changes add up to zero, or a shut-in condition.

PRINCIPLE OF SUPERPOSITION — ALTERNATE METHOD

If a well begins producing at time zero with a constant rate of q_o, the pressure in the well will decline as shown in the early time portion of Part II of Figure 16-5. After producing for a time t_1, the pressure drop is Δp_1. Similarly, after producing for time t_2, the pressure drop is Δp_2. If the producing rate, q_o, is continued to time t, then the Δp is equal to Δp_t. At this point in time (producing time equal to t), if the producing rate is reduced to zero (shut-in), then the well pressure begins to rise.

To predict the well pressure versus shut-in time, it should be remembered that the earlier production period set up pressure transients in the reservoir that will still be affecting the pressure response at the well. This is handled mathematically by letting the producing rate continue through the shut-in period. Notice that on the pressure versus time graph, the dashed line illustrates the well pressure that would have existed if the well had been allowed to continue producing with the constant rate q_o.

Then at the time of the shut-in (producing time = t), a second producing rate, equal to $-q_o$, is added which continues through the shut-in period. So, there are two rates that are affecting the pressure solution after shut-in: $+q_o$ and $-q_o$. Notice that their sum is equal to zero or a net shut-in condition. The $-q_o$ pressure solution is added directly to the $+q_o$ pressure solution. However, it should be noted that two different time references are used. Shut-in time is represented as Δt, while the total producing time at shut-in was t. Therefore, a shut-in time of Δt_1 would correspond to a producing time of $t + \Delta t_1$.

So, to determine the total effect of the producing period and the shut-in at a shut-in time of Δt_1, the $-q_o$ pressure solution at a time of Δt_1 is added to the $+q_o$ pressure solution at a time of $t + \Delta t_1$. Note that the pressure change due to the $-q_o$ rate at a shut-in time of Δt_1 is Δp_1 (Figure 16-5). Thus, this pressure change is added from the dashed line which represents the continuing $+q_o$ pressure solution. Similarly, at a shut-in time of Δt_2, the $-q_o$ pressure change is Δp_2. Once again this is added to the $+q_o$ pressure solution that corresponds to a time of $t + \Delta t_2$.

It should also be noted that at shut-in time $\Delta t_1 = t_1$, the two "Δp_1's" in Figure 16-5 are equal. That is, the pressure drop due to producing the well at $+q_o$ is equal to the pressure rise due to injecting in the well at a rate of $-q_o$, assuming that the times of production and injection are equal. Similarly, assuming that $\Delta t_2 = t_2$, the two "Δp_2's" in Figure 16-5 are equal.

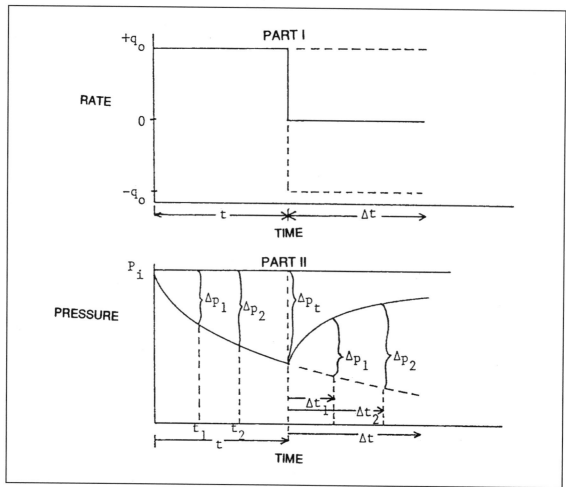

Fig. 16-5. Superposition for a buildup test.

It should be evident that the actual path of the pressure-time response during shut-in depends on the pressure-time behavior at the end of the producing period. If the slope at the end of the producing period is near zero (stabilized flow), then the shut-in pressure response will only minimally be affected by the producing time. Indeed, in this case, if the Δp's during shut-in are plotted downward instead of upward (see Figure 16-5), then the shut-in pressure change - time curve would look almost exactly like the producing curve.

However, if the slope of the pressure - time curve at the end of the producing period is not negligible, then the shape of the pressure buildup curve is a function of the producing time.

Thus, pressure buildup analysis methods must be divided into two categories: (1) short producing time buildup tests and (2) long producing time buildup tests. The short producing time buildup methods utilize the principle of superposition and consider the effect of the period of production prior to shut-in. The long producing time buildup methods assume that the well was stabilized at the time of shut-in. Thus, the time of production is no longer affecting the shut-in pressure response, and superposition is not needed. So, the long producing time buildup analysis techniques are usually simpler.

DISCUSSION OF THE SKIN EFFECT (PERMEABILITY CHANGE NEAR WELLBORE)

Experience has shown that most wells have a different average permeability near the well than that existing at a remote distance from the well. This permeability alteration can be caused by a number of different processes during the drilling, completion, production, and stimulation of a well.

There is cause for concern when the formation has been damaged or a high apparent resistance to flow exists in the vicinity of the wellbore. Some of the factors contributing to such a condition are:

(1) Invasion of drilling fluids during the drilling operation.

(2) A higher gas or water saturation existing near an oil-producing well. This is sometimes called a gas or water block and results from producing excess free gas or water. A similar condition can result with the production of water from a gas well.

(3) A concentration of solid fines which tend to flow toward the well with the produced fluids will reduce the formation conductivity close to the wellbore. This is particularly true of wells that produce water from unconsolidated or only partially consolidated sand reservoirs.

(4) Partial penetration of the total pay zone or insufficient perforations cause extra resistance to flow at the wellbore. To derive Equation 16-7, the Continuity Equation, the assumption of constant thickness was made. Thus, the effect of partial penetration or insufficient perforation density is to cause vertically convergent flow resulting in a greater pressure drop near the wellbore than theory would predict.

In 1953, van Everdingen[6] defined the "skin effect" (also called "skin factor"). He suggested that the phenomenon of local permeability alteration could be handled by adding a term to the pressure solution that would act as a pure resistance felt right at the wellbore. With this approach, the effect of fluid expansion in the damaged zone has been neglected, but the affected reservoir volume is negligible. With the van Everdingen approach, the "skin effect" is represented by a single additional term, s, added to the basic dimensionless pressure drop; i.e.:

$$(\Delta P_{dw})_{Revised} = (\Delta P_{dw})_{Previous} + s$$

The revised equation for pressure change at the well is:

$$\Delta p_w = \left[\frac{q \mu B}{2 \pi h k} \right] \left[\Delta P_{dw} (t_{dw}) + s \right]$$

or

$$(p_i - p_{wf}) = \left[\frac{q \mu B}{2 \pi h k} \right] \left[\Delta P_{dw} (t_{dw}) + s \right] \tag{16-17}$$

The interpretation of this skin effect term, s, is as follows:

> s > 0 damaged condition,
>
> s = 0 uniform permeability (no altered zone),
>
> s < 0 improved condition (stimulated).

The wise well test analyst who has just calculated a positive skin for an oil well should realize, before recommending a costly work over, that the total skin seen on a well test is actually comprised of two parts: true skin plus pseudoskin. Pseudoskin is also sometimes referred to as apparent skin.

$$s_{total} = s_{true} + s_{pseudo} \tag{16-18}$$

True skin is that skin that is due to actual formation damage; e.g., that due to formation incompatibility with the drilling fluid filtrate. So, skin that is caused by an actual change in formation permeability is referred to as true skin. Pseudoskin is that skin which results from partial penetration or insufficient perforation density. Pseudoskin could also result from a poorly designed wire screen or gravel pack. Obviously, if the total skin seen on a well test is mostly due to perforating only the top five feet of a zone that is 50 feet thick, then a workover will be ineffective in reducing this skin. It pays to characterize the nature of the skin before attempting to alleviate it.

Figure 16-6 is a result of the work of Brons and Marting.[8] It relates the effects of partial penetration in a single zone assuming single phase flow. So, to find the skin attributable to damage, subtract the pseudoskin factor from the total skin seen on a well test.

Equation 16-18 is actually an oil well equation. Most gas wells have turbulence around the wellbore which causes extra pressure drop. This appears as skin on a well test and is rate-dependent. This situation will be discussed under gas wells later in this chapter.

According to Reynolds *et al.*[9], the correlation of Brons and Marting does not adequately predict the pseudoskin factor caused by partial penetration when multi-phase flow exists (for instance, water or gas coning). Under these conditions, the procedures given by Reynolds *et al.*[9] are suggested to estimate pseudoskin factor.

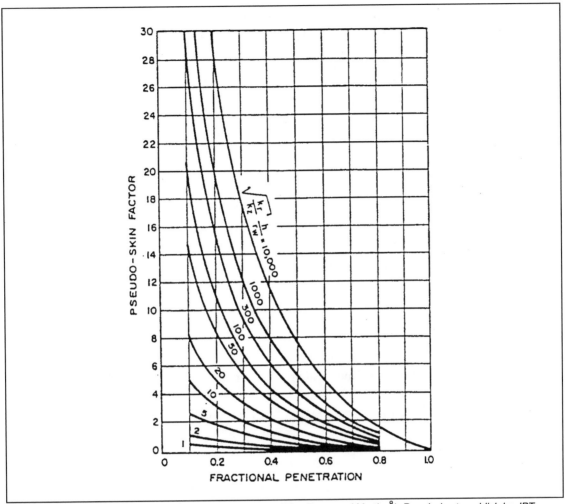

Fig. 16-6. Pseudoskin factor for partially penetrating wells. (After Brons and Marting[8]). Permission to publish by JPT.

PRESSURE BUILDUP THEORY

The basic principles underlying the nature of pressure buildup in producing wells during a shut-in period are as follows:

   (1) A two-rate situation where the second rate is zero, and

   (2) the principle of superposition applies.

According to the principle of superposition, for a situation involving two different constant rates, the pressure change seen at the well as a function of time is:

$$p_i - p_w(t_t) = \frac{q_1 B \mu}{2 \pi h k} \Delta P_{dw}(t_{tdw}) + \frac{(q_2 - q_1) B \mu}{2 \pi h k} \Delta P_{dw}(t_{tdw} - t_{dw})$$ (16-19)

where:

t_t = the total actual elapsed time since the well was put on production to the current time of interest,

t_{tdw} = dimensionless time referred to the wellbore radius, based on total elapsed time since the well began producing to the current time of interest, and

t_{dw} = dimensionless producing time before shut-in, and

t = actual producing time before shut-in.

If $q_2 = 0$, then $p_w(t_t)$ for $t_t > t$ will be p_{ws}.

$$t_{tdw} = (t + \Delta t)_{dw}$$

$$t_{tdw} - t_{dw} = (\Delta t)_{dw}$$

With the above substitutions, Equation 16-19 becomes:

$$p_i - p_{ws} = \frac{q_1 B \mu}{2 \pi h k} \left\{ \Delta P_{dw} \left[(t + \Delta t)_{dw} \right] - \Delta P_{dw} (\Delta t_{dw}) \right\} \tag{16-20}$$

For a radial-cylindrical flow system, using the natural logarithmic approximation (16-12) to the solution to the Continuity Equation (and including the skin effect):

$$\Delta P_{dw} (t_{tdw}) = 0.5 [Ln(t_{tdw}) + 0.80907 + 2s] \tag{16-21}$$

Therefore,

$$\Delta P_{dw}[(t + \Delta t)_{dw}] = 0.5 [Ln(t + \Delta t)_{dw} + 0.80907 + 2s]$$

and:

$$\Delta P_{dw}[(\Delta t)_{dw}] = 0.5 [Ln(\Delta t)_{dw} + 0.80907 + 2s]$$

Hence:

$$\Delta P_{dw}[(t + \Delta t)_{dw}] - \Delta P_{dw}[(\Delta t)_{dw}] = 0.5 \ Ln\left[\frac{t + \Delta t}{\Delta t}\right]$$

Then, letting $q = q_1$, Equation 16-20 transforms into:

$$p_i - p_{ws} = \left[\frac{q B \mu}{2 \pi h k}\right] \left[\frac{1}{2}\right] Ln\left[\frac{t + \Delta t}{\Delta t}\right]$$

$$\tag{16-22}$$

Equation 16-22 has been developed using Darcy units. If this equation is changed into conventional practical oil field units, then, the following equation results:

$$p_i - p_{ws} = 162.6 \left[\frac{q\mu B}{k\,h} \right] Log_{10} \left[\frac{t + \Delta t}{\Delta t} \right] \tag{16-23}$$

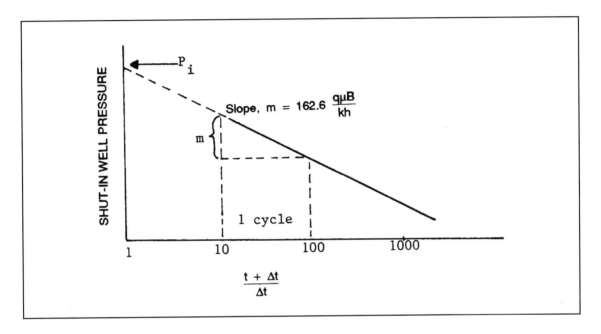

This plot has through the years come to be called the Horner[12] plot.

If the cumulative production from the test well has been at varied rates over its life, then the evaluation of producing time, t, is accomplished as follows:

$$t = \frac{(\text{Cumulative Production, STB})\,(24)}{(\text{Rate Prior to S.I., STB}/D)} \quad hours \tag{16-24}$$

For the situation where $t \gg \Delta t$, then $Ln\,(t + \Delta t) \approx Ln\,t$; so:

$$p_{ws} = p_i - \frac{qB\mu}{4\pi k h} [\,Ln\,t_{dw} + 0.80907 + 2s\,] + \frac{qB\mu}{4\pi h k} [\,Ln\,(\Delta t)_{dw} + 0.80907 + 2s\,]$$

But,

$$p_i - p_{wf} = \frac{qB\mu}{4\pi h k} [\,Ln\,t_{dw} + 0.80907 + 2s\,]$$

Therefore,

$$p_{ws} = p_{wf} + \frac{qB\mu}{4\pi h k}[Ln\,(\Delta t)_{dw} + 0.80907 + 2s]\qquad(16\text{-}25)$$

Equation 16-25 is given in Darcy units. Transforming to practical units:

$$p_{ws} = p_{wf} + \frac{162.6\,qB\mu}{kh}\left[Log\,(\Delta t) + Log\left[\frac{k}{\phi\mu c r_w^2}\right] - 3.23 + 0.87s\right]$$
$$(16\text{-}26)$$

This equation suggests that a plot of p_{ws} vs. Log Δt should yield a straight line.

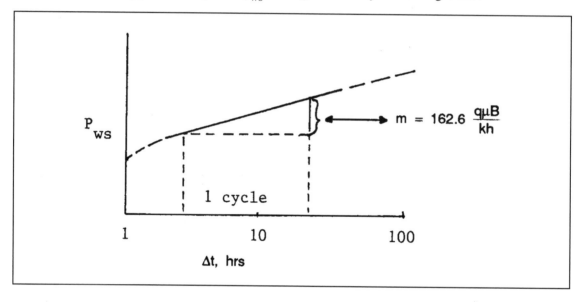

So, if the producing time is long enough (to reach stabilization), then this log Δt approximation plot can be used for the analysis. The originators of this semi-log graph are Miller, Dyes, and Hutchinson[13], and it is normally called simply an MDH plot.

A guide to the required producing time to reach stabilization (semi-steady state) is:

$$t \geq \frac{381\,\phi\mu c A}{k} = t_s\qquad(16\text{-}27)$$

where:

   t = producing time, hours,
   ϕ = porosity, fraction,
   μ = viscosity, cp.,
   A = drainage area of test well, ft.^{2},
   k = formation permeability, md.,
   c = compressibility, 1/psi, and
   t_s = producing time required to reach stabilization, hrs.

If the actual producing time (Eq. 16-24) is longer than the time to stabilization (16-27), then the simpler log Δt semi-log plot can be used. Quantitatively, this is what is meant by a long producing-time pressure buildup: one where the calculated producing time is greater than the time to reach stabilization.

The Horner plot must be used for a short producing-time buildup test analysis. It can also be used to consider a long producing-time buildup. That is, the long producing-time buildup can be analyzed with either the Horner method or with the MDH plot.

A typical Horner Plot is shown in Figure 16-7. Note that it has been divided into three different regions.

Region I contains pressure data collected during very early shut-in times. Thus, these data reflect conditions in the wellbore and near the well. In this plot, wellbore storage (to be discussed later) is assumed to be negligible. Thus, only the near-wellbore conditions are considered. Illustrated in Figure 16-7 are three different possibilities for the region I pressure data depending on the values of the near-wellbore permeability, k_1, and the interwell (out away from the wellbore) permeability, k_2. These distinct curve shapes illustrated in the plot suggest that curve shape alone could be used as a guide to the contrast between near-well and remote permeabilities. Unfortunately, wellbore storage also affects the shut-in pressure response sufficiently to disguise the true contrast.

Regardless of the shape of the region I data, the region II data plot as a straight line. Near-well effects of damage (or stimulation) and wellbore storage have dissipated during this time interval, and the effects of the drainage area boundaries have not yet been felt. Thus, the region II data are collected when the well is infinite acting. Further, the slope of this straight line can be used to calculate the value of the interwell formation flow capacity, kh.

Region III contains the late time shut-in pressure data that are being affected by the nature of the boundaries around the test well. In this region, the curve tends to flatten with shut-in time. These data can be used to determine the static well pressure over the test well drainage area.

Unfortunately, in many cases, the well is not shut-in long enough to see any of the flattened portion of the data. In this situation, the latest shut-in data still fall on the straight line (region II). When this happens, the static well pressure must be determined by extrapolating the region II straight line to the point where:

$$[\,(\,t\ +\ \Delta t\,)\,/\,\Delta t\,]\ =\ 1.0$$

which corresponds to infinite shut-in time. This extrapolated pressure is the true original pressure (p_i) if the original pressure still exists somewhere within the drainage area of the well. Of course, this would be the case for a new well. If the shut-in test well is a mature well, such that the original pressure no longer exists anywhere within the drainage area, then the extrapolated pressure is called p* ("p-star"). It is also called the "false pressure." This is not a real pressure near this well. Normally, it is a pressure higher than any pressure within the drainage area of the well. However, unless the formation permeability is very low; e.g., a few millidarcies or less, this value of p* is often used as the static well pressure.

The MDH plot is represented in Figure 16-8. It has many of the same features of the Horner plot; however, it is not possible to graphically extrapolate to infinite shut-in time to determine the original pressure or p*. It is emphasized that the MDH plot can be used only after the producing time (before shut-in) has exceeded that necessary to reach stabilization. Hence, the original pressure no longer exists within the drainage area of the well. Notice that the same three regions discussed under the

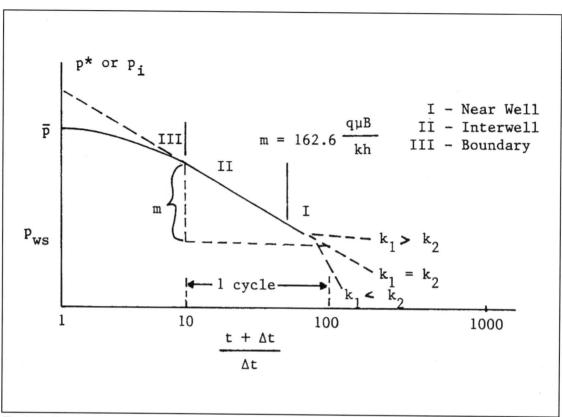

Fig. 16-7. Horner Representation.

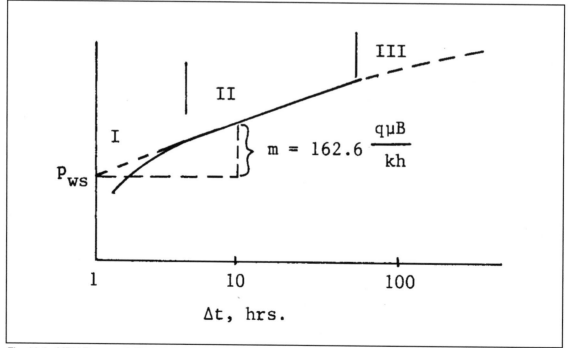

Fig. 16-8. MDH Representation.

Horner plot also exist here. The flow capacity is calculated with the region II straight line slope using the same equation that is used with a Horner analysis. The MDH plot has application especially for conditions where the producing time is difficult, if not impossible, to determine.

EVALUATION OF TOTAL MOBILITY, $(k/\mu)_t$

For situations where more than one fluid is flowing through the reservoir, modifications must be made in the well test analysis. The theory called for single phase flow. To extend the single-phase analysis techniques to multiphase flow situations, the effects of relative permeability must be taken into account.

Thus, the value of (k/μ) found in various transient flow equations must be the total mobility, $(k/\mu)_t$, not just the major hydrocarbon-phase mobility.

$$\left(\frac{k}{\mu}\right)_t = \left(\frac{k}{\mu}\right)_o + \left(\frac{k}{\mu}\right)_g + \left(\frac{k}{\mu}\right)_w \quad md/cp \qquad (16\text{-}28)$$

For situations where more than one fluid is being produced from the test well, assuming that there is a negligible saturation gradient in the vicinity of the well, then the same pressure gradient which controls one phase controls all phases. Since the slope of the buildup curve is related to the pressure gradient in the neighborhood of the producing well, then the semi-log plot slope, m, can be used to evaluate the mobility of each phase.

The individual phase equations are then:

$$\frac{k_o}{\mu_o} = \frac{162.6 \; q_o \; B_o}{m \, h}$$

$$\frac{k_w}{\mu_w} = \frac{162.6 \, q_w \, B_w}{m \, h}$$

$$\frac{k_g}{\mu_g} = \frac{162.6 \, (q_g - q_o R_s) \, B_g}{m \, h}$$

Notice that in the gas-phase equation, only that part of the total gas seen at the surface that existed as free gas in the reservoir is considered in the equation. The gas that was in solution in the reservoir, $q_o R_s$), is subtracted off because it was actually part of the oil while in the reservoir.

Therefore, the total mobility can be computed as:

$$(\frac{k}{\mu})_t = \frac{162.6}{m \, h} [\, q_o B_o + q_w B_w + (q_g - q_o R_s) B_g \,] \qquad (16\text{-}29)$$

where:
$q_o B_o$, $q_w B_w$, $q_g B_g$, and $q_o R_s B_g$ have units of reservoir barrels per day.

EVALUATION OF EFFECTIVE DRAINAGE RADIUS, r_e

The magnitude of the drainage radius of a producing or an injection well is a function of the producing or injection rate of the tested well as compared to the rates from the wells that offset the test well.

It can be shown that when producing a reservoir at semi-steady state, each well drains a volume that is proportional to its rate. Therefore, if we assume that the net reservoir thickness is constant over a given region of the reservoir, then for stabilized flow conditions, the area drained by each well is proportional to its rate.

To implement this principle in a practical easy-to-manage method, only the direct offsets to the test well are considered. To illustrate the computational procedure, consider a tested well in the middle of eight wells.

Problem 16-1:

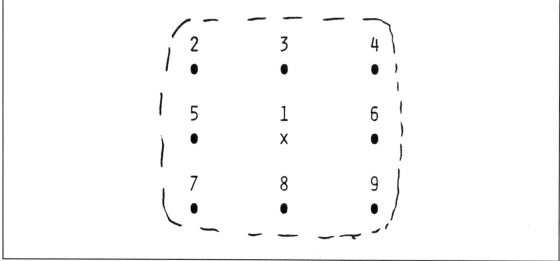

Assume that the area drained by all nine wells is 400 acres. The producing rates of the various wells prior to shut-in of well no. 1, the test well, are:

Well No.	Producing Rate B/D
1	200
2	100
3	150
4	150
5	190
6	250
7	210
8	250
9	200
Total = 1,700 B/D	

Area drained by Well No. 1:

$$= (200/1700)(400)$$

$$= 47.06 \quad \text{acres}$$

Assuming that this area is circular-shaped, an effective drainage radius can be calculated:

$$r_e = \begin{bmatrix} \textit{Effective} \\ \textit{Drainage} \\ \textit{Radius} \end{bmatrix} = \sqrt{\frac{(47.06)(43,560)}{\pi}}$$

$$r_e = 807.8 \ \textit{feet}$$

The above method is applicable for a well in the vicinity of other wells which affect the drainage radius of the test well due to mutual interference.

On the other hand, if the problem is to find the drainage radius of a well that has no other wells near it, then a different approach is needed. In this case, for a new well, the drainage radius is strictly a function of time, reservoir characteristics, and distance to the nearest barrier. The drainage radius can best be determined from pressure drawdown data which will be discussed at a later point in this chapter.

For a new well, the radius of drainage is equal to the radius of investigation (r_i) until the time at which the pressure transients (caused by production) reach the drainage area boundaries. The equation relating the growing radius of drainage (with time) is:

$$r_i = 0.0328 \ \sqrt{\frac{[(k/\mu)_t] \ t}{\phi \ c_e}} \tag{16-30}$$

where:

r_i = the distance the pressure transient has moved from the wellbore in producing time "t", ft.,

t = producing time, hours,

$(k/\mu)_t$ = total mobility, md/cp,

ϕ = formation porosity, fraction,

c_e = the effective compressibility of the formation-fluid system, psi^{-1}.

TOTAL COMPRESSIBILITY IN OIL RESERVOIRS

When a well test equation calls for compressibility, the total system compressibility should be used. To use oil compressibility for total compressibility is to assume that (1) oil is the only fluid within the reservoir, and (2) the rock is incompressible. Both of these assumptions are always wrong.

When more than one phase exists in the reservoir (always), the total compressibility must include the compressibilities of all the fluids present (saturation-weighted average) plus the pore volume compressibility.

UNDERSATURATED OIL RESERVOIRS

$$c_e = c_o S_o + c_w S_w + c_f \qquad\qquad (16\text{-}31)$$

where:

c_e = effective (total) compressibility, psi^{-1},
c_o = compressibility of oil, psi^{-1},
c_w = water compressibility, psi^{-1},
c_f = formation compressibility (actually pore volume compressibility), psi^{-1},
S_o = oil saturation, fraction of pore space, and
S_w = water saturation, fraction of pore space.

For undersaturated oil reservoirs, the value of oil compressibility will normally fall between 5×10^{-6} and 30×10^{-6} psi^{-1}. If the oil formation volume factor, B_o, versus pressure relationship is available, then the value of c_o can be calculated. Using data above the bubblepoint, plot $\ln B_o$ versus pressure. Determine the slope in psi^{-1}. Then,

$$c_o = -\ (slope) \quad or \quad c_o = -\ \frac{1}{B_o}\ \frac{\partial B_o}{\partial p} \qquad\qquad (16\text{-}32)$$

The compressibility of water is influenced by pressure, temperature, gas in solution, and by dissolved solids or salinity content. The value of c_w will normally be between 2.5×10^{-6} and 4.0×10^{-6} psi^{-1} with an average value of about 3.5×10^{-6} psi^{-1}.

The formation compressibility called for in the equation is pore volume compressibility, not the rock grain or matrix compressibility. For a consolidated formation, if c_f has not been measured in the laboratory, then many people use the Hall correlation given in the Rock Properties chapter. This correlation relates c_f versus porosity. Newman and van der Knaap, working separately, have worked in this area with results summarized in Earlougher[3], Appendix D. In addition to consolidated rocks, Newman considered friable and unconsolidated formation compressibilities. Little correlation with porosity was found.

For consolidated rocks, the value of c_f will normally fall within the range of 2×10^{-6} and 10×10^{-6} psi^{-1}. However, the friable and unconsolidated formation compressibilities can be much higher.

SATURATED OIL RESERVOIRS

In this case, the reservoir static pressure is below the bubblepoint pressure, and a free gas saturation exists in the reservoir. Therefore, the total or effective compressibility is:

$$c_e = c_o S_o + c_w S_w + c_g S_g + c_f \qquad (16\text{-}33)$$

where:

c_g = compressibility of gas, psi^{-1},
S_g = the gas saturation, fraction, and
c_o, S_o, c_w, S_w, c_f, and c_e are defined as before.

Because the reservoir static pressure is below the bubblepoint, a decrease in pressure due to production is associated with gas coming out of solution in the oil. Of course, this did not happen above the bubblepoint pressure. Therefore, the oil compressibility equation must be altered to account for the changing solution gas with pressure that occurs below the bubblepoint:

$$c_o = -\frac{1}{B_o}\left[\frac{\partial B_o}{\partial p}\right] + \left[\frac{B_g}{B_o}\right]\left[\frac{\partial R_s}{\partial p}\right] \qquad (16\text{-}34)$$

The term $\left[-\dfrac{1}{B_o}\left(\dfrac{\partial B_o}{\partial p}\right)\right]$ represents the shrinkage of the oil phase with decreasing pressure below the bubblepoint.

The term $\left[\dfrac{B_g}{B_o}\right]\left[\dfrac{\partial R_s}{\partial p}\right]$ represents the change in relative oil volume due to the liberated solution gas as pressure is decreased. As mentioned earlier, the water compressibility is influenced by pressure and gas in solution, but its change with changing pressure is usually small. So, water compressibility is normally regarded as constant for a given reservoir.

The gas compressibility may be computed as follows:

$$c_g = -\left[\frac{1}{B_g}\right]\left[\frac{\partial B_g}{\partial p}\right] = \frac{1}{p} - \left[\frac{1}{z}\right]\left[\frac{\partial z}{\partial p}\right] \qquad (16\text{-}35)$$

where:

B_g = the gas formation volume factor, and
z = the gas-deviation factor.

Generally speaking, for pressures less than 3000 psia, the change in z-factor with pressure is quite small. Therefore, the value of the term:

$$(1 / z) (\partial z / \partial p)$$

is normally negligible when compared to (1/p). So, for pressures less than 3000 psia, c_g is often calculated as:

$$c_g \approx 1 / p \ \ psi^{-1}$$

However, for pressures above 3000 psia, the z-factor is normally changing almost linearly with pressure, and the ($\partial z / \partial p$) term cannot be neglected. So, above 3000 psia, the complete gas compressibility formula is used:

$$c_g = (1 / p) - (1 / z) (\partial z / \partial p) \ \ psi^{-1}$$

Problem 16-2:

Example Calculation of c_e in a Saturated Oil Reservoir:

Data:

p	2100	2000	1900	PSIA
B_o	1.285	1.271	1.257	BBL/STB
R_s	0.543	0.511	0.479	MCF/STB
B_g		1.238		B/MCF
S_o		0.756		
S_g		0.044		
S_w		0.200		
c_w		3.5×10^{-6}		PSI^{-1}
c_f		3.8×10^{-6}		PSI^{-1}

Problem:

Calculate total system compressibility @ 2000 psia.

Calculations:

$$c_o \cong - \frac{1}{B_o} \frac{\Delta B_o}{\Delta p} + \frac{B_g}{B_o} \frac{\Delta R_s}{\Delta p}$$

$$= - \left[\frac{1}{1.271} \right] \left[\frac{1.285 - 1.257}{200} \right] + \left[\frac{1.238}{1.271} \right] \left[\frac{0.543 - 0.479}{200} \right]$$

$$= 201.5 \times 10^{-6} \ psi^{-1}$$

$$c_g \cong 1/p = 1/2000 = 500 \times 10^{-6} \ psi^{-1}$$

$$c_e = c_o S_o + c_g S_g + c_w S_w + c_f$$

$$= [(201.5)(0.756) + (500)(0.044) + (3.5)(0.2) + 3.8] \times 10^{-6} \ psi^{-1}$$

$$c_e = 178.8 \times 10^{-6} \ psi^{-1}$$

TOTAL COMPRESSIBILITY IN GAS RESERVOIRS

Assuming only two phases to be present, free gas and connate water, then:

$$c_e = S_g c_g + S_w c_w + c_f \qquad (16\text{-}36)$$

Free gas compressibility is normally hundreds of times more compressible than either water or pore volume. Therefore, the last two terms are typically neglected (except for overpressured reservoirs), and we have:

$$c_e = S_g c_g = S_g [(1/p) - (1/z)(\partial z / \partial p)] \qquad (16\text{-}37)$$

For pressures less than 3000 psia, we can further approximate the total compressibility as:

$$c_e = S_g c_g = S_g / p \qquad (16\text{-}38)$$

For pressures above 3000 psia, the z-factor contribution cannot be neglected in the gas compressibility term, so Equation 16-37 is used.

EVALUATION OF WELLBORE DAMAGE (SKIN EFFECT)

If the Horner time function, $(t + \Delta t) / \Delta t$, is inverted, then the Horner plot for a pressure buildup will look similar to:

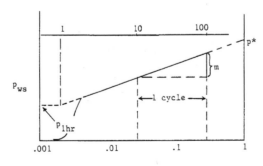

For a Horner analysis of a pressure buildup, it makes no difference whether the Horner diagram is plotted as above or as shown in Figure 16-7.

Over 30 years ago, van Everdingen[6] presented the concept of the skin effect. According to the theory, the bottomhole pressure just prior to the shut-in (i.e., the flowing bottomhole pressure) is needed for comparison with a pressure taken from the semilog straight line. By convention, and because the skin factor equation simplifies somewhat, the semilog straight line pressure is usually taken at a time of one hour.

Unfortunately, it is common for early measured pressures not to fall on the semilog straight line due to "afterflow" and skin. This will be discussed in detail later in this chapter. If the measured pressure at a shut-in time of one hour does not fall on the semilog straight line, then the semilog straight line must be extrapolated to a time of one hour. Then, the pressure, p_{1hr}, is read from the semilog straight line at a shut-in time of one hour for use in the following skin factor equation:

$$s = 1.151 \left\{ \frac{p_{1hr} - p_{wf}}{m} - Log \left[\frac{k}{\phi \mu c_e r_w^2} \right] + 3.23 \right\} \tag{16-39}$$

where:

   m = the Horner or MDH slope, psi/cycle, and producing time, t, is assumed to be greater than five hours.

The skin effect (or skin factor), s, is a dimensionless quantity which is interpreted as follows:

   s > 0, Damaged condition

   s < 0, Improved condition

   s = 0, Uniform permeability (no damage or stimulation)

Because the skin effect is a dimensionless quantity, perhaps a more convenient parameter to judge the magnitude of permeability alteration near the wellbore is the pressure drop due to skin. This pressure drop due to skin is not being used to flow fluids to the wellbore. It is merely extra pressure drop (in the case of damage) that is caused by the reduced permeability in the vicinity of the well.

$$\Delta p_{skin} = (m s / 1.151) = 0.87 m s \tag{16-40}$$

Some analysts have found the condition ratio convenient to use:

$$Condition\, Ratio = CR = (p^* - p_{wf} - \Delta p_{skin}) / (p^* - p_{wf}) \tag{16-41}$$

This is approximately the ratio of the well's actual capacity to the theoretical capacity with no permeability alteration near the wellbore. Therefore, the condition ratio may be interpreted as follows:

CR < 1.0, Damaged Condition,

CR > 1.0, Stimulated or Improved Condition,

CR = 1.0, Uniform Permeability.

A slightly more rigorous (but less convenient) measure of a well's relative capacity is the "flow efficiency" which uses $\bar{p}$ (static pressure) instead of p*:

$$Flow\ Efficiency\ =\ FE\ =\ (\ \bar{p}\ -\ p_{wf}\ -\ \Delta p_{skin}\ )\ /\ (\bar{p}\ -\ p_{wf}) \qquad (16\text{-}42)$$

Methods for the determination of $(\bar{p})$ will be covered later in this chapter. The interpretation of flow efficiency value is the same as for the condition ratio.

TRANSIENT FLOW OF GASES

The basic theory used to describe gas flow under transient conditions is quite similar to that for liquid flow. It includes:

1. the equation of continuity,

2. an equation of state, and

3. Darcy's law

EQUATION OF CONTINUITY:

$$-\ \frac{\partial}{\partial r}\ (\rho q)\ =\ 2\pi h \phi r\ \frac{\partial \rho}{\partial t} \qquad (16\text{-}1)$$

where:

ρ = gas density (weight per volume),
q = gas flow rate,
r = radial distance,
t = time,
h = pay thickness, and
ϕ = gas porosity.

EQUATION OF STATE:

$$\rho = \frac{W}{V} = \frac{p\,M}{zRT} \qquad (16\text{-}43)$$

where:

    ρ = gas density,
    W = weight of gas,
    V = volume of gas,
    p = pressure,
    M = molecular weight of gas,
    z = gas deviation factor,
    R = universal gas constant (numerical value dependent on units),
    T = reservoir absolute temperature, $°R = °F + 460$

DARCY'S LAW:

$$q = -\frac{2\pi h k}{\mu}\left[r\,\frac{\partial p}{\partial r}\right] \qquad (16\text{-}44)$$

where:

    q = gas flow rate,
    h = pay thickness,
    k = formation permeability,
    μ = gas viscosity,
    r = radial distance, and
    p = pressure.

Introducing Equations 16-43 and 16-44 into Equation 16-1 and simplifying, the following can be obtained:

$$\frac{1}{r}\frac{\partial}{\partial r}\left\{\left[\frac{k}{\mu}\right]\left[\frac{p}{z}\right]r\left[\frac{\partial p}{\partial r}\right]\right\} = \phi\,\frac{\partial}{\partial t}\left[\frac{p}{z}\right] \qquad (16\text{-}45)$$

Recalling the equation relating gas compressibility to pressure and z-factor:

$$c_g = \left[\frac{1}{p} - \frac{1}{z}\frac{\partial z}{\partial p}\right] \text{ at constant temperature} \qquad (16\text{-}35)$$

By using this gas compressibility relation, Equation 16-45 can be transformed to:

$$\frac{1}{r}\frac{\partial}{\partial r}\left\{\left[\frac{p}{\mu z}\right]r\,\frac{\partial p}{\partial r}\right\} = \frac{\phi}{k}\left[\frac{p}{z}\right]\left[c_g\right]\frac{\partial p}{\partial t} \qquad (16\text{-}46)$$

Equation 16-46, which describes gas transient flow, is a much more difficult form of the Continuity Equation than resulted for liquid transient flow. Consequently, the simple liquid analysis techniques cannot be used to obtain a solution to Equation 16-46. Al-Hussainy *et al.*[15] developed a transformation which may be used to convert Equation 16-46 into a simpler form comparable to the liquid Continuity Equation. This potential function is known as the "Real Gas Potential" or "Gas Pseudopressure:"

$$\psi(p) = 2 \int_{a}^{p} \frac{p}{\mu z} \, dp \qquad (16\text{-}47)$$

where:

$\psi(p)$ = the Real Gas Potential,
a = reference pressure lower than or equal to the lowest pressure of the system (atmospheric pressure is often used).

If Equation 16-47 is introduced into Equation 16-46, the following can be obtained:

$$\frac{1}{r} \frac{\partial}{\partial r} \left[r \frac{\partial \psi}{\partial r} \right] = \left[\frac{\phi \mu c_g}{k} \right] \frac{\partial \psi}{\partial t} \qquad (16\text{-}48)$$

Notice that Equation 16-48, the gas Continuity Equation in terms of the real gas potential, has exactly the same form as the liquid Continuity Equation that is in terms of pressure:

$$\frac{1}{r} \frac{\partial}{\partial r} \left[r \frac{\partial p}{\partial r} \right] = \left[\frac{\phi \mu c}{k} \right] \frac{\partial p}{\partial t} \qquad (16\text{-}6)$$

If Equation 16-48, the gas equation, is compared with Equation 16-6, the liquid equation; the similarity should be apparent. The dependent variable in the liquid Equation 16-6 is pressure; whereas, the dependent variable in the gas Equation 16-48 is the real gas potential or gas pseudopressure.

In the development of Equation 16-48, it was assumed that c_g and μ, which are both functions of pressure, do not change enough to preclude treating them as constant over a limited range of pressure values. This is a limitation of the gas well test analysis methods that are based on Equation 16-48. In extreme cases, such as the test of a gas well in a microdarcy formation where a range of pressures on the order of thousands of psi are seen during the test, it may be appropriate to solve Equation 16-45 numerically (reservoir simulation), or to use pseudo-time which will be discussed shortly.

Often, over low pressure ranges, say for pressures less than 3000 psia, the product of μ and z is reasonably constant. Assuming this to be the case, then the real gas potential Equation 16-47 simplifies:

$$\psi(p) = 2 \int_a^p \left[\frac{p}{\mu z} \right] dp = \frac{2}{\overline{\mu}\,\overline{z}} \int_a^p p\; dp$$

$$\psi(p) = \frac{1}{\overline{\mu}\,\overline{z}} \left(p^2 - a^2 \right) \tag{16-49}$$

By inspection of Equation 16-49, it should be apparent that if the variations in the $(\mu)(z)$ product are minor over the pressure range considered, then changes in the real gas potential are directly related to changes in p^2. In fact, if Equation 16-49 is substituted into Equation 16-48, then an equation results that is exactly like Equation 16-6, except that the pressures are squared instead of linear:

$$\frac{1}{r} \frac{\partial}{\partial r} \left[r \frac{\partial p^2}{\partial r} \right] = \left[\frac{\phi \mu c_g}{k} \right] \frac{\partial p^2}{\partial t} \tag{16-50}$$

Thus, the analysis of pressure buildups in low pressure gas wells, where $(\mu)(z)$ is approximately a constant, can be done quite similarly to the analysis of oil well pressure buildups. It will be shown that in this situation, a semilog plot can be made, plotting p^2 instead of p.

For high pressure gas wells (above 3000 psia), quite often the quantity $p/(\mu z)$ is nearly a constant. If this is true, then a second approximation for the real gas potential can be developed:

$$\psi(p) = 2 \left[\frac{p}{\mu z} \right]_{avg} \int_a^p dp$$

$$\psi(p) = \left\{ 2 \left[\frac{p}{\mu z} \right]_{avg} \right\} (p - a) \tag{16-51}$$

Notice that at elevated pressures, Equation 16-51 suggests that (assuming $[p/(\mu z)]$ is a constant) the real gas potential varies linearly with pressure. This is an important result, because when Equation 16-51 is substituted into Equation 16-48, then an equation results that is exactly like the liquid Continuity Equation:

$$\frac{1}{r} \frac{\partial}{\partial r} \left[r \frac{\partial p}{\partial r} \right] = \left[\frac{\phi \mu c_g}{k} \right] \frac{\partial p}{\partial t} \tag{16-52}$$

Therefore, high pressure gas wells can very often be analyzed with oil well techniques such as the Horner and MDH plots and equations that have already been considered.

EVALUATION OF THE REAL GAS POTENTIAL VERSUS PRESSURE

Since the real gas potential, $\psi(p)$, is defined by an integral expression, the practical evaluation involves plotting $2(p/\mu z)$ versus pressure. The beginning pressure is the reference pressure, a; and the real gas potential associated with a particular reservoir pressure is equal to the area under the curve out to the pressure of interest. As long as the pressure increment (distance between plotted points) is not too large, say 25 psi, then a simple quadrature scheme, such as the trapezoidal rule, can be used to calculate the area (Figure 16-9).

Considering Figure 16-9, the value of the real gas potential at pressure p_1 is simply the area under the curve out to p_1. Similarly, the real gas potential at p_2 is the area under the curve to p_2. Problem 5-7, "Gas Reservoirs" chapter, illustrates the calculation procedure.

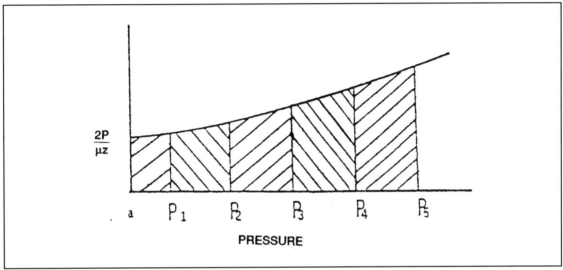

Fig. 16-9. Evaluation of Real Gas Potential versus pressure.

FLOW CAPACITY DETERMINATION

The low-pressure, gas well pressure-buildup Horner analysis [assuming that $(\mu)(z)$ is approximately a constant] is illustrated on the left in Figure 16-10. The only difference between this plot and the oil well Horner plot is that the gas plot has p^2 on the vertical axis, instead of just p (oil well).

The real gas potential can be used over any pressure range, and that Horner plot is illustrated on the right in Figure 16-10.

SKIN EFFECT DETERMINATION

The calculation of skin for a gas well is addressed in Figure 16-11. The skin computed by the equations given in Figure 16-11 actually is s_{total}.

Recall that for an oil well:

$$s_{total} = s_{true} + s_{pseudo} \tag{16-18}$$

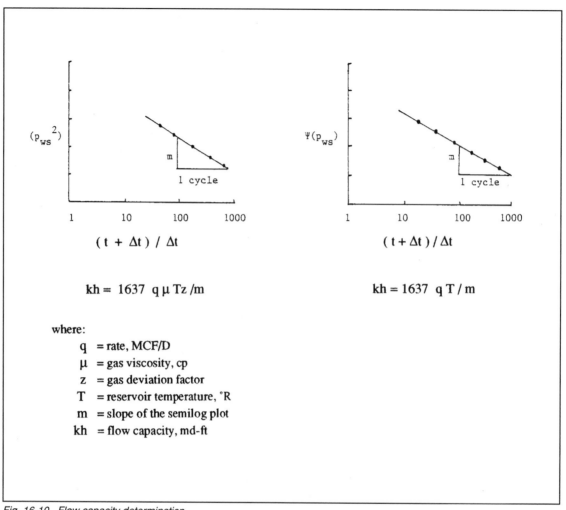

Fig. 16-10. Flow capacity determination.

For a gas well, the calculated skin is often seen to be a function of rate. This happens because the theory assumed laminar flow conditions; whereas, near-well gas flow is often turbulent. The character of the turbulence is a function of the well rate. Hence, the total skin gas-well equation needs a term relating to rate-dependent turbulence. For a gas well:

$$s_{\text{total}} = s_{\text{true}} + s_{\text{pseudo}} + (D)(q)$$ (16-53)

where:
   s_{true} and s_{pseudo} were defined under Equation 16-18,
      D = the non-Darcy flow coefficient, and
      q = the surface well flow rate.

Notice that Equation 16-53 can be written as:

$$s_{\text{total}} = s_{\text{fixed}} + (D)(q)$$ (16-54)

Real Gas Potential Analysis:

$$s = 1.151\left[\frac{\psi(p)_1 - \psi(p_{wf})}{m} - Log\left[\frac{k}{\phi\mu c_e r_w^2}\right] + 3.23\right]$$

$$\Delta\psi_s = 0.87\,m\,s; \qquad \psi(p)_{ns} = \psi(p_{wf}) + 0.87\,m\,s$$

$$C.R. = \frac{\psi^* - \psi_{wf} - \Delta\psi_s}{\psi^* - \psi_{wf}}$$

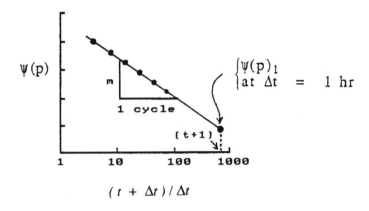

Low Pressure (p^2) Analysis:

$$s = 1.151\left\{\left[\frac{(p_w^2)_1 - (p_{wf}^2)}{m}\right] - Log\left[\frac{k}{\phi\mu c_e r_w^2}\right] + 3.23\right\}$$

$$\Delta p_s = \left[\sqrt{p_{wf}^2 + (0.87\,m\,s)}\right] - p_{wf}$$

$$C.R. = \frac{(p^*)^2 - p_{wf}^2 - (\Delta p_s)^2}{(p^*)^2 - p_{wf}^2}$$

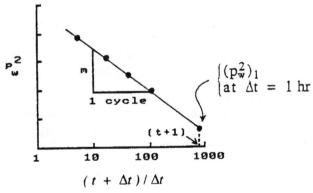

Fig. 16-11. Skin effect determination.

where:

$$s_{fixed} = s_{true} + s_{pseudo}$$

The implication of (16-54) is that if skin is calculated from several different well tests associated with different rates, then these skins (s_{total}) can be plotted versus rate. If a linear trend is seen, then s_{fixed} and the non-Darcy flow coefficient, D, can be determined:

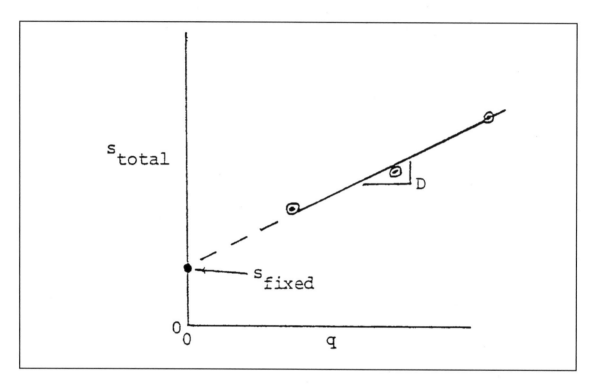

DEFINITION AND USE OF PSEUDOTIME[47]

Recall the gas-well continuity equation in terms of the real gas potential:

$$\frac{1}{r} \frac{\partial}{\partial r} \left[r \frac{\partial \psi}{\partial r} \right] = \left[\frac{\phi \mu c_g}{k} \right] \frac{\partial \psi}{\partial t}$$

(16-48)

To use the solution methods depicted in Figure 16-10 and 16-11, Equation 16-48 was assumed to be linear; i.e., ($\phi \mu c_g / k$) should not change significantly. This is normally the case.

However, if there is a large pressure change seen on a test, then Equation 16-48 may actually represent a nonlinear equation. In this event, superposition is not necessarily valid, and the solution techniques discussed thus far may not yield accurate results.

If porosity and connate water are allowed to change with pressure, then the gas well continuity equation can be written as:

$$\frac{1}{r}\frac{\partial}{\partial r}\left[r\ \frac{\partial \psi}{\partial r}\right] = \left[\frac{\phi_i\ \mu\ c_e}{k}\right]\frac{\partial \psi}{\partial t}$$

(16-55)

where:

c_e, the total compressibility, was defined with Equation 16-36; and
i = subscript indicating initial conditions.

Agarwal[47] attempted to further linearize the continuity equation by using pseudotime:

$$\tau = \int_{o}^{t}\frac{dt}{\mu\ c_e}$$

(16-56)

Another way to define pseudotime which has the benefit of retaining units of hours is[48]:

$$\tau_n = (\mu\ c_e)_i \int_{o}^{t}\frac{dt}{\mu\ c_e}$$

(16-57)

where:

τ_n = normalized pseudotime, hrs.

Substituting (57) into (55):

$$\frac{1}{r}\frac{\partial}{\partial r}\left[r\ \frac{\partial \psi}{\partial r}\right] = \frac{(\phi\mu\ c_e)_i}{k}\frac{\partial \psi}{\partial \tau_n}$$

(16-58)

Equation 16-58 is the gas continuity equation in terms of the real gas potential and normalized pseudotime. The pressure dependence of the porosity-viscosity-compressibility product has been collected within the normalized pseudotime variable. So, if test time is converted to normalized pseudotime with Equation 16-57, then the analysis methods in Figures 16-10 and 16-11 (using normalized pseudotime instead of real time) can be used without changing the equation constants. Interestingly, Reynolds *et al.*[49] have shown that pseudotime should *not* be used when analyzing drawdown data with semilog techniques.

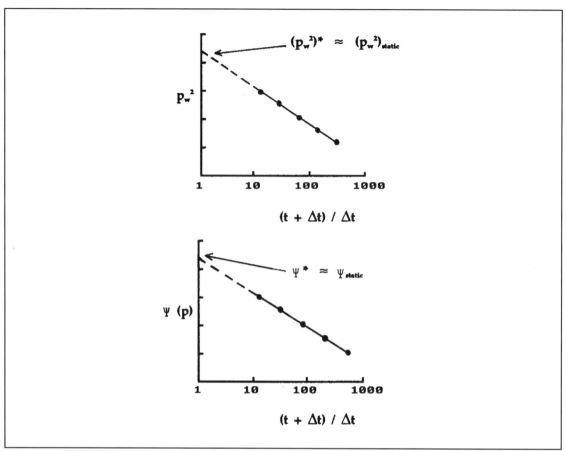

Fig. 16-12. Static pressure determination for gas wells.

CALCULATION OF STATIC PRESSURE — OIL WELLS

When discussing the calculation of static pressure over the drainage area of a test well, two different flow situations must be considered. During the production period, first, the test well is considered to be draining from an ever expanding drainage radius (infinite drainage area), and secondly, the test well is draining from a finite drainage area.

INFINITE DRAINAGE AREA

A plot of p_{ws} vs. log $[(t + \Delta t)/\Delta t]$ should yield a straight line; i.e., a Horner plot should have all of the data points, except possibly the early time data points, falling on a straight line.

Of course, there are no true infinite drainage areas. On a drillstem test or on the initial production test, before the pressure transients due to production propagate out to the drainage boundaries, the well will be infinite-acting and will not feel its boundaries. In this case, the initial system pressure still exists beyond the point to which the pressure transients have propagated.

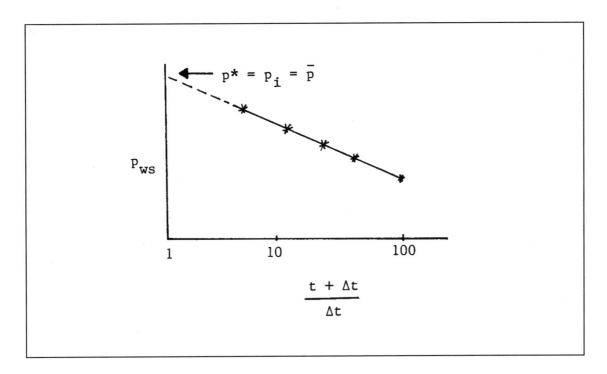

In this system, the initial pressure can be obtained by extrapolating the Horner plot straight line to the point where $[(t + \Delta t)/\Delta t] = 1.0$. Notice that this value of the Horner time function corresponds to infinite shut-in time.

So, as long as the test well is draining from a regular area in which the initial pressure still exists, that value will be the indicated pressure obtained by the extrapolation of the Horner straight line to infinite shut-in time.

FINITE OR LIMITED DRAINAGE AREA

When discussing a well producing from a finite drainage area, what is really meant is a well that has produced long enough to feel its boundaries.

Four different methods will be discussed for the calculation of static pressure for this type of well: (1) Matthews, Brons, and Hazebroek;[19] (2) Dietz;[21] (3) Miller, Dyes, and Hutchinson;[13] and (4) Muskat.[23]

MATTHEWS, BRONS, AND HAZEBROEK (MBH)[19]

This method uses the Horner plot of the pressure buildup data. However, the extrapolation of the straight line to:

$$(t + \Delta t) / \Delta t = 1.0$$

yields p*. p* is sometimes called the "false pressure" because in a mature well it is a pressure that does not really exist anywhere. Actually, p* is the pressure that would exist at the outer boundary if the pressure distribution were a steady state pressure distribution.

Figures 16-13 through 16-16 are from Earlougher.[20] These curves were originally developed by Matthews, *et al.*[19] to be used for the determination of static pressure. The charts are for different drainage shapes and different relative positions of the well within the drainage area. (See Figures 16-13 through 16-16).

Figure 16-13 is used in the majority of cases and handles a well in the center of a: (1) circle, (2) hexagon, (3) square, (4) equilateral triangle, (5) rhombus, or (6) right triangle.

In the MBH figures, the ordinate is the Matthews-Brons-Hazebroek dimensionless pressure which is defined as:

$$p_{DMBH} = 2.303 \ (p* - \bar{p}) \ / \ m \qquad (16\text{-}59)$$

where:

$p*$ = pressure indicated by extrapolating the Horner plot straight line to $[(t + \Delta t) \ / \ \Delta t] = 1.0$
$\bar{p}$ = static well pressure, psia
m = slope of the Horner plot straight line
t = producing time, hours
Δt = shut-in time, hours

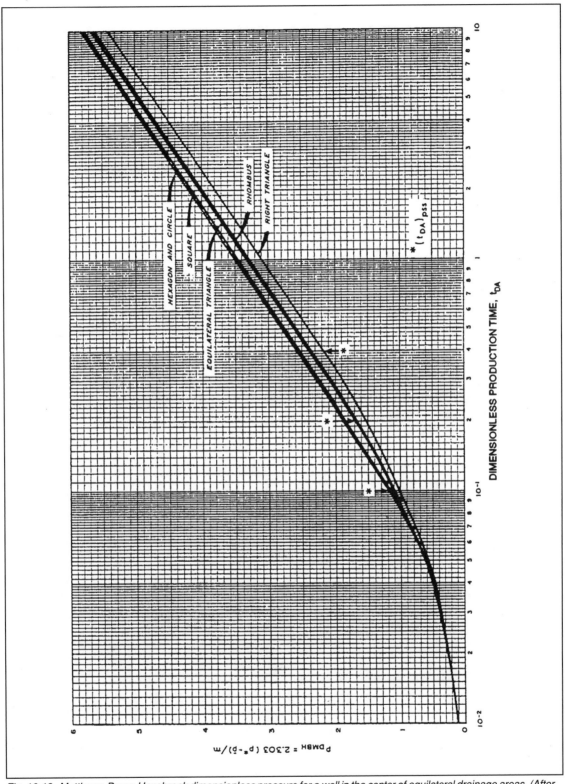

Fig. 16-13. Matthews-Brons-Hazebroek dimensionless pressure for a well in the center of equilateral drainage areas. (After Matthews, Brons, and Hazebroek[19]). Permission to publish by SPE.

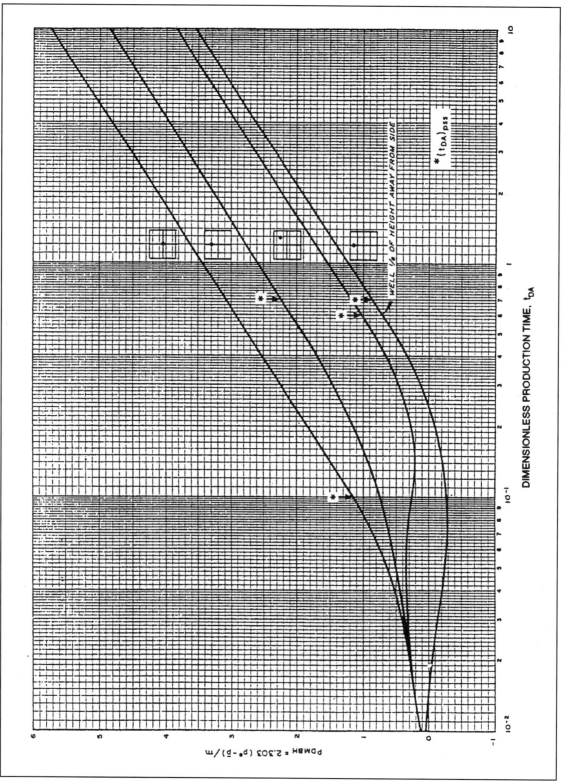

Fig. 16-14. Matthews-Brons-Hazebroek dimensionless pressure for different well locations in a square drainage area. (After Matthews, Brons, and Hazebroek[19]). Permission to publish by SPE.

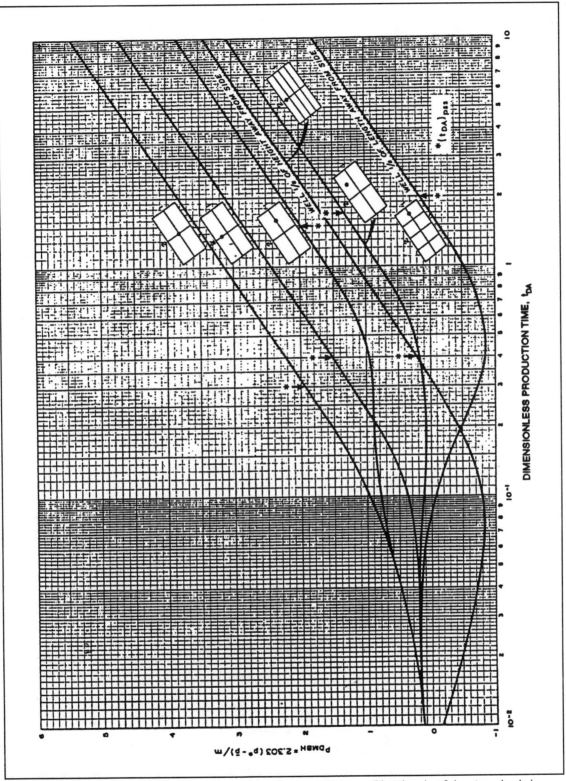

Fig. 16-15. Matthews-Brons-Hazebroek dimensionless pressure for different well locations in a 2:1 rectangular drainage area. (After Matthews, Brons, and Hazebroek[19]). Permission to publish by SPE.

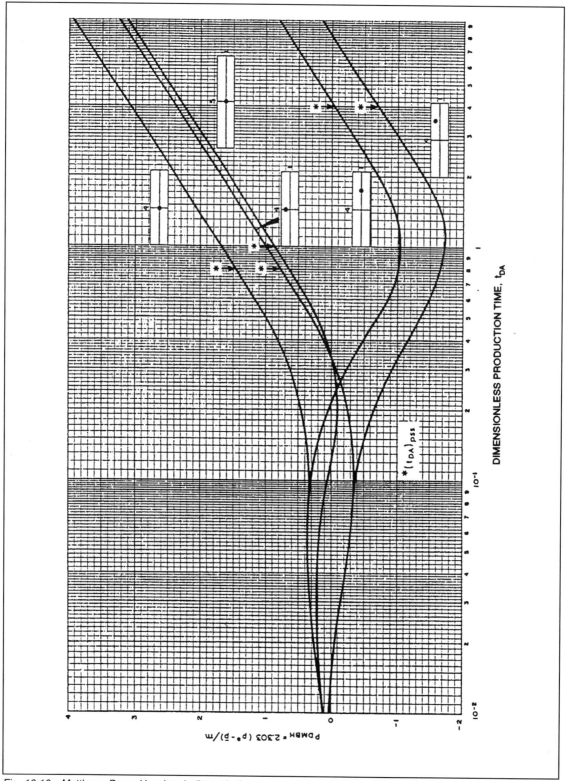

Fig. 16-16. Matthews-Brons-Hazebroek dimensionless pressure for different well locations in 4:1 and 5:1 rectangular drainage areas. (After Matthews, Brons, and Hazebroek[19]). Permission to publish by SPE.

The abscissa is the dimensionless producing time based on drainage area:

$$t_{DA} = \frac{2.64 \times 10^{-4}}{\phi\, c_e\, A} \left[\frac{k}{\mu} \right]_t (t)$$ (16-60)

where:

t = producing time, hours,

ϕ = porosity, fraction,

c_e = total or effective compressibility, psi^{-1}

A = drainage area of the well, ft^2,

$[k/\mu]_t$ = total mobility, md. / cp.

CALCULATION PROCEDURE

1. Plot the pressure buildup data on a Horner plot:

$$p_{ws} \; or \; \Delta p_{ws} \; vs. \; Log \; [(t + \Delta t) \,/\, \Delta t]$$

(Semilog graph paper is normally used.)

2. Determine the slope, m, of the straight line data; and extrapolate the straight line to:

$$[(t + \Delta t) \,/\, \Delta t] = 1.0.$$

This pressure is p*.

3. Choose from Figures 16-13 through 16.16 the drainage shape that most nearly corresponds to the actual well drainage shape.

4. Calculate the dimensionless producing time:

$$t_{DA} = \frac{(2.64)(10^{-4})(t)}{\phi\, c_e\, A} \left[\frac{k}{\mu} \right]_t$$

5. Using the appropriate figure and the dimensionless producing time from step 4, determine dimensionless pressure:

$$p_{DMBH} = 2.303 \left[\frac{p^* - \bar{p}}{m} \right]$$

6. Calculate the well static pressure:

$$\bar{p} = p^* - \left[\frac{m}{2.303} \right] \left[p_{DMBH} \right]$$

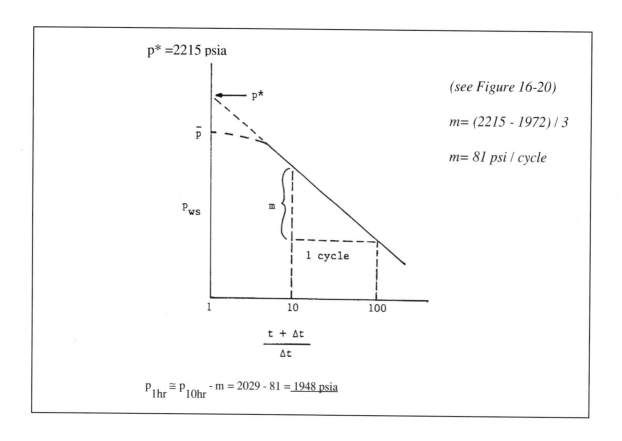

$$p_{1hr} \cong p_{10hr} - m = 2029 - 81 = \underline{1948\ psia}$$

DIETZ METHOD[21]

This method uses the results of the work of Matthews *et al.*[19] If one were to inspect Figures 16-13 through 16-16, it would be evident that the late time portion of each of these curves are straight lines with a common slope. Dietz[21] used this characteristic to develop a simpler method of evaluating the static pressure.

The Dietz method is only valid after the dimensionless producing time has reached a certain value, i.e., the well should be producing at semi-steady state before shut-in. The amount of needed producing time (before shut-in) is dependent on drainage shape. The more irregular the drainage shape or the more off-center the well is, then the longer is the required producing time.

The amount of dimensionless producing time (based on drainage area) that is needed may be found in Column 4 of Table 16-1. Column 4 is headed "Exact for $t_{DA} > .$" These values indicate the minimum value of dimensionless producing time that is needed to use the Dietz method for exact answers. However, if the dimensionless producing time is less than the value in Column 4 but greater than the value in Column 5, then the Dietz method may still be used with less than one percent error.

CALCULATION PROCEDURE — DIETZ METHOD

1. Plot the pressure buildup data using either the Horner representation or the MDH plot (p_{ws} vs. log Δt).

2. Determine the slope of the straight line portion of the pressure buildup data, m, psi/cycle.

3. Calculate the value of dimensionless producing time (based on drainage area):

$$t_{DA} = \frac{(2.64)(10^{-4})(t)}{\phi\, c_e\, A} \left[\frac{k}{\mu} \right]_t$$

4. Compare the value of t_{DA} with the limiting value for the drainage shape used. If the value calculated in step 3 is larger than the value given in Column 4 or 5 (depending on the accuracy required) of Table 16-1, then the Dietz method can be used.

5. Calculate the value of shut-in time at which to read the static pressure from the extrapolated straight line of the semilog plot.

$$\Delta t_s = 3792\ \phi \left[\frac{\mu}{k} \right]_t \left[\frac{c_e\, A}{C_A} \right] \tag{16-61}$$

where:

C_A = the shape factor for the drainage area and well geometry, taken from Col. 1 of Table 16-1, and

Δt_s = the shut-in time to which the straight line of the semilog plot is extrapolated to read static pressure

6. If the Horner plot has been used in the analysis, then the Horner time function must be calculated that corresponds to Δt_s:

$(t + \Delta t_s)\ /\ \Delta t_s$

Then, extrapolate the Horner plot straight line to this value of the Horner time function to read the static pressure.

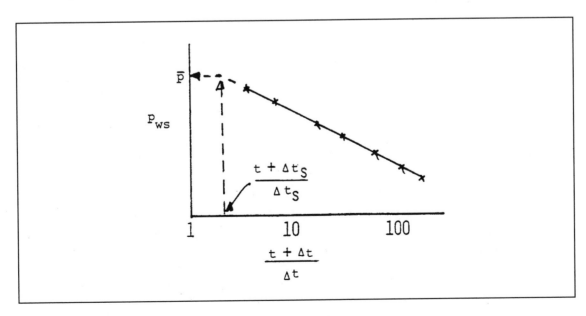

Table 16-1
Shape factors for various closed single-well drainage areas. (From Earlougher[20]).

IN BOUNDED RESERVOIRS	C_A	$\ln C_A$	$\frac{1}{2}\ln\left(\frac{2.2458}{C_A}\right)$	EXACT FOR $t_{DA}>$	LESS THAN 1% ERROR FOR $t_{DA}>$	USE INFINITE SYSTEM SOLUTION WITH LESS THAN 1% ERROR FOR $t_{DA}<$
(circle)	31.62	3.4538	−1.3224	0.1	0.06	0.10
(hexagon)	31.6	3.4532	−1.3220	0.1	0.06	0.10
(triangle)	27.6	3.3178	−1.2544	0.2	0.07	0.09
(parallelogram 60°)	27.1	3.2995	−1.2452	0.2	0.07	0.09
(right triangle 1/3)	21.9	3.0865	−1.1387	0.4	0.12	0.08
(triangle 3,4)	0.098	−2.3227	+1.5659	0.9	0.60	0.015
(square)	30.8828	3.4302	−1.3106	0.1	0.05	0.09
(square 2×2 centered)	12.9851	2.5638	−0.8774	0.7	0.25	0.03
(square 2×2 offset)	4.5132	1.5070	−0.3490	0.6	0.30	0.025
(square 3×2 offset)	3.3351	1.2045	−0.1977	0.7	0.25	0.01
(rectangle 1×2 centered)	21.8369	3.0836	−1.1373	0.3	0.15	0.025
(rectangle 1×2)	10.8374	2.3830	−0.7870	0.4	0.15	0.025
(rectangle 1×2 offset)	4.5141	1.5072	−0.3491	1.5	0.50	0.06
(rectangle 1×2 offset)	2.0769	0.7309	+0.0391	1.7	0.50	0.02
(rectangle 1×2 offset)	3.1573	1.1497	−0.1703	0.4	0.15	0.005

Table 16-1 (Continued)

	C_A	$\ln C_A$	$\frac{1}{2}\ln\left(\frac{2.2458}{C_A}\right)$	EXACT FOR $t_{DA}>$	LESS THAN 1% ERROR FOR $t_{DA}>$	USE INFINITE SYSTEM SOLUTION WITH LESS THAN 1% ERROR FOR $t_{DA}<$
	0.5813	−0.5425	+0.6758	2.0	0.60	0.02
	0.1109	−2.1991	+1.5041	3.0	0.60	0.005
	5.3790	1.6825	−0.4367	0.8	0.30	0.01
	2.6896	0.9894	−0.0902	0.8	0.30	0.01
	0.2318	−1.4619	+1.1355	4.0	2.00	0.03
	0.1155	−2.1585	+1.4838	4.0	2.00	0.01
	2.3606	0.8589	−0.0249	1.0	0.40	0.025

IN VERTICALLY-FRACTURED RESERVOIRS   USE $(x_e/x_f)^2$ IN PLACE OF A/r_w^2 FOR FRACTURED SYSTEMS

	C_A	$\ln C_A$		EXACT FOR $t_{DA}>$	LESS THAN 1% ERROR FOR $t_{DA}>$	USE INFINITE... FOR $t_{DA}<$
0.1 = x_f/x_e	2.6541	0.9761	−0.0835	0.175	0.08	CANNOT USE
0.2	2.0348	0.7104	+0.0493	0.175	0.09	CANNOT USE
0.3	1.9986	0.6924	+0.0583	0.175	0.09	CANNOT USE
0.5	1.6620	0.5080	+0.1505	0.175	0.09	CANNOT USE
0.7	1.3127	0.2721	+0.2685	0.175	0.09	CANNOT USE
1.0	0.7887	−0.2374	+0.5232	0.175	0.09	CANNOT USE

IN WATER-DRIVE RESERVOIRS

	C_A	$\ln C_A$				
	19.1	2.95	−1.07	—	—	—

IN RESERVOIRS OF UNKNOWN PRODUCTION CHARACTER

	C_A	$\ln C_A$				
	25.0	3.22	−1.20	—	—	—

MILLER, DYES, AND HUTCHINSON (MDH) METHOD[13]

This method is similar to the Dietz method in its application. A shut-in time will be calculated to which the semilog plot straight line is extrapolated to determine the static pressure.

The principal drawback of this method is that it is limited to either a square or circular drainage shape. Further, the producing time must have been sufficient to have reached stabilization. If these two restrictions are met, then the MDH and Dietz methods produce essentially the same results.

Although the MDH method has extra inherent restrictions, there are also extra benefits to be derived. The method not only has application to wells with no-flow boundaries, but also to wells that are affected by constant pressure boundary conditions. Notice that this would be the case for a well surrounded by injectors. A pseudo-constant pressure boundary condition (on one side, anyway) would also be generated by a gas cap near the test well. Tracy[38] has extended this method to include the situation where the test well has one half no-flow boundaries and one half constant pressure boundaries.

CALCULATIONS OF SHUT-IN TIME TO DETERMINE STATIC PRESSURE

Case 1: No-flow boundary or test well offset by producers.

$$\Delta t_s = 120.7 \left[\frac{\phi \mu c_e A}{k} \right] \; hours \tag{16-62}$$

Case 2: 50% no-flow boundary and 50% constant pressure boundary.

(Example: Test well offset by producers on one side and injectors or a gas cap on the other side.)

$$\Delta t_s = 253.5 \left[\frac{\phi \mu c_e A}{k} \right] \; hours \tag{16-63}$$

Case 3: Constant pressure boundary. (Example: Well surrounded by injectors on all sides)

$$\Delta t_s = 537.1 \left[\frac{\phi \mu c_e A}{k} \right] \; hours \tag{16-64}$$

If an MDH plot is being used in the analysis, then the appropriate extrapolation shut-in time is calculated. Next, the semilog straight line is extrapolated to this time to read the static pressure as illustrated in Figure 16-17.

The case 1 extrapolation (to Δt_1) yields $\bar{p}$, while cases 2 and 3 do not. With a constant pressure boundary, extrapolating to Δt_3 yields p_e, the pressure at the boundary. The case 2 result[38] is a line integral average pressure (approximately $[p_e + \bar{p}] / 2$.

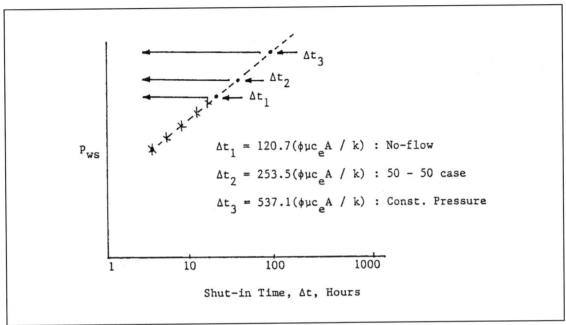

$\Delta t_1 = 120.7(\phi\mu c_e A \,/\, k)$: No-flow

$\Delta t_2 = 253.5(\phi\mu c_e A \,/\, k)$: 50 - 50 case

$\Delta t_3 = 537.1(\phi\mu c_e A \,/\, k)$: Const. Pressure

Fig. 16-17. Miller, Dyes and Hutchinson method.[13] Permission to publish by SPE.

Figure 16-18 contains some of the results of the Miller *et al.*[13] work. Dimensionless pressure is plotted versus dimensionless shut-in time for the constant pressure boundary case and for the no-flow boundary situation. By inspection of the no-flow boundary curve, it can be seen that the end of the straight line portion occurs at an approximate value of $\Delta t_{DA} \approx 0.009$. This result can be used to determine the real time at which the semilog plot straight line should end.

$$\Delta t_{DA} = 0.009 = \frac{(2.64)(10^{-4})(k)(\Delta t_{end})}{\phi\,\mu\,c_e\,A}$$

$$\Delta t_{end} = 34.1\,\frac{\phi\,\mu\,c_e\,A}{k}\quad hours \tag{16-65}$$

Problem 16-3:

To illustrate the previous MDH equation for the determination of the end of the semilog straight line data, the following data will be used:

Permeability = 50 md.

Porosity = 0.15

Viscosity = 1.5 cp.

Compressibility = 25.0×10^{-6} psi^{-1}

Area = 80 acres = 3.48×10^{6} ft^{2}

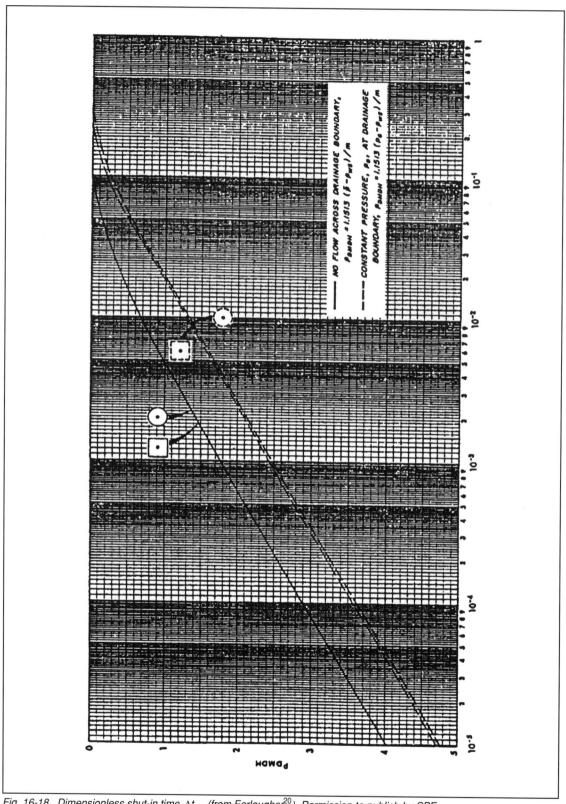

Fig. 16-18. Dimensionless shut-in time, Δt_{DA} (from Earlougher[20]). Permission to publish by SPE.

Therefore:

$$\Delta t_{end} = (34.1) \frac{\phi \mu c_e A}{k}$$

$$= (34.1)(.15)(1.5)(25)(10^{-6})(3.48)(10^6)/50$$

$$= 13.5 \text{ hours} = \text{end of the straight line data.}$$

Pressures after this time will tend to flatten.

MUSKAT METHOD[23]

This method is the only one considered here that uses late transient (or late time) pressure buildup data. In fact, the method requires that the data be at times later than the end of the semilog straight line data. Only pressures collected during the flattening part of the semilog buildup plot are applicable. The Muskat method data are in the transition period between infinite-acting and fully-equalized pressures. These pressures are being affected by the drainage boundaries.

The Muskat method yields results that are more approximate than those achieved by the classical semilog methods. More shut-in time is required to reach the Muskat data. However, there are times when the late transient buildup data arrive fairly quickly, and the classical semilog straight line can be completely obscured by wellbore storage and skin effects. In this case — particularly in high permeability systems with small total compressibility, and/or small well spacing — the Muskat method finds use. Normally, economic factors preclude the use of the Muskat method due to the long shut-in times usually needed.

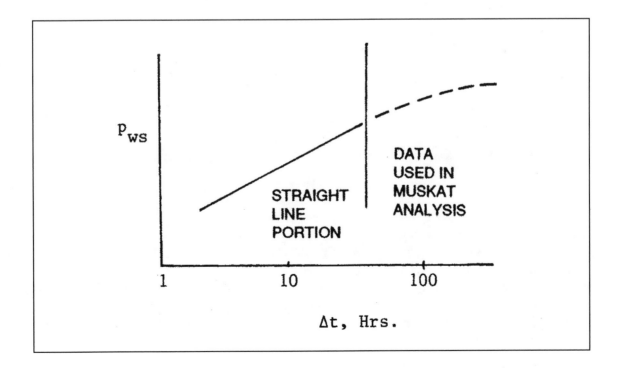

The Muskat method is only valid for a regular drainage area with the well in the center. The circular shape flow capacity equation gives a result that is about 25% higher than the square drainage area equation. Because Muskat flow capacity results are normally lower then those calculated by the more accurate semilog methods, the closed circular drainage area equation is reported here. Not only can flow capacity be calculated with the Muskat method, but also an indication of the well's drainage pore volume can be obtained. The computed pore volume is likely to be even more approximate than the flow capacity result. Finally, a good approximation of static pressure results from this method.

The following is the equation that describes the well pressure behavior during the late transient period of time:

$$Log\,(\bar{p} - p_{ws}) = Log\,\frac{118.6\,q\,\mu\,B}{k\,h} - 0.00528\,\frac{k\,\Delta t}{\phi\,\mu\,c_e\,A} \qquad (16\text{-}66)$$

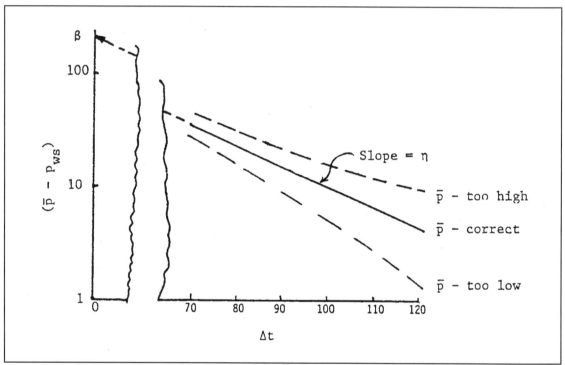

Fig. 16-19. Static pressure determination — Muskat method.[23] Permission to publish by SPE.

The Muskat procedure is a trial and error one. Normally, the procedure is to assume a static pressure ($\bar{p}$) that is a few psi higher than the last measured buildup pressure. (Remember that the Muskat method uses late transient data only.) Then, a value of ($\bar{p}$ - p_{ws}) is calculated for each pressure point, p_{ws}, recorded during the late transient period. Then, a plot, such as the one in Figure 16-19 is made. Note that the vertical ($\bar{p}$ - p_{ws}) scale is a log scale, and the horizontal shut-in time scale is linear. The correct value of $\bar{p}$ will result in a straight line plot. If the assumed $\bar{p}$ is too high, then the resulting curve will be concave upward. A curve that is concave downward indicates that the assumed $\bar{p}$ is too low.

The extrapolated y-intercept of the Muskat plot, β, is related to the flow capacity.

Flow Capacity Determination:

$$Intercept = \beta = 118.6 \left[(q \mu B) / (k h) \right]$$

$$k h = 118.6 (q \mu B) / \beta \tag{16-67}$$

The slope, η (cycles per hour), of the Muskat straight line is related to the drainage area as:

$$\eta = \frac{0.00528 \, k}{\phi \mu c_e A} \tag{16-68}$$

This result may be combined with the flow capacity equation to approximate the well's drainage pore volume.

Pore Volume Approximation:

$$Pore \ Volume = \frac{\phi A h}{5.615} \ bbls$$

$$\beta = 118.6 \, \frac{q \mu B}{k h}$$

$$h = 118.6 \, \frac{q \mu B}{k \beta}$$

$$\eta = Slope = \frac{0.00528 \, k}{\phi \mu c_e A}$$

$$\phi A = \frac{0.00528 \, k}{\eta \mu c_e}$$

$$Pore \ Volume = PV = \left[\frac{0.00528 \, k}{\eta \mu c_e} \right] \left[118.6 \, \frac{q \mu B}{k \beta} \right] \left(\frac{1}{5.615} \right) bbls$$

$$PV = \frac{0.1115 \, (q B)}{\eta \beta c_e} \ bbls \tag{16-69}$$

STATIC PRESSURE DETERMINATION — GAS WELLS

Gas wells are normally completed with large spacing, typically at least 160 acres/well. This results in $\bar{p}$ and p^* being quite close together. (This can be proved by using a large "A" with the MBH method.) Thus, the normal practice is simply to use p^* for $\bar{p}$ as indicated in Figure 16-12.

For those desiring a more accurate approximation for $\bar{p}$, the MBH method can be used. However, when the liquid MBH method is applied to gas wells with long producing times before shut-in, the results may not be reliable. Reynolds *et al.*[49] have suggested a more rigorous procedure.

Problem 16-4:

Pressure Buildup in a Well Producing From a Limited Drainage Area

A well is completed in a reservoir where the well spacing is 40 acres per well and is known to have the following information and fluid properties:

Initial pressure, p_i	2400 psia
Producing rate prior to shut-in	50 STB/D
Pay thickness, h	20 feet
Porosity, ϕ	15%
Wellbore radius, r_w	0.25 feet
Oil viscosity, μ_0	0.8 cp
Oil formation volume factor, B_o	1.25 Bbl/STB
Effective compressibility, c_e	19.8×10^{-6} psi^{-1}
Cumulative production	4167 STBO
Drainage area = (40) (43,560)	1.742×10^6 ft^2

Calculation of effective producing time:

$$t = [(4167)(24)] / 50 = 2000 \ hr.$$

Pressure Buildup Data:

Shut-in Time, Δt, hrs	p_{ws}, psia	$(t + \Delta t)/\Delta t$
0	1565	—
1	1949	2001
3	1988	667.7
5	2006	401
7	2017	286.7
9	2026	223.2
10	2029	201
15	2043	134
20	2052	101
25	2060	81
30	2066	67.7
35	2070	58.1
40	2074	51.0
45	2078	45.5
50	2081	41.0
60	2086	34.4
70	2090	29.6
80	2093	26.0
90	2096	23.2
100	2098	21.0
110	2100	19.2
120	2102	17.7
130	2103	16.4

Determine the following:

1. kh, s, Δp_{skin}

2. Static pressure by:

   a. Matthews, Brons, and Hazebroek (MBH),

   b. Dietz,

   c. Miller, Dyes, and Hutchinson (MDH), and

   d. Muskat method.

3. Determine kh and approximate pore volume using the Muskat analysis.

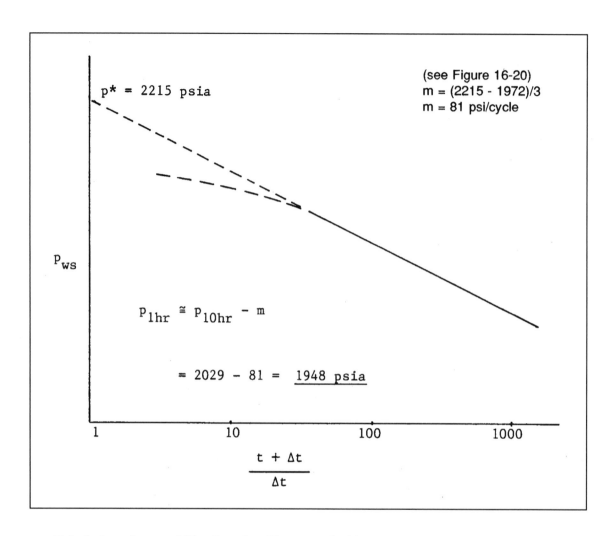

Calculation of permeability (based on Horner analysis):

$$kh = \frac{162.6 \, q \, \mu \, B}{m} = \frac{(162.6)(50)(0.8)(1.25)}{81} = 100.4 \; md\text{–}ft$$

$$k = kh / h = \frac{100.4}{20} = 5.02 \; md$$

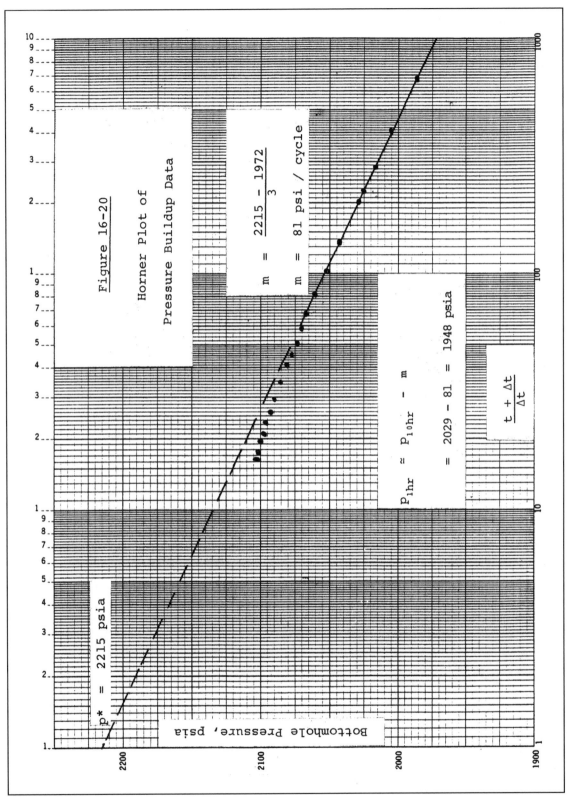

Fig. 16-20. Horner type plot of pressure buildup data.

Calculation of the end of the semilog straight line data:

$$\Delta t_{end} = (34.1) \frac{\phi \mu c_e A}{k}$$

$$= \frac{(34.1)(0.15)(0.8)(19.8 \times 10^{-6})(1.74 \times 10^{+6})}{5.02} = 28.1 \; hrs$$

Calculation of skin effect, s:

$$s = 1.151 \left\{ \frac{p_{1hr} - p_{wf}}{m} - Log \left[\frac{k}{\phi \mu c_e r_w^2} \right] + 3.23 \right\}$$

$$= (1.151) \left\{ \frac{1948 - 1565}{81} - Log \left[\frac{5.02}{(.15)(.8)(19.8 \times 10^{-6})(.25)^2} \right] + 3.23 \right\}$$

$$= 0.494$$

So

$$\Delta p_{skin} = 0.87(m)(s) = (0.87)(81)(0.494) = 35 \; psi$$

Skin Damage is positive, but how bad?

Last measured pressure = 2103

Flowing pressure = 1565

Approximate Δp = 538

Skin Ratio = 35/538 = 0.065

So, 6.5% of the total drawdown is not being used for flow of fluids to the wellbore. This portion of the total drawdown is the additional amount of pressure drop needed just to get the existing fluid production through the damaged zone.

STATIC PRESSURE DETERMINATION

MATTHEWS, BRONS, AND HAZEBROEK METHOD:

Calculate dimensionless producing time:

$$t_{DA} = \frac{2.64 \times 10^{-4} \, k \, t}{\phi \, \mu \, c_e \, A}$$

$$= \frac{(2.64)(10^{-4})(5.02)(2000)}{(0.15)(0.8)(19.8)(10^{-6})(1.742)(10^6)}$$

$$= 0.640$$

From Figure 16-13 (square drainage area):

pDMBH = 3.03

$$\bar{p} = p^* - (m \, / \, 2.303)(\text{pDMBH})$$

$$= 2215 - [(81)(3.03)/2.303] = 2215 - 107$$

$$= 2108 \text{ psia}$$

DIETZ METHOD:

Since the dimensionless producing time, $t_{DA} = 0.64$, is greater than the value given in Column 4 of Table 16-1 (square with well in the center), the Dietz method can be used.

From Column 1 of Table 16-1, the shape factor, $C_A = 30.9$ for a square drainage area.

$$\Delta t_s = \frac{3792 \, \phi \, c_e \, A}{C_A} \left[\frac{\mu}{k} \right]_t$$

$$= \frac{(3792)(0.15)(19.8)(10^{-6})(1.742)(10^6)(0.8)}{(30.9)(5.02)}$$

$$= 101 \text{ hrs}$$

To convert this shut-in time to the corresponding Horner time function:

$$(t + \Delta t_s) \, / \, \Delta t_s = (2000 + 101) \, / \, 101 = 20.8$$

Extrapolating the Horner plot straight line to a time ratio of 20.8 yields a static pressure of 2109 psia (Figure 16-20).

MILLER, DYES, AND HUTCHINSON METHOD:

$$\Delta t_s = 120.7 \frac{\phi \, \mu \, c_e \, A}{k} \quad \textit{(no-flow boundary)}$$

$$= \frac{(120.7)(0.15)(0.8)(19.8)(10^{-6})(1.742)(10^6)}{5.02}$$

$$= 99.5 \; hrs$$

Evaluating the corresponding Horner time function:

$$(t + \Delta t_s) \, / \, \Delta t_s = (2000 + 99.5) \, / \, 99.5 = 21.1$$

Extrapolating the Horner plot straight line to a time ratio of 21.1 yields a static pressure of 2108 psia (Figure 16-20).

MUSKAT METHOD:

The very late time pressure buildup data work best in the Muskat method. As the end of the semilog data was calculated to be almost 30 hours, therefore, only data collected at 60 hours and after will be considered.

To start, guess $\bar{p}$ to be slightly higher than the last measured pressure.

Next, calculate ($\bar{p}$ - p_{ws}) for each pressure point from 60 hours to the last measured pressure point. The calculations are shown below for three investigated pressures: 2106, 2112 and 2118 psia.

Δt	$\bar{p} = 2106$ $\bar{p}$ - p_{ws}	$\bar{p} = 2112$ $\bar{p}$ - p_{ws}	$\bar{p} = 2118$ $\bar{p}$ - p_{ws}
60	20	26	32
70	16	22	28
80	13	19	25
90	10	16	22
100	8	14	20
110	6	12	18
120	4	10	16
130	3	9	15

From Figure 16-21, the pressure that produced the best straight line on the Muskat plot is:

$$\bar{p} = 2112 \; psia$$

With this pressure, the y-intercept is:

$$\beta = 63 \; psi$$

The slope:

$$\eta = 0.00653 \; cycles/hr$$

Flow Capacity (Muskat Method):

$$kh = 118.6 \; \frac{(q \mu B)}{\beta}$$

$$kh = [\,(\,118.6\,)(\,50\,)(\,0.8\,)(\,1.25\,)\,]\,/\,63 \;=\; 94.1 \; md\text{-}ft.$$

Pore Volume Determination (Muskat Method):

$$PV = \frac{0.1115 \, q \, B}{\eta \, \beta \, c_e} = \frac{[\,(\,0.1115\,)(\,50\,)(\,1.25\,)\,]}{(\,0.00653\,)(\,63\,)(\,19.8\,)(\,10^{-6}\,)} = 856{,}000 \; bbls.$$

Pore volume determination from basic data (for the sake of comparison):

$$PV = \frac{A \, \phi \, h}{5.615} = \frac{(\,1.742\,)(\,10^{6}\,)(\,0.15)\,(20)}{5.615} = 931{,}000 \; bbls$$

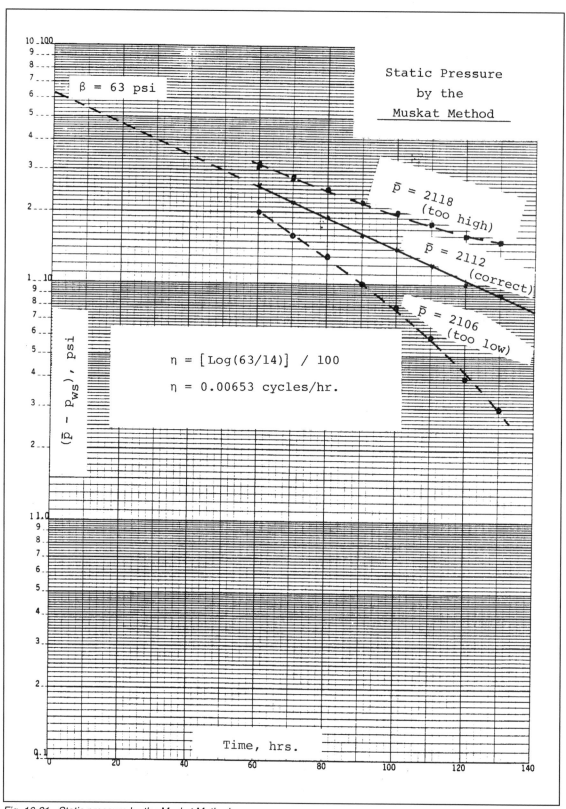

Fig. 16-21. Static pressure by the Muskat Method.

TYPE CURVE MATCHING

The theory governing pressure buildup analysis assumes that the well is shut-in at the formation face. Normally, except in the case of drillstem tests, a well is shut-in at the surface, not downhole. Therefore, fluids continue to enter the wellbore through much, if not all, of the shut-in period. However, the rate of fluid entry generally drops to a negligible value after enough shut-in time. In some cases the shut-in time required to reach the point where fluid entry does not adversely affect the pressure buildup data is a major part of the allocated shut-in time.

Type curve analysis was developed to analyze the data collected from wells where there are significant "afterflow" effects for most of the shut-in time. There have been numerous type curves developed. A type curve technique is merely a graphical method of solving a difficult analytical problem.

Typically, if we plot data from a pressure buildup test, such as using an MDH analysis, the plot might look like this:

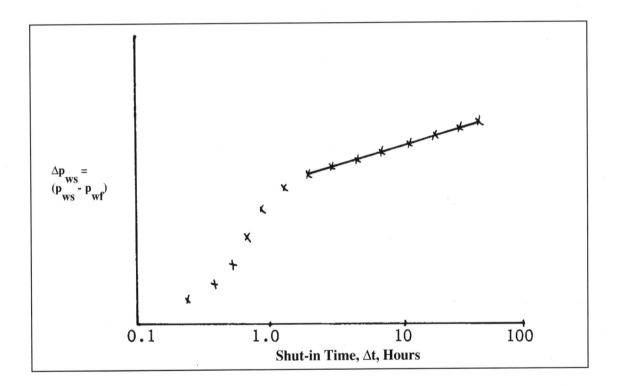

Note that the vertical pressure scale is not bottomhole shut-in pressure. Instead, the difference between the bottomhole shut-in pressure and the flowing pressure (just before shut-in) has been plotted. This is still a valid MDH plot.

Notice that there is usually some early time pressure data that do not fall on the extrapolated semilog straight line. These early points are being affected by "afterflow" or "wellbore storage." Early time data typically plot below the extrapolated straight line. If these data are replotted on log-log paper, the shape usually looks like:

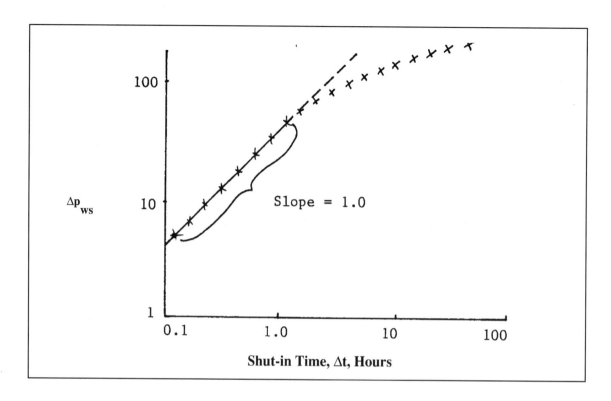

Notice that the first group of data points is linear with a slope of one. These data represent the "afterflow" period. It can be shown that the volumetric rate being delivered to the well has not changed substantially from that which existed just before shut-in. It is possible for this "afterflow" period to exist for a significant length of time only if the pressure drop within the damaged zone is large compared to that over the rest of the drainage area. This means that if there is a large positive skin effect, the unit slope above can last for a while; but without much skin, then the "afterflow" period is short.

A so-called "afterflow-dominated" buildup is one where all of the buildup data, when graphed on the log-log plot, deviates only slightly from the unit slope line. In this situation, if analysis is attempted, different combinations of permeability and skin can give rise to the same afterflow-dominated pressure response.

Two rules of thumb are used in the industry for graphically determining the time to the beginning of well test data that can be analyzed by classical semilog techniques. The first one is illustrated in Figure 16-22 and Equation 16-70.

Notice that:

Δt_1 = the amount of shut-in time to the end of the log-log unit slope line, and

Δt_2 = the amount of shut-in time needed to reach the end of wellbore storage effects. Note that this is also the beginning of the data that can be analyzed by semilog techniques.

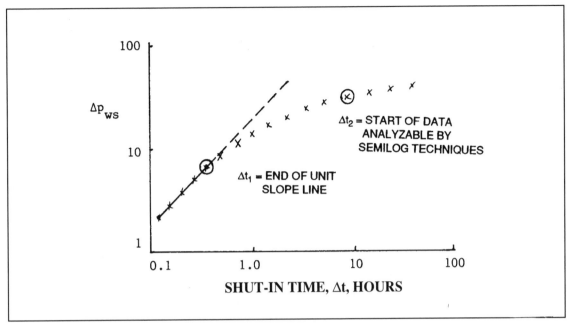

Fig. 16-22. Determining the end of wellbore storage.

$$\Delta t_2 \approx (31 \text{ to } 50) (\Delta t_1) \qquad\qquad (16\text{-}70)$$

The previous equation indicates that the semilog data begins somewhere between 31 and 50 times the amount of time it took to reach the end of the log-log straight line. The second rule of thumb[3] also uses the log-log plot of Δp vs. Δt. The semilog analysis data begins between 1 and 1-1/2 cycles in time beyond the end of the unit slope data. Notice that the first rule of thumb suggests that if the unit slope line lasts 2 hours or more, there is a very good chance that the buildup will not last long enough to obtain any pressure data that can be used for semilog analysis. However, type curve analysis can still be used.

Three different type curve methods will be considered in detail: Agarwal *et al.*, Bourdet and Gringarten, and McKinley. The first two use type curve solutions that were developed for a drawdown test. These can be used effectively for buildups as long as the producing time is long enough. If not, then superposition should be used before applying the method. The McKinley type curve discussed here was developed for build-ups, but does not take into account the length of the producing time before the shut-in.

AGARWAL, *et al.* TYPE CURVES[26]

Four different dimensionless variables are used:

      (1) Dimensionless pressure drop, P_d,

      (2) Dimensionless time, t_d,

      (3) Skin effect, s, and

      (4) Dimensionless wellbore storage coefficient, $\overline{C}$.

The definition of dimensionless pressure drop is dependent on whether the test well is an oil well or a gas well:

$$P_d = \frac{k\,h\,(\Delta p)}{141.2\,q\,\mu\,B} \qquad oil\ wells \qquad\qquad (16\text{-}71)$$

$$P_d = \frac{k\,h\,(\Delta p^2)}{1424\,q\,\mu\,T\,z} \qquad gas\ wells \qquad\qquad (16\text{-}72)$$

where:

T = temperature, °R

It should be noted that in both of these relationships, the right-hand side is made up of terms that are regarded as constants for a given well test, except for the pressure drop factor.

The dimensionless time is defined on the basis of the wellbore radius:

$$t_d = \frac{2.64 \times 10^{-4}\,k\,t}{\phi\,\mu\,c_e\,r_w^2} \qquad\qquad (16\text{-}73)$$

Only the time (hours) varies on the right hand side.

The skin effect, s, is the same as previously discussed and is a measure of the amount of flow resistance near the wellbore (due to damage or stimulation).

The dimensionless storage coefficient is defined as:

$$\overline{C} = \frac{5.615\,C}{2\,\pi\,h\,\phi\,r_w^2\,c_e} \qquad\qquad (16\text{-}74)$$

where:

C = the wellbore storage coefficient, bbls/psi,
h = the net pay thickness, ft,
ϕ = the porosity, fraction,
r_w = wellbore radius, ft, and
c_e = total reservoir compressibility, psi^{-1}

The wellbore storage coefficient is evaluated differently depending on well conditions. For a full wellbore, this term represents the product of the average compressibility of the fluids in the wellbore and the available wellbore volume:

$$C = (V_w)(c_{avg}) \qquad\qquad (16\text{-}75)$$

Therefore, for a gas well, the value of C can be calculated as:

$$C = (V_w)(c_g) \approx (V_w)\left[\frac{2}{p_{BHP} + p_{SURF}}\right] \qquad\qquad (16\text{-}76)$$

For oil wells producing significant quantities of water and/or free gas, the value of C can be calculated by using a producing-rate weighted average of the compressibilities of the different fluids:

$$c_{avg} = \frac{B_o c_o + (GOR - R_s) B_g c_g + (WOR) c_w}{B_o + (GOR - R_s) B_g + WOR} \qquad (16\text{-}77)$$

Then, the wellbore storage coefficient is:

$$C = (c_{avg})(V_w)$$

For clean oil wells, the wellbore storage coefficient is normally calculated from data collected during the log-log plot unit slope. If the correlation of the pressure change (p_{ws} - p_{wf}) is linear with shut-in time, Δt, then a plot of log Δp vs. log Δt will yield a unit slope. Therefore, any corroborated points along the unit slope line will yield the same quotient, ($\Delta p/\Delta t$). Consequently, any point on the unit slope line may be used in the following formula for wellbore storage coefficient:

$$C = \frac{q_o B_o}{(24)(\Delta p / \Delta t)_{unit\ slope}} \qquad (16\text{-}78)$$

where:

q_o = oil rate just prior to shut-in, STB/D,
B_o = oil-formation volume factor, Bbl/STB,
($\Delta p/\Delta t$) = ratio of the pressure change to the shut-in time from some point on the unit slope line, psi/hr.

Consider the Agarwal *et al.* type curve, Figure 16-23. Dimensionless pressure is plotted against dimensionless time. Notice that there are a number of different curves corresponding to different skin factors (ranging from -5 to +20) and different dimensionless wellbore storage coefficients. To analyze buildup data using the Agarwal *et al.* type curve method, the log-log plot of Δp versus Δt, using actual well data, must be on log-log paper with axes the same size as those on the Agarwal *et al.* type curve.

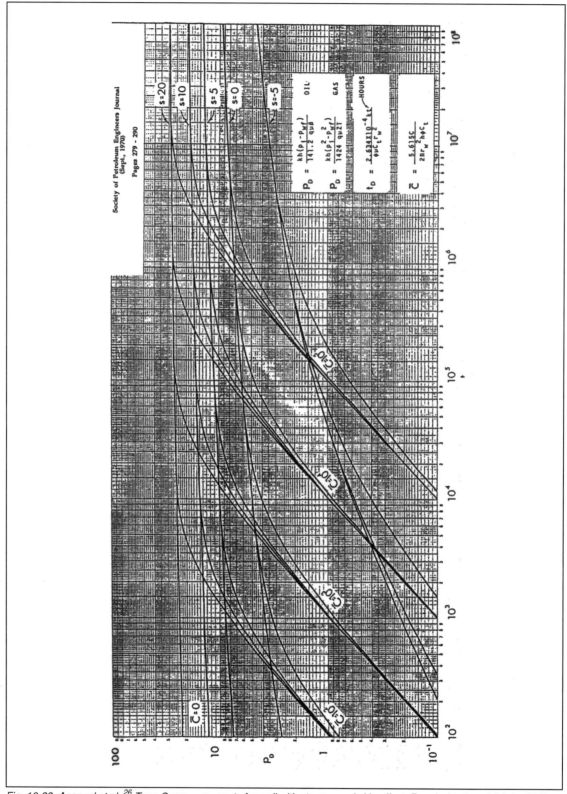

Fig. 16-23. Agarwal et al.,[26] Type Curve — p_D vs. t_D for well with storage and skin effect. Permission to publish by SPE.

AGARWAL *et al.* ANALYSIS PROCEDURE — OIL WELL

1. Prepare a table of data of Δp versus Δt, where for an oil well buildup,

$$\Delta p = p_{ws} - p_{wf}$$

2. Obtain a transparent sheet of paper suitable for overlaying on the Agarwal *et al.* type curves. Overlay the transparency onto the Agarwal graph and draw the major axes on the transparency. This ensures that your graph has the same size log cycles as the type curve. In other words, draw the major grids such as:

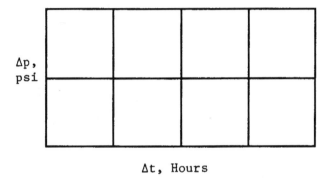

Δt, Hours

3. Label the vertical log axis on the transparency as Δp (psi). Ignore the scale on the type curve and use whatever scale is convenient for your data. Label the horizontal log axis as shut-in time in hours. Once again, use whatever scale is convenient for your data.

4. Plot the test data (Δp vs. Δt) on the log-log transparency.

5. Calculate the dimensionless storage coefficient, $\overline{C}$. (This allows you to know which bundle of curves on the Agarwal type curve to try to match.) The first task is to compute the wellbore storage coefficient. Depending on whether the well is an oil well or gas well, one of the equations discussed earlier is used. For a clean oil well, the recommended equation uses data from the unit slope line:

$$C = \frac{q_o B_o}{(24)(\Delta p / \Delta t)_{unit\ slope}}$$

Then, the dimensionless wellbore storage coefficient can be calculated:

$$\overline{C} = \frac{0.894\ C}{\phi\ h\ c_e\ r_w^2}$$

where: the terms in these two equations were defined earlier.

6. The matching process is performed. It is important that during the matching process, the vertical axis on the transparency must remain parallel to the vertical axis on the type curve. Noting the value of $\overline{C}$ obtained in the last step (to indicate which group of curves should be attempted to be matched), slide the transparent paper around until a match is obtained between the well test data and one of the type curves. This is sometimes a difficult procedure, and a "perfect" match is seldom obtained. Typically, more horizontal shifting (than vertical) is required. Remember that in the moving process the transparency and type curve axes must remain parallel.

7. After making a successful match, and the curves are overlaying each other, note the skin effect labeled on the matched type curve. Then, pick any point from your plot on the transparency such as the intersection of two major grid lines, (Δp, Δt). This point will be referred to as the match point. Then, note the corresponding point on the type curve grid, (P_d, t_d). This is also called a match point.

8. Calculate the flow capacity.

$$(kh)_o = 141.2 \, q_o \, \mu_o \, B_o \left[\frac{P_d}{\Delta p} \right]_{match\ point} \tag{16-79}$$

This result is usually of better quality than any of the other answers obtained from this analysis. However, the value of skin, s, is also a fair estimate.

9. Calculate the porosity-thickness product:

$$(h\phi) = \frac{2.64 \times 10^{-4} \, (k h)_o}{\mu_o \, c_e \, r_w^2} \left[\frac{\Delta t}{t_d} \right]_{match\ point} \tag{16-80}$$

This is usually the poorest quality result from the Agarwal *et al.* type curve analysis.

10. Calculate pressure drop due to skin:

$$\Delta p_{skin} = (s) \left[\frac{\Delta p}{P_d} \right]_{match\ point} \tag{16-81}$$

For gas wells, the Agarwal type curve procedure is similar. The differences usually are in:

(1) the log-log plot (Δp^2 vs. Δt),

(2) calculation of C (with Equation 16-76),

(3) the kh product is found by rearranging Equation 16-72, and

(4) $\Delta p_{skin}^2 = [(s) (\Delta p^2 / P_d)_{m.p.}]$.

SHORT PRODUCING TIME BUILDUPS

The Agarwal *et al.* type curves were constructed for drawdown tests. The reason for this is simple: superposition is not needed. Thus, the effect of the length of producing time has not been considered when using the Agarwal *et al.* type curves to analyze buildup data. For buildups with a long producing time before the shut-in, there is no problem in using the curves as suggested previously. However, for short producing time buildup tests, the length of the producing time needs to be taken into account. Agarwal[34] suggests using the superposition principle and converting shut-in time to an equivalent producing (drawdown) time:

$$\Delta t_e = \frac{(t)(\Delta t)}{t + \Delta t} \tag{16-82}$$

where:

t = producing time, hrs, and
Δt = shut-in time, hrs.

When using this equivalent drawdown time, instead of the actual shut-in time in the Agarwal type curve analysis method, the effects of producing time, skin, and buildup wellbore storage are properly taken into account. The assumption is made that the producing time lasted long enough to be beyond producing wellbore storage effects.

Thus, considering the producing time before the buildup test, a dimensionless producing time is calculated:

$$t_d = \frac{0.000264 \, k \, t}{\phi \, \mu \, c_e \, r_w^2} \tag{16-73}$$

For the "equivalent drawdown time" method to be valid, the dimensionless producing time must satisfy:

$$t_d \geq (60 + 3.5 \, s)(\overline{C}) \tag{16-83}$$

where:

$$\overline{C} = \frac{5.615 \, C}{2 \, \pi \, \phi \, h \, c_e \, r_w^2}$$

Thus, by using Δt_e, instead of Δt, the Agarwal *et al.* type curve analysis may be performed.

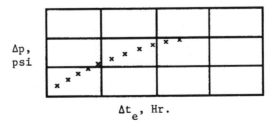

Further, instead of using a Horner plot for a short producing-time buildup test, an MDH type semilog plot can be made with shut-in pressure plotted on the linear vertical axis and the equivalent drawdown time (hours) plotted on the log horizontal scale (see figure 16-24). Notice that at infinite shut-in time, the value of equivalent drawdown time is equal to the producing time before shut-in. Thus, the semilog straight line can still be extrapolated to obtain p*.

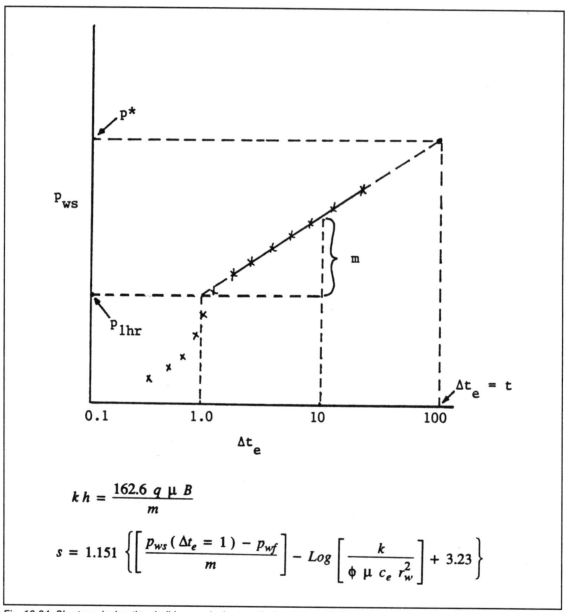

$$k\,h = \frac{162.6\ q\ \mu\ B}{m}$$

$$s = 1.151 \left\{ \left[\frac{p_{ws}\,(\Delta t_e = 1) - p_{wf}}{m} \right] - Log \left[\frac{k}{\phi\ \mu\ c_e\ r_w^2} \right] + 3.23 \right\}$$

Fig. 16-24. Short producing-time buildup, equivalent producing-time analysis.

BOURDET AND GRINGARTEN TYPE CURVES

These curves were developed for the analysis of double porosity systems, so comments concerning these types of reservoirs will be made.

DOUBLE POROSITY RESERVOIRS

A double porosity system, such as a naturally fractured reservoir, has two different types of porosity. Typically, there is one system of high conductivity (fractures) in communication with another system of fairly low conductivity (matrix blocks). The pressure response obtained from such a system is complicated. It depends on the storativities (volume-compressibility product) of the two different porous systems as well as their conductivities.

Warren and Root[30] have defined the two different porosities to be (1) primary (intergranular or matrix block) porosity and (2) secondary (fractures, vugs, etc.) porosity. It was assumed that pseudosteady state flow existed between the primary porous elements and the secondary porosity system, or:

$$\phi_m\, c_m\, \frac{\partial p_m}{\partial t} = \frac{\alpha\, k_m}{\mu}\, (p_f - p_m) = q_{inter} \tag{16-84}$$

where:

$subscript\ ``m" =$ refers to primary porosity,

$subscript\ ``f" =$ refers to secondary porosity,

$\alpha =$ a shape factor with dimensions of reciprocal area, which reflects the geometry of the matrix elements and controls flow between the porous element "m" and porous element "f," and

$q_{inter} =$ interporosity flow rate.

The structure of the resulting pressure-time equations for double porosity behavior made it convenient to define the following parameters[30]:

$$\lambda = \alpha\, \frac{k_m\, r_w^2}{k_f} \tag{16-85}$$

$$\omega = \frac{\phi_f\, c_f}{\phi_m\, c_m + \phi_f\, c_f} \tag{16-86}$$

where:

ω = a parameter relating the fluid capacitance (storativity) of the secondary porosity to that of the total system, dimensionless,

λ = a parameter governing interporosity flow, dimensionless,

α = shape factor discussed previously,

k_m = permeability of primary porosity elements,

r_w = wellbore radius,

$\overline{k}_f = \sqrt{k_{fx} k_{fy}}$ = geometric mean of anistropic secondary (fracture) porosity elements,

ϕ_f = bulk porosity of the secondary porosity elements (fractures); i.e., pore volume of the fractures divided by the total bulk volume of the reservoir,

ϕ_m = bulk porosity of the primary porosity elements; i.e., pore volume of the matrix blocks divided by total reservoir bulk volume,

c_m = total compressibility associated with the primary porosity elements, and

c_f = total compressibility associated with the secondary porosity elements.

In 1969, Kazemi[32] published a numerical procedure allowing the prediction of double porosity system behavior. His method was significant in that transient flow of fluids from the matrix blocks into the fractures was allowed.

In 1976, de Swaan[33] treated the same problem: complete unsteady state flow including flow from the matrix blocks into the fracture system, and he developed an analytical solution. Although a significant achievement, the resulting expressions are so complex that little benefit can be derived for the purpose of rapid hand calculation techniques.

BOURDET AND GRINGARTEN[27]

In 1980, Bourdet and Gringarten published type curves for analysis of double porosity reservoirs based on the work of Mavor and Cinco.[31] These curves can also be used for single porosity systems (Figure 16-25).

The method features some of the same dimensionless variables as used by Agarwal *et al.* However, where the Agarwal *et al.* curves use dimensionless time as the abscissa, the Gringarten *et al.* curves use the ratio of dimensionless time to dimensionless wellbore storage coefficient. By using this approach, the value of the dimensionless wellbore storage coefficient does not have to be calculated before the matching process as it was in the Agarwal *et al.* procedure. When using the Gringarten curves, the wellbore storage coefficient is calculated after the match is obtained. On a practical note, this aspect of the procedure can make the matching process more difficult.

Bourdet and Gringarten chose to characterize double porosity systems with two parameters originally defined by Warren and Root: λ and ω.

$$\lambda = \alpha \, r_w^2 \, \frac{k_m}{k_f}$$

(16-87)

$$\omega = \frac{(\phi \, v \, c_e)_f}{(\phi \, v \, c_e)_{f+m}}$$

(16-88)

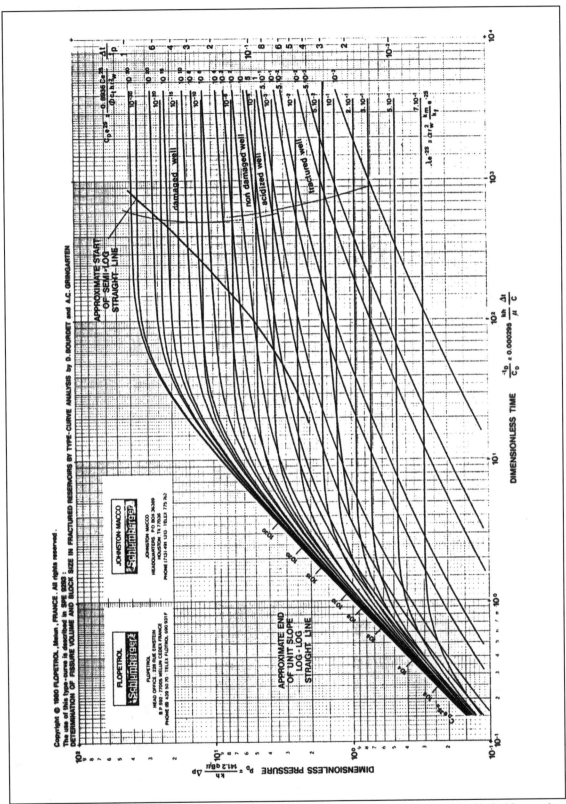

Fig. 16-25. Bourdet - Gringarten[27] Type curve — for a well with wellbore storage and skin (reservoir with double porosity behavior). Permission to publish by SPE.

where:

   λ = the interporosity flow coefficient, dimensionless (indicates how easily fluid can flow from the blocks into the fissures),

   α = a shape factor relating to the geometry of the matrix blocks and fracture system, dimensions of (1/area),

   k_m = permeability associated with the matrix blocks,

   k_f = permeability associated with the fracture system,

   r_w = radius of the wellbore,

   ω = ratio of the storativity of the fissure system to that of the total reservoir (fissures plus matrix blocks),

   ϕ_f = intrinsic porosity of the fracture system (fracture pore volume divided by fracture system bulk volume),

   v_f = bulk volume of the fracture system divided by the total reservoir bulk volume (fractures plus matrix blocks),

   $(c_e)_f$ = total compressibility associated with the fracture system,

   ϕ_m = intrinsic porosity of the matrix blocks; i.e., total pore volume of the matrix blocks divided by the total bulk volume of the matrix blocks,

   v_m = bulk volume of the matrix blocks divided by the total reservoir bulk volume (fractures plus matrix blocks), and

   $(c_e)_m$ = total compressibility associated with the matrix blocks.

Notice that although Gringarten and Bourdet chose to define ω in terms of intrinsic porosity values (as opposed to the bulk values that Warren and Root used), the actual value of ω is the same as that of Warren and Root. Similarly, the Gringarten *et al.*, λ is the same as the Warren and Root λ.

BOURDET AND GRINGARTEN DIMENSIONLESS VARIABLES:[27]

Dimensionless pressure (similar to Agarwal et al.[26]):

$$P_d = \frac{k_f h}{141.2\, q\, B\, \mu}\, \Delta p \tag{16-89}$$

Dimensionless wellbore storage coefficient (same as Agarwal *et al.*):

$$\overline{C} = C_d = \frac{0.8936\, C}{\phi\, c_e\, h\, r_w^2}$$

except:

$$(\phi)(c_e) = [(\phi)(c_e)(v)]_f + [(\phi)(c_e)(v)]_m$$

Skin factor: (dimensionless, same as that introduced to the industry by van Everdingen and used by Agarwal *et al.*)

Dimensionless time (similar to Agarwal *et al.*):

$$t_d = \frac{0.000264 \, k_f}{\phi \, c_e \, \mu \, r_w^2} \, \Delta t \qquad\qquad (16\text{-}90)$$

except:

$$(\phi)(c_e) = [\,(\phi)(c_e)(v)\,]_f + [\,(\phi)(c_e)(v)\,]_m$$

Dimensionless time is not found explicitly on the Gringarten type curves as defined above. Instead, the type curve abscissa is equal to (t_d/C_d).

Dimensionless interporosity flow coefficient: (as defined previously)

$$\lambda = \alpha \, r_w^2 \, \frac{k_m}{k_f}$$

Note that this parameter is a measure of the flow of fluids from the matrix rock to the more permeable portions (fractures or vugs) of the pay zone. The value of λ is typically in the range:

$$10^{-9} \le \lambda \le 10^{-3}$$

However, there are some double porosity reservoirs where λ approaches 1.0. In this instance, the flow behavior is similar to that of a single porosity homogeneous reservoir.

Dimensionless storativity ratio:

(as defined previously)

$$\omega = \frac{(\phi \, v \, c_e)_f}{(\phi \, v \, c_e)_{f+m}}$$

The normal double porosity reservoir range is:

$$10^{-4} \le \omega \le 10^{-1}$$

STEPWISE PROCEDURE FOR GRINGARTEN TYPE CURVE USE

1. Obtain a piece of unlined transparent paper.

2. Overlay the transparency on the Gringarten type curves and draw in the major grids (axes). This step ensures that the transparency log axes and the type curve log axes are the same size.

3. (a) Label the vertical axis of the transparency as Δp, or pressure change during the test using a convenient scale for the actual test data.

   For a buildup: $\Delta p = p_{ws} - p_{wf}$

   For a drawdown: $\Delta p = p_i - p_{wf}$

   (b) Label the horizontal axis of the transparency as elapsed test time, using a convenient scale. For a buildup, this is shut-in time; and for a drawdown, this is production time.

4. Plot the test data on the log-log axes of the transparency.

5. The matching process is performed. During the matching process, the vertical axis on the transparency must remain parallel to the vertical axis on the type curve.

   For a single porosity reservoir, slide the transparent paper around until a match is obtained between the well test data and one of the $[C_d e^{2s}]$ type curves.

   For a double porosity reservoir, the procedure is more involved. The actual well test data on the transparency should be considered in three parts: early time, middle time, and late time. The early time data is controlled by wellbore storage, skin effects, and then flow in the fractures. These early data points should have a positive slope and are to be matched with one of the $[C_d e^{2s}]$ curves. The middle time data represents a transition period. During this time the character of flow changes from only flow in the fractures to a combination of fracture flow together with flow from the matrix blocks into the fractures. During this transition period, the data points typically flatten, and these points are to be matched with one of the $[\lambda e^{-2s}]$ curves. The late time data represents pressure equilibrium between the fractures and the matrix blocks. Flow is occurring both in the fracture system and from the matrix blocks into the fractures. This late time data should be matched with a $[C_d e^{2s}]$ curve, different from that of the early time data. The analyst must match all portions of the well test data simultaneously.

   The matching procedure discussed in the last paragraph for a double porosity system may be summarized as follows. Overlay the transparency onto the type curve and match simultaneously:

   (a) the early time data to a $[C_d e^{2s}]$ curve,

   (b) the middle time data (flat) to a $[\lambda e^{-2s}]$ curve, and

   (c) the late data to another $[C_d e^{2s}]$ curve.

The matching process is sometimes a difficult procedure (particularly with a double porosity system), and a "perfect" match is seldom obtained. Remember that in the moving process the transparency and type curve axes must remain parallel.

6. Once an actual data and type curve match have been obtained, then choose any value of Δp from the transparency (often taken as the value of one of the major grids, such as 10 or 100) and the corresponding point from the type curve (value of P_d right underneath the chosen Δp on the transparency). These two points are called match points, and from them (kh/μ) can be calculated:

$$\frac{kh}{\mu} = 141.2 \, q \, B \left[\frac{P_d}{\Delta p} \right]_{match \, point}$$

7. Pick a value of shut-in time from the transparency, and the corresponding value of t_d/C_d from the type curve. These are two more match points.

(a) Calculate the value of the wellbore storage coefficient as:

$$C = 0.000295 \left(\frac{k \, h}{\mu} \right) \left[\frac{\Delta t}{t_d / C_d} \right]_{match \, point} \qquad (16\text{-}91)$$

(b) Calculate the dimensionless wellbore storage coefficient:

$$C_d = \frac{0.8936 \, C}{\phi \, c_e \, h \, r_w^2}$$

For double porosity, recall that:

$$\phi \, c_e = [\phi \, c_e \, v]_f + [\phi \, c_e \, v]_m$$

8. For a single porosity system, note the value of $[C_d e^{2s}]$ that corresponds to the matched type curve.

For a double porosity system, note the value of $[C_d e^{2s}]$ that corresponds to the matched type curve for the late data.

Then, the skin factor, s, can be calculated in two steps:

$$e^{2s} = \frac{(C_d \, e^{2s})_{match \, point, \, late \, data}}{C_d}$$

$$s = 0.5 \, Ln \, e^{2s} \qquad (16\text{-}92)$$

For a single porosity system, the analysis is finished.

9. For a double porosity system, note the value of the matched curve corresponding to the early time data, $[C_d e^{2s}]$. Then the storativity ratio, ω, may be calculated (assuming that both early time and late time $[C_d e^{2s}]$ matches were obtained:

$$\omega = \frac{(C_d e^{2s})_{match\ point,\ late\ data}}{(C_d e^{2s})_{match\ point,\ early\ data}} \qquad (16\text{-}93)$$

10. Finally, considering double porosity systems, the interporosity flow coefficient can be calculated using the matched curve, $[\lambda\, e^{-2s}]$, for the middle time data:

$$\lambda = [(\lambda\, e^{-2s})_{match\ point}]\ e^{2s} \qquad (16\text{-}94)$$

As with the Agarwal *et al.* type curves, the Gringarten type curves were really constructed for drawdowns. However, they are used more often for buildup analysis. There is no problem with this approach if the shut-in time is small compared to the producing time.

Bourdet and Gringarten have placed a vertical scale on the extreme right hand side of their type curves headed: $\Delta t/t_p$. This scale is used to decide whether these drawdown type curves can be used for a buildup analysis with sufficient accuracy (within five percent). Using the last buildup point matched, proceed horizontally over to the $\Delta t/t_p$ scale and read the required ratio. Then take the length of shut-in time associated with the last matched buildup point and divide this shut-in time by the $\Delta t/t_p$ ratio. This calculation yields the required amount of producing time for the match to be valid.

If the producing time was not sufficient, then the match is not correct. The transparency plot $\log \Delta p$ vs. $\log \Delta t$ will have to be redone using a different Δp. In such case, it is best to calculate the Δp associated with each shut-in point as:

$$\Delta p = p_{ws}(\Delta t) - p_{wf,ext}(\Delta t) \qquad (16\text{-}95)$$

where:

$p_{ws}(\Delta t)$ = shut-in bottomhole pressure at shut-in time Δt,

$p_{wf,ext}(\Delta t)$ = the drawdown pressure extrapolated to shut-in time Δt. Of course, this corresponds to a producing time of $t + \Delta t$.

MCKINLEY TYPE CURVE[28]

This type curve was designed to obtain quick answers in single porosity reservoirs. Less variables are involved, and some limiting assumptions have been made that render the results of a McKinley analysis less rigorous than those of an Agarwal or Gringarten analysis.

The McKinley type curve (Figure 16-26) has actual time (not dimensionless time) plotted on the log vertical scale. Further, the time units are specified in minutes. This simplifies the matching process, because the actual data is plotted with the same time scale (both axis size and units of minutes). Therefore, data collected at a shut-in time of 10 minutes will be correlated on the data plot and on the type curve with the 10 minute line. So, during the matching process, no vertical shifting is allowed (10 minutes on the data plot must stay right on top of 10 minutes on the type curve); only horizontal shifting of the data plot is permitted to obtain a match.

The two variable groups used are:

the Ratio of Transmissibility to Storage Factor, T/F,

the Pressure Buildup group, $\Delta pF/q$

The term "T" is the transmissibility, kh/μ. Its units are md-ft/cp. "F" is the storage factor with units of cubic ft/psi. "F" is roughly equivalent to the wellbore storage coefficient (C) of Agarwal, but 5.615 times as large. "F" is not directly evaluated as was "C" in the Agarwal method. Instead, the storage factor is determined after the matching process with the early time data.

The "Δp" term indicates the pressure change (p_{ws} - p_{wf}), and "q" is the reservoir flow rate in barrels per day.

As was the case in the Agarwal, *et al.* type curve analysis, a piece of transparent paper is placed over the type curve, and the major axes are drawn on the transparent paper. The vertical log scale on the transparent paper is the time scale and must be in minutes as is the type curve. The horizontal log scale on the transparent paper is pressure change. The units are psi, and the scale should be whatever best fits the data.

Once the data are plotted, the matching process begins. The McKinley type curve was built on the assumption that s = 0. Therefore, it is necessary to match the data in two parts: early time and late time. The early time data is largely affected by skin and wellbore storage; while, the late time data is assumed not to be. Quite frequently the split between early time data and late time data is taken to be 30 minutes. The data collected during the first 30 minutes are often dominated by wellbore storage; therefore, it is from these data that the storage factor, F, is determined.

During the matching process, the transparency is shifted horizontally (to the side) only. Once a match of the early data (up to 30 minutes shut-in time) is made, the value of (T/F) is read from the matched type curve.

With the actual data plot (transparency) still overlaying the type curve with the early time data matched, read any convenient value of Δp from the transparency and the corresponding value of ($\Delta pF/q$) from the type curve. Thus, using the early time data, three match points are to be determined: (T/F), (Δp), and ($\Delta pF/q$).

Using the early time data match points, calculate the wellbore storage factor:

$$F = \left[\left(\frac{\Delta p \, F}{q} \right)_{curve} \right] \left[\frac{q_o \, B_o}{(\Delta p)_{match \, point, \, early \, data}} \right]$$

The value of transmissibility near the well is computed:

$$(T)_{near \, well} = \left[\left(\frac{T}{F} \right)_{early \, match} \right] [\, F \,] = (k \, h / \mu)_{near \, well}$$

Next, the late time data (after 30 minutes) is matched. The transparency should be shifted to the left or right only with no vertical "slippage" relative to the type curve. Move the transparency until the upper portion of the plot (late time data) matches one of the type curves. Using the matched curve's value of (T/F), together with the value of the storage factor (F) computed earlier, the value of the transmissibility away from the wellbore can be computed.

$$(T)_{interwell} = \left[\left(\frac{T}{F} \right)_{late \, match} \right] [\, F \,] = (k \, h / \mu)_{interwell}$$

Comparing the calculated transmissibility away from the wellbore with that computed near the wellbore will give a qualitative indication of positive or negative skin effect:

$$(T)_{near \, well} \; < \; (T)_{interwell} : \; s = +$$

$$(T)_{near \, well} \; > \; (T)_{interwell} : \; s = -$$

As a practical note, for damaged wells, the interwell transmissibility calculated with the McKinley method will usually be on the low side. Higher transmissibilities (than actual) will likely result in the presence of negative skin.

To calculate the approximate condition ratio,[58] two additional quantities must be noted from the data plot: Δp^* and Δp_d. Recall that the condition ratio is:

$$C.R. = \frac{p^* - p_{wf} - \Delta p_s}{p^* - p_{wf}} = \frac{\Delta p^* - \Delta p_s}{\Delta p^*} \tag{16-41}$$

Note that Δp^* is the vertical asymptote approached by the late time data on the McKinley plot, while Δp_d is the pressure change associated with the departure of the actual data from the *early* time type curve.

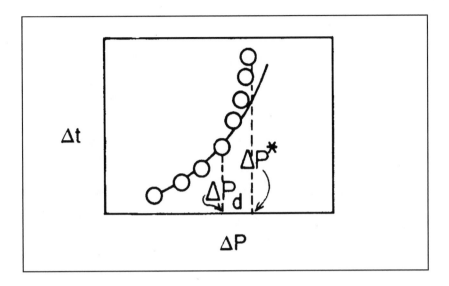

Then,

$$\Delta p_s = \Delta p_{skin} \approx \Delta p_d \left[1 - \frac{T_{near\,well}}{T_{interwell}} \right]$$

And,

$$C.R. = \frac{\Delta p^* - \Delta p_s}{\Delta p^*}$$

The approximate skin factor is:

$$s \approx \frac{(\Delta p_s)(T_{interwell})}{141.2\,q\,B}$$

According to Agarwal,[34] the McKinley type curves were published for buildup data in a radial flow system, and they include the inherent assumption of long producing times. Therefore, these type curves are unsuitable for short producing time buildups.

In 1977, Crawford, *et al.*[50] extended the McKinley method and published type curves suitable for drillstem test analysis. These curves were specifically constructed for flow times between 15 and 120 minutes.

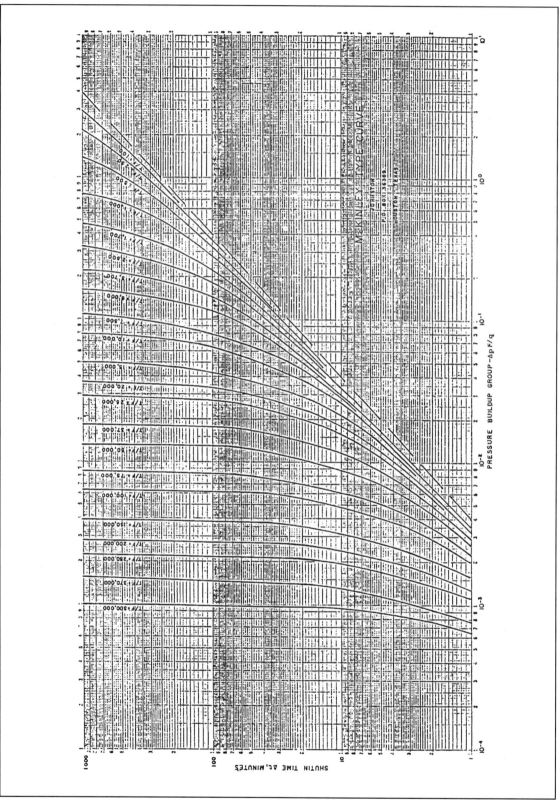

Fig. 16-26. McKinley Type Curve.

DERIVATIVE TYPE CURVES

With conventional type curves, such as Figure 16-25, for large values of wellbore storage and/or damage, the curves have quite similar curvature. The curves all go from the unit slope line to a flat response quickly. Hence, the correct curve to be matched is often ambiguous.

When only mechanical pressure gauges were available, well test analysis had to be based solely on actual pressure data versus time. The analysis of transient well test data involves the solution of the continuity equation which includes directly the derivative of pressure with respect to time. With the advent of electronic downhole pressure gauges came the ability to measure pressure points that are accurate and close together in time. Both the accuracy and the closely spaced points are requirements if meaningful pressure derivatives are to be calculated.

Even with an expensive electronic gauge, there is still noise present in both the pressure and time scales. With pressure data obtained from an actual well test, some sort of smoothing process is needed before the derivative values can be used in well test analysis.[54,56,57]

One importance of the pressure derivative is in the diagnosis of flow regime. From wellbore storage to pure radial (or linear) flow, proper use of the derivative allows determination of the flow state. When well test equipment with surface readout is available (together with simultaneous analysis), the derivative can be used to determine when a test should be terminated; i.e., when enough data has been collected. This is especially important when an expensive drilling rig is tied up for test time.

For most of the derivative type curves in use today, the term "pressure derivative" is misleading. It is actually the slope of the appropriate semilog plot. Some examples are:

$$\frac{d\,(\Delta p)}{d\,(Ln\,\Delta t)} = \Delta t \,\frac{d\,(\Delta p)}{d\,(\Delta t)} \tag{16-96}$$

$$\frac{d\,(\Delta p)}{d\left[Ln\,\dfrac{t+\Delta t}{\Delta t}\right]} = \frac{(\Delta t)\,(t+\Delta t)}{t}\,\frac{d\,(\Delta p)}{d\,(\Delta t)} \tag{16-97}$$

$$\frac{d\,[\,\psi\,(p)\,]}{d\,[\,Ln\,\Delta t\,]} = \frac{2p}{\mu z}\,(\Delta t)\,\frac{d\,(\Delta p)}{d\,(\Delta t)} \tag{16-98}$$

where:

     ψ = real gas potential (Equation 16-47).

The "derivative" usually is not simply $d(\Delta p)/d(\Delta t)$. For the type curve itself, dimensionless pressure and the dimensionless pressure derivative are often plotted as a function of: t_d/C_d. The radial flow, single porosity system, derivative type curve of Bourdet *et al.*[51] is shown in Figure 16-27.

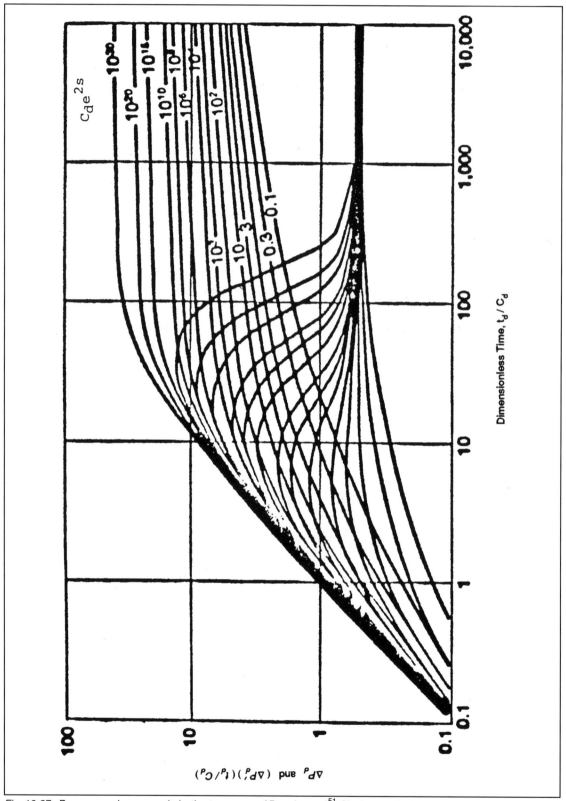

Fig. 16-27. *Pressure and pressure derivative type curve of Bourdet et al.*[51] *Single porosity formation, radial flow geometry.*
    Permission to publish by World Oil.

Recall that during pure wellbore storage, pressure change with time has a unit slope on the log-log plot. This implies that during pure wellbore storage:

$$\Delta P_d = \frac{t_d}{C_d} \quad \text{or} \quad \frac{d(\Delta P_d)}{d(t_d/C_d)} = 1.0 = \Delta P_d' \tag{16-99}$$

which further implies that,

$$\frac{d(\Delta P_d)}{d Ln(t_d/C_d)} = (\Delta P_d')(t_d/C_d) = t_d/C_d \tag{16-100}$$

Thus, as shown in Figure 16-27, with ΔP_d and the derivative, $d\Delta P_d / d[Ln(t_d/C_d)]$, both plotted on the same graph versus t_d/C_d, during pure wellbore storage, they plot as the same unit slope straight line.

The radial flow, logarithmic approximation solution to the continuity equation is:

$$\Delta P_d = 0.5(Ln\ t_d + 0.80907 + 2s)$$

Hence, for infinite-acting radial flow:

$$\frac{d(\Delta P_d)}{d[Ln(t_d/C_d)]} = 0.5 \tag{16-101}$$

Thus, these derivative data will also plot as a straight line; but, this time it is horizontal.

When considering actual well test pressure-derivative data, on the log-log plot of derivative versus time, two straight lines will occur (unit slope and horizontal) if both wellbore storage and infinite-acting radial flow have occurred. Therefore, matching the two straight lines onto the asymptotes of the type curve yields only one possible match.

The actual well test derivative horizontal data is placed on top of the type curve 0.5 flat line, and moved sideways until a match is obtained. Unfortunately, with real data, the late time portion of the curve is not always well defined. Thus the well test pressure change data is also imposed on the same graph. When a match is obtained with the derivative data, a simultaneous match should also occur with the pressure data. In some instances, this eliminates the need to do a Horner analysis to confirm the match. After a match has been obtained, the calculations are similar to those described under the Bourdet and Gringarten type curve.

For buildups, it is recommended, especially for short prior producing times, that the derivative be based on the Horner time function, rather than simply test time (use Equation 16-97). Because of the difficulty in getting a constant rate, it is best not to analyze drawdown data by type curve matching.[52]

The derivative curves are also quite useful if infinite acting data is not reached or if there is very little wellbore storage. Between these two regimes, the curvature of the derivative curves is significantly greater than that of the pressure curves. Thus, a unique match is much easier to obtain.

Derivative type curves[52,53,56] have also been developed to analyze well tests in dual porosity reservoirs. The type curve shown in Figure 16-28 assumes pseudosteady state flow from the matrix blocks to the fissures. The use of the derivative leads to quick and clear recognition of heterogeneity, which may not be seen on the log-log pressure plot. Actually, the derivative pressure response in a naturally fractured reservoir is similar to that of a homogeneous reservoir except for the addition of an extra "dip." The derivative data still starts off on a unit slope line (pure wellbore storage) and at late times goes to a horizontal line (total system infinite-acting radial flow). But just before the horizontal straight line data, the "dip" occurs which corresponds to the transition period. Figure 16-29 illustrates the transient inter-porosity naturally fractured reservoir type curve.[53] For details of the calculation procedures, see references 52 and 53.

Based on the model of Serra *et al.*[55] a later set of dual porosity reservoir type curves were published by Petak *et al.*[56] There are 10 different type curves, each for a different $C_d e^{2s}$ value, ranging from 0.5 to 10^{10}. An example is shown in Figure 16-30. Use of the Petak curves concentrates on well response during the transition period. A reported advantage of these curves is that they represent the dependence of the transition period on the radial flow conditions; and they require less data (than other curves) to do a complete analysis.

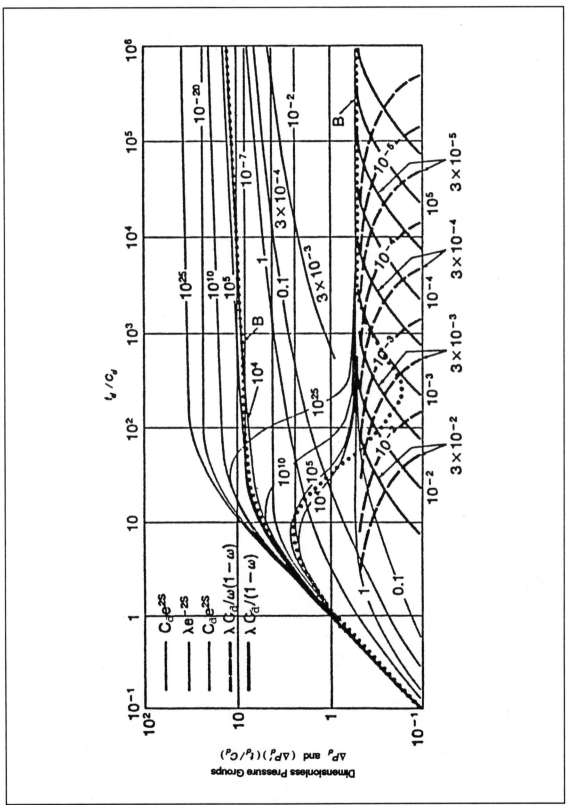

Fig. 16-28. Pressure and pressure derivative type curve of Bourdet et al.[51] Double porosity formation, radial flow geometry, pseudosteady inter-porosity flow. Permission to publish by World Oil.

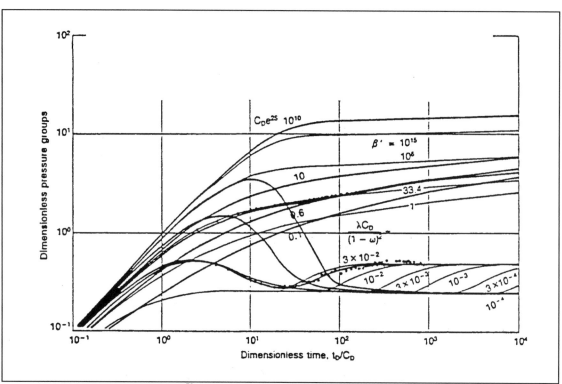

Fig. 16-29. Type curve of Bourdet et al.[53] Double porosity formation, transient inter-porosity flow. Permission to publish by World Oil.

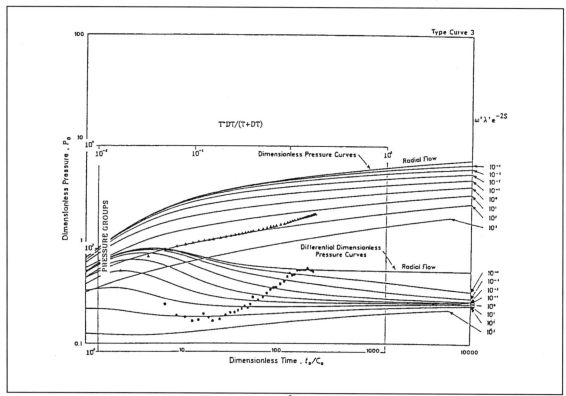

Fig. 16-30. Unsteady state natural fracture type curve for $C_D e^{2S}$ = 10.0. Naturally fractured formation, transient flow type curve example from Petak et al.[56] Permission to publish by SPE.

DRAWDOWN TEST THEORY

When a well is placed on production, the pressure declines steadily when the rate is maintained at or near a constant level. The radius of investigation or current drainage radius moves outward from the well during the time period referred to as the "transient phase." This period of time lasts until a boundary affects the pressure data. A boundary can be an actual physical entity such as a fault or permeability barrier, or it can be interference effects from offset wells.

Dimensionless producing time is defined as:

$$t_{de} = \frac{2.64 \times 10^{-4}\, k\, t}{\phi\, \mu\, c_e\, r_e^2}$$

(16-102)

where:

t_{de} = dimensionless time referred to the external radius, r_e,
k = formation permeability, md.,
t = producing time, hours
ϕ = formation porosity, fraction,
μ = fluid viscosity, cp
c_e = total compressibility of the system, psi^{-1}, and
r_e = external radius of drainage (when producing at semi-steady state), ft.

Between dimensionless time (t_{de}) of 0 and 0.25, transient flow is occurring. During this time the well is "infinite-acting;" i.e., the well's pressure response is exactly the same as if it truly had no boundaries. At a dimensionless producing time of 0.25, for a well with a radial drainage area, the pressure transients (due to production) have just reached the external radius of drainage, r_e.

Between dimensionless time values (t_{de}) of 0.25 and 0.30 (transition period), a modification of the pressure distribution occurs. At t_{de} = 0.25, when the pressure transients due to producing have just reached the boundary, the pressure distribution across the drainage area is the same as a steady state pressure distribution. So, for an instant (at t_{de} = 0.25), the external drainage radius boundary condition can be represented as a constant pressure boundary. By the time that a dimensionless producing time of 0.30 is reached, the pressure condition at r_e has changed to a no-flow boundary condition (dp/dr = 0) reflecting the fixed boundary.

Following the transition period, the flow regime becomes semi-steady state. With this type of flow, the pressure declines at a rate proportional to the reservoir withdrawal rate and inversely proportional to the (drainage pore volume)(total compressibility) product. The following plot of flowing bottomhole pressure versus producing time depicts the various phases of pressure behavior during an extended drawdown test:

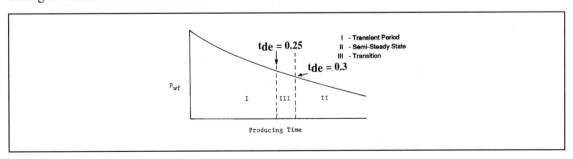

TRANSIENT FLOW PERIOD

During the transient flow period; i.e., when the radius of drainage is expanding, a plot of the well's bottomhole pressure versus the log of producing time will yield a straight line, after wellbore effects have died out and before the drainage boundaries are felt.

Figure 16-31 is an illustrative plot of well pressure versus the log of producing time in hours. The early time region (A) is affected by two departures from theory that occur right at the wellbore: skin effect and wellbore storage. The middle time region (B) is not affected by near-wellbore conditions or by the drainage boundaries. Thus the expression of an infinite acting well is seen here: the semilog straight line. The late time region (C) is being affected by drainage boundaries, and the pressure is falling off more rapidly than if the reservoir were still infinite acting.

The slope of the infinite acting region straight line can be used to calculate the flow capacity in the same way as was done for a buildup analysis. The equations which relate pressure and time for a drawdown test during the transient flow phase are quite similar to those for a pressure buildup test where the producing time was long compared to the shut-in time.

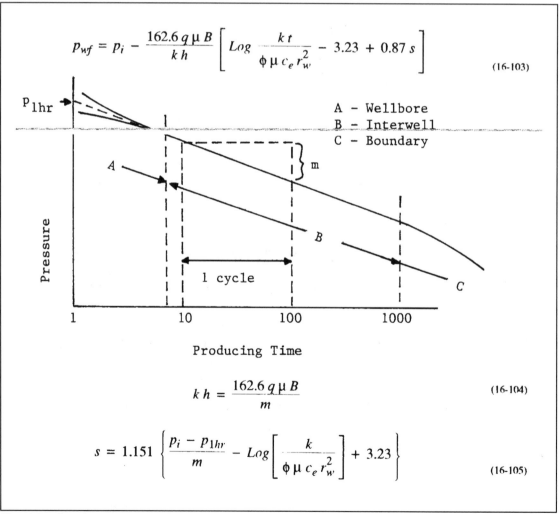

$$p_{wf} = p_i - \frac{162.6\,q\,\mu\,B}{k\,h}\left[Log\,\frac{k\,t}{\phi\,\mu\,c_e\,r_w^2} - 3.23 + 0.87\,s\right] \qquad (16\text{-}103)$$

A — Wellbore
B — Interwell
C — Boundary

Producing Time

$$k\,h = \frac{162.6\,q\,\mu\,B}{m} \qquad (16\text{-}104)$$

$$s = 1.151\left\{\frac{p_i - p_{1hr}}{m} - Log\left[\frac{k}{\phi\,\mu\,c_e\,r_w^2}\right] + 3.23\right\} \qquad (16\text{-}105)$$

Fig. 16-31. Drawdown: transient radial flow.

SEMI-STEADY STATE FLOW

If a well is produced at a constant rate for an extended period time, the rate of pressure decline with time will approach a constant. This constant rate of pressure change can be used to calculate the pore volume within the drainage area of the well. The theory behind the application of the semi-steady state portion of the drawdown test is exactly the same as that used to develop the reservoir limit test (next topic).

$$p_{wf} = p_i - \left[\frac{0.0417\, qBt}{V_p\, c_e} \right] - \frac{141.2\, q\mu B}{kh} \left\{ Ln \left[\frac{r_e}{r_w} \right] - \frac{3}{4} + s \right\} \qquad (16\text{-}106)$$

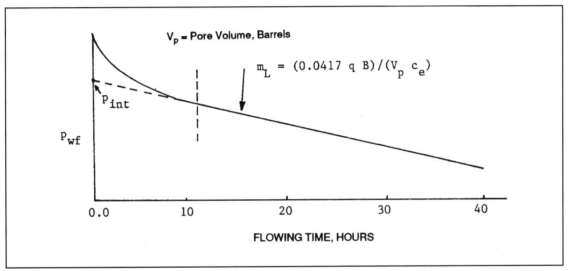

Fig. 16-32. Extended drawdown test.

$$V_p = \frac{0.0417\, q\, B}{m_L\, c_e} \quad bbls \qquad (16\text{-}107)$$

where:

qB = total reservoir flow rate, BPD
m_L = slope of semi-steady state straight line, psi/hr
V_p = pore volume, bbls

If the transient drawdown analysis is also available, according to Earlougher,[7] the drainage area shape factor can be calculated:

$$C_A = 5.456\, \frac{m}{m_L} \exp \left[-2.303\, (p_{1hr} - p_{int})\, / \, |m| \right] \qquad (16\text{-}108)$$

The calculated shape factor can then be compared with the closest shape factor in Table 16-1 to determine drainage configuration.

Problem 16-5:

Drawdown Test for a Well in an Undersaturated Reservoir:

An extended drawdown test was conducted on a new well to determine the interwell permeability, skin factor, and pore volume of the drainage area associated with the well. The following basic data are known.

p_i = 4000 psia	ϕ = 10 %
q_o = 500 STBO/D	μ_o = 1.0 cp
h = 10 ft	B_o = 1.25 BBL/STB
r_w = 3 in	c_e = 20 x 10^{-6} psi^{-1}

The following pressure data were collected during the test.

TIME, HRS.	P_{wf}, PSIA	TIME, HRS.	P_{wf}, PSIA
0.	4000	25.	3149
0.1	3641	30.	3133
0.25	3559	35.	3118
0.5	3498	40.	3105
1.	3438	45.	3092
2.	3377	50.	3080
5.	3295	60.	3055
7.	3265	70.	3032
10.	3234	80.	3008
15.	3198	90.	2986
20.	3169	100.	2963

Problem Solution:

Figure 16-33 shows the transient pressure data plot. The straight line slope is determined to be 203 psi/cycle. Based on this slope, the permeability was calculated to be 50.1 md.

Skin effect calculation:

$$s = 1.151 \left\{ \left[\frac{p_i - p_{1hr}}{m} \right] - Log \left[\frac{k}{\phi \, \mu \, c_e \, r_w^2} \right] + 3.23 \right\}$$

$$s = 1.151 \left\{ \left[\frac{4000 - 3438}{203} \right] - Log \left[\frac{50.1}{(\,0.1\,)(\,1\,)(\,20\,)(\,10^{-6}\,)(\,0.0625\,)} \right] + 3.23 \right\}$$

$$s = -3.0$$

$$\Delta p_{skin} = (\,0.87\,)(\,m\,)(\,s\,) = (\,0.87\,)(\,203\,)(\,-3.0\,) = -530 \ psi$$

Figure 16-34 presents the analysis of the late time or semi-steady state data. This is the flow regime that exists when the Cartesian plot of well pressure versus producing time becomes

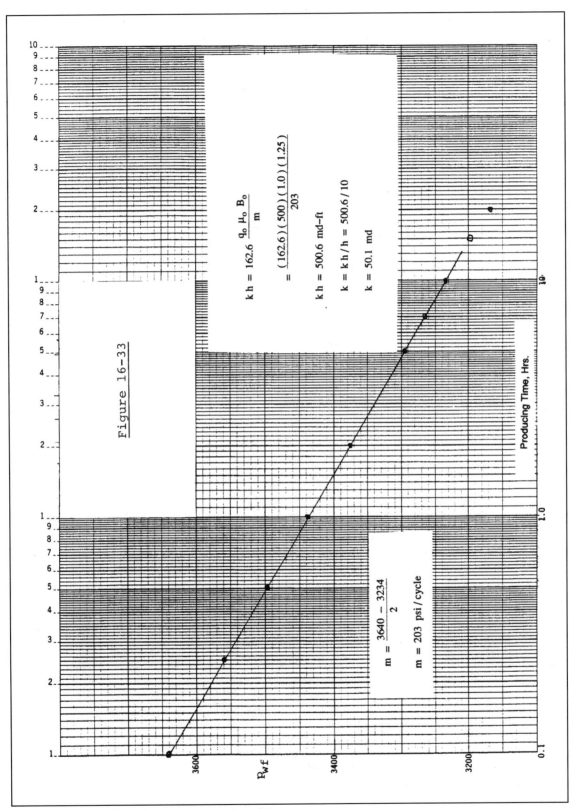

Figure 16-33

$$k\,h = 162.6\,\frac{q_o\,\mu_o\,B_o}{m}$$

$$= \frac{(162.6)\,(500)\,(1.0)\,(1.25)}{203}$$

$$k\,h = 500.6\ \text{md-ft}$$

$$k = k\,h\,/\,h = 500.6\,/\,10$$

$$k = 50.1\ \text{md}$$

$$m = \frac{3640 - 3234}{2}$$

$$m = 203\ \text{psi / cycle}$$

Producing Time, Hrs.

P_{wf}

Fig. 16-33. Analysis of transient pressure drawdown data.

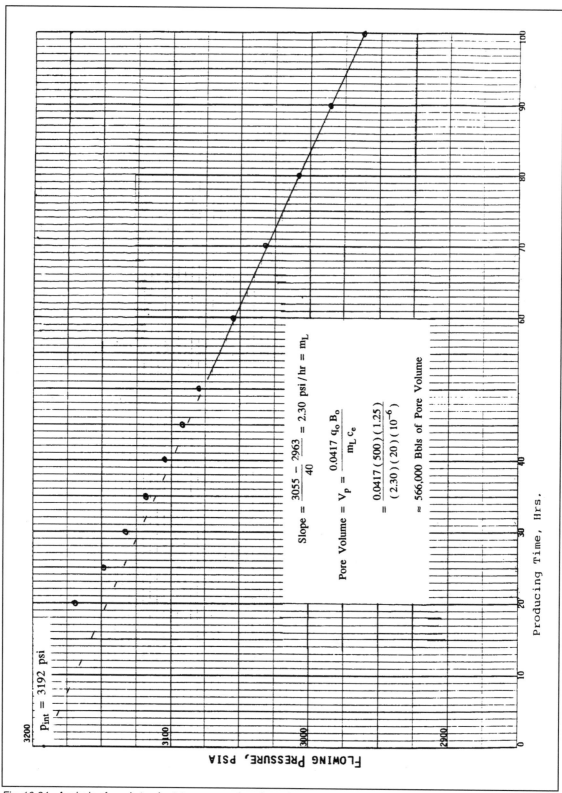

Fig. 16-34. Analysis of semi steady state pressure drawdown data.

linear. The semi-steady state pressure decline at the well is seen to be substantial: 2.30 psi/hr. On this basis, the pore volume calculation indicates that this well is in contact with a very small drainage area: only about 566,000 barrels of pore volume.

Shape factor calculation:

From Figure 16-33: $m = 203$ psi/cycle, $p_{1hr} = 3438$ psi

From Figure 16-34: $m_L = 2.30$ psi/hr, $p_{int} = 3192$ psi

Using Equation 16-108:

$$C_A = 5.456 \left(\frac{203}{2.3} \right) \exp \left[\frac{(-2.303)(3438 - 3192)}{203} \right]$$

$$= (5.456)(88.261)(0.0614)$$

$$C_A = 29.6$$

The closest shape factor in Table 16-1 is 30.9, which is for a square with a centered well. However, the other regular shapes (circle, hexagon, triangle) are also close fits.

RESERVOIR LIMIT TEST

This test is an adaptation of the semi-steady state pressure drawdown theory. The primary difference between this analysis and that discussed earlier is the method of plotting or representing the data. This method was developed by Park Jones[36] who suggested that a log-log plot be prepared for the function:

$$Y = - \frac{1}{q_t} \frac{\partial p_{wf}}{\partial t} \tag{16-109}$$

versus producing time. Note that in this function, q_t is the total reservoir flow rate:

$$q_t = q_o B_o + q_w B_w + (q_g - q_o R_s) B_g \tag{16-110}$$

During the transient phase of drawdown data, the relationship of Y vs. producing time will be linear with a slope if (-1), or an angle of 45 degrees below the horizontal.

During the semi-steady state phase of the drawdown data, the relationship of Y vs. producing time approaches a constant value (Y_{Stab}).

Then:

$$\frac{\partial p_{wf}}{\partial t} = \frac{0.0417 \ q_t}{V_p \ c_e} \tag{16-111}$$

where:

q_t = total reservoir flow rate, reservoir B/D
V_p = pore volume, reservoir barrels,
p_{wf} = flowing bottomhole pressure, psia,
t = flowing time, hrs, and
c_e = total system compressibility, psi^{-1},

$$c_e = c_o S_o + c_w S_w + c_g S_g + c_f$$

Then,

$$Y_{Stab} = -\frac{1}{q_t}\frac{\partial p_{wf}}{\partial t} = \frac{0.0417}{V_p c_e} \qquad (16\text{-}112)$$

So,

$$V_p = \frac{0.0417}{(Y_{Stab})(c_e)} \ bbls$$

Best results are obtained with this method when applying it to an undersaturated oil reservoir. This is due to the difficulty in determining the factors in the c_e relationship. Fluid saturations are difficult to determine in a reservoir with multiphase flow.

Problem 16-6:

DETERMINATION OF ORIGINAL OIL IN PLACE (Park Jones Method)

A well test is conducted in a well remote from other production. The objective of this test was to evaluate the original oil in place so as to determine if a second well could be drilled. It takes at least 500,000 barrels per well to justify a new well.

Basic data:

Producing Rate	100 BOPD
Oil formation volume factor	1.25 bbl/STB
Oil viscosity	0.8 cp
Pay thickness	20 Ft
Effective compressibility	20 x 10^{-6} psi^{-1}
Connate water saturation	20%

Test pressure - time data:

Flowing Time Hours	Incremental Time Hours	Bottomhole Pressure Psia	Incremental Pressure Change
0		2500	
2	2	2409	91
5	3	2403	6
10	5	2398	5
20	10	2393	5
30	10	2391	2
50	20	2387	4
100	50	2380	7
150	50	2373	7
200	50	2366	7
250	50	2359	7
300	50	2352	7

Inspection of these data indicate an approximate linear trend of p_{wf} vs. time for the last 200 hours of flowing time. A finite difference approximation was used to calculate the pressure derivative needed in the Y function:

$$Y = -\frac{1}{q_t}\frac{\partial p_{wf}}{\partial t}$$

$$\frac{\partial p_{wf}}{\partial t} \approx \frac{\Delta p_{wf}}{\Delta t}$$

Figure 16-35 illustrates the Park Jones[36] Y plot for this problem.

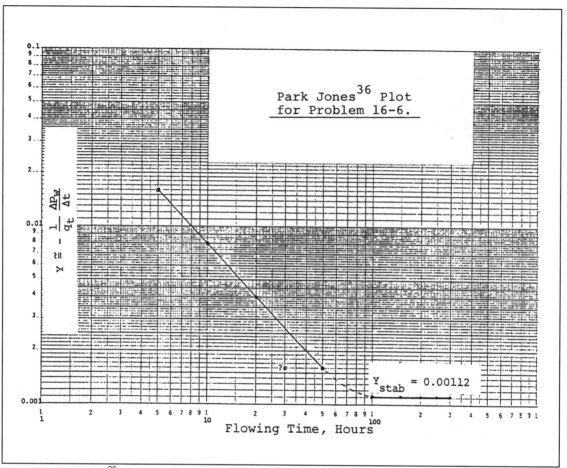

Fig. 16-35. Park Jones[36] plot to find Y_stab. Permission to publish by Oil & Gas Journal.

$$\left[\frac{\Delta p}{\Delta t} \right]_{stab} = \frac{2352 - 2380}{300 - 100} = -0.14 \; psi / hr$$

$$Y_{stab} = -\frac{1}{qB}\left[\frac{\Delta p}{\Delta t} \right]_{stab} = -\frac{-0.14}{(100)(1.25)} = 0.00112$$

$$V_p = \frac{0.0417}{Y_{stab} \, c_e} = \frac{0.0417}{(0.00112)(20)(10^{-6})}$$

$$= 1.862 \times 10^6 \; Reservoir \; barrels$$

$$N = \frac{(V_p)(1 - S_w)}{B_{oi}} = \frac{(1.862)(10^6)(0.8)}{1.25}$$

$$= 1.19 \times 10^6 \; STB$$

Another well is probably justified.

RESERVOIR LIMIT TEST FOR GAS WELLS

The same general theory that applies to oil wells also applies to gas wells[37]. The equations describing the relation between well pressure and cumulative production can be divided into two periods: transient period and stabilized period.

TRANSIENT PERIOD:

$$Ln \left[- \frac{\Delta(p_{wf})^2}{\Delta G_p} \right] = Constant - Ln\, G_p$$

(16-113)

STABILIZED PERIOD (SEMI-STEADY STATE):

$$- \frac{\Delta(p_{wf})^2}{\Delta G_p} = \frac{2\, p_i^2}{G} - \left[\frac{2\, p_i^2}{G^2} \right] G_p$$

(16-114)

The idea is to perform an extended drawdown test at a constant rate, with the well pressure beginning at p_i, to determine the original gas in place. Unfortunately, this theory can only be practically applied to wells in small gas reservoirs. The amount of time required to reach semi-steady state in a large reservoir can easily be years. And a significantly changing skin factor during this time would invalidate the analysis.

To begin, a Cartesian plot of p_{wf}^2 versus cumulative gas production during the drawdown test is prepared:

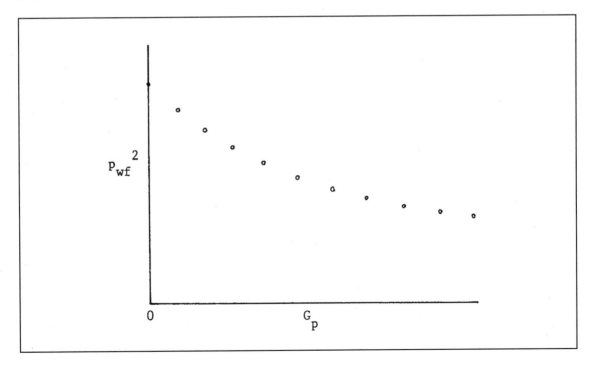

The slope, $\Delta(p_{wf})^2/\Delta G_p$, is determined at a number of different points (G_p's) on this graph. Then, the following log-log plot is made.

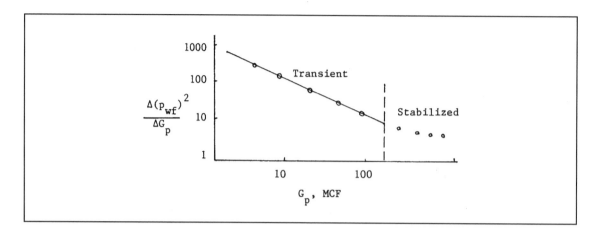

Notice that the transient flow period shows up as a straight line on the log-log plot. Then when the curve begins to flatten, the boundaries are starting to be felt. The data points, after the transient flow period, are plotted on Cartesian coordinate paper.

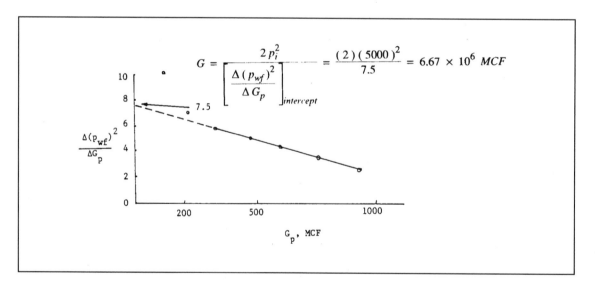

Semi-steady state flow is illustrated on this last plot as a straight line. Extrapolate the straight line back to the y-axis to determine the y-intercept. Then,

$$G = \frac{2p_i^2}{(y - intercept)} \tag{16-115}$$

For example, in the previous illustration, $p_i = 5000$ psia, and the y-intercept is 7.5 psi^2/MCF. Hence, G = 6.67 MMMCF.

GAS WELL RESERVOIR LIMIT TEST—SECOND METHOD[49]

According to Reynolds *et al.*[49] gas wells with long flow times can be analyzed for pore volume similar to the oil well analysis if flowing pseudopressure is used. For an oil well at semi-steady state, there is a linear change in well pressure with flowing time (Figure 16-32). For gas wells, a similar plot can be made, but the flowing well pressures must be converted to pseudopressures using Equation 16-47. The method is valid for dimensionless flowing times within:

$$0.07 \leq t_{DA} \leq 0.3$$

$$t_{DA} = \frac{0.000264 \ k \ t}{\phi \ (c_e \ \mu)_i \ A} \tag{16-116}$$

and:

k = permeability, md,
t = flowing time, hr,
ϕ = fractional porosity,
A = drainage area, ft^2,
$(c_e)_i$ = total compressibility at original pressure, psi^{-1},
μ_i = gas viscosity at the original pressure, cp.

If a flowing pseudopressure vs. time plot is made and becomes linear at large flow times, then the pore volume can be estimated with:

$$V_p = \frac{2.358 \ q \ T}{(c_e \mu)_i \ m_L} \ ft^3 \tag{16-117}$$

where:

V_p = drainage pore volume, ft^3
q = producing rate, MSCF/day,
m_L = slope of the linear part of the pseudopressure vs. time plot, (psi^2/cp)/hr, and
T = reservoir temperature, °R.

Note that the pseudopressure vs. time plot should use time in hours.

DETECTING FAULTS OR PERMEABILITY BARRIERS

Pressure transient well test data collected from a well that is in the vicinity of a linear barrier such as a fault will indicate the presence of this barrier. This situation can be modeled mathematically by neglecting the barrier (that is a distance "x" from the test well) and including an "image" well (that is a distance "2x" from the test well) as illustrated below. Note the isopotential lines (constant pressure contours) that are caused by placing the test well (and image well) on production. At early test times the isopotential lines are close to the well and radial in shape. The lines further away from the well correspond to later times.

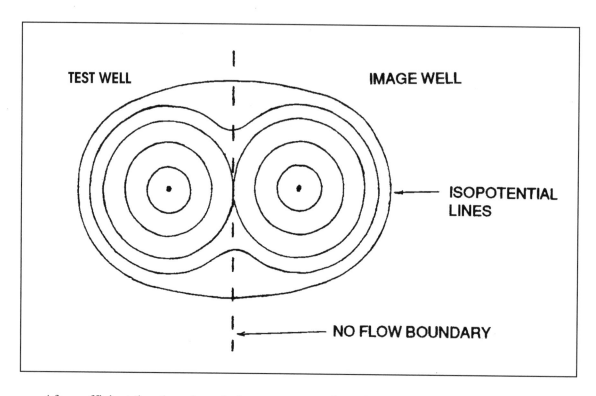

After sufficient time has elapsed, the pressure transients from the real well and the image well intersect (pressure interference). There is a brief transition period where the isopotential lines from the test and image wells merge. At the end of the transition period, the combined real well and image well system is acting like a single well with twice the actual rate. Thus, the pressure change seen on a semilog plot doubles. An equivalent response is obtained when a test well is produced long enough to feel a linear barrier. The semilog pressure response doubles. Physically, this may be reasoned that the doubling in pressure decline occurs because the expanding circular drainage area, after contacting the barrier, assumes more of a semi-circular shape (draining one-half of a circle instead of a whole circle).

Similarly, on pressure buildup test semilog plots (Horner or MDH) a doubling in slope occurs when a physical barrier (or image well interference) is encountered.

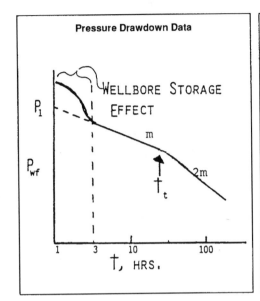

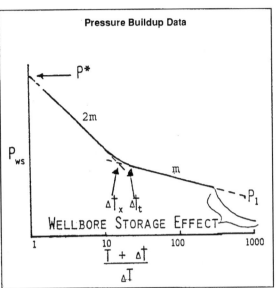

If the linear barrier is encountered after the effects of wellbore storage have ended, then two different straight line segments are seen, with slopes having the approximate ratio: 2 to 1. The flow capacity and skin effect should be calculated on the basis of the first line — the one that is demonstrated before the barrier is felt. In the previous diagram, this is the one with slope = m. For either a buildup or drawdown:

$$kh = 162.6 \, \frac{q \, \mu \, B}{m}$$

The skin effect is calculated as before:

Pressure Drawdown Data:

$$s = 1.151 \left[\frac{p_i - p_{1hr}}{m} - Log \left(\frac{k}{\phi \, \mu \, c_e \, r_w^2} \right) + 3.23 \right]$$

Pressure Buildup Data:

$$s = 1.151 \left[\frac{p_{1hr} - p_{wf}}{m} - Log \left(\frac{k}{\phi \, \mu \, c_e \, r_w^2} \right) + 3.23 \right]$$

Then, for either case:

$$\Delta p_{skin} = 0.87 \, m \, s$$

For a buildup, note from the diagram that p* can be determined (with a Horner plot) by extrapolating the second straight line to infinite shut-in time; i.e., where the time ratio:

$$(t + \Delta t) / (\Delta t) = 1.0$$

Normally, p* is a good approximation to the regional (or static) pressure near the well. If the exact drainage area and well geometry are known, then the Matthews-Brons-Hazebroek method can be used to calculate $\bar{p}$ using the second line.

Assuming a constant transmissibility between the well and the barrier, then the distance from the test well to the barrier should be related to the test time when the semilog plot slope doubles. Gray[39] has suggested that the distance to the fault or barrier can be approximated using the following equation:

$$D = 0.0328 \sqrt{\frac{k \, \Delta t_t}{\phi \, \mu \, c_e}} \qquad (16\text{-}118)$$

where:

D = distance to the barrier or fault, ft,
k = formation permeability, md,
ϕ = porosity, fraction,
μ = fluid viscosity, cp,
c_e = total system compressibility, psi^{-1}
Δt_t = end of the first straight line segment, hr.

A second equation has been suggested by Davis and Hawkins.[40] The required time, Δt_x or t_x, represents the time of intersection of the two straight lines seen on the semilog plot.

$$D = 0.0122 \sqrt{\frac{k \, \Delta t_x}{\phi \, \mu \, c_e}} \qquad (16\text{-}119)$$

For the Gray equation and the Davis-Hawkins equations to yield similar results, the ratio:

$$\Delta t_x / \Delta t_t \approx 7$$

However, in practice, this ratio is seldom seen to be more than 2.5 to 3.0. The Davis-Hawkins equation has a better theoretical basis. According to Tracy,[38] however, in cases where he was able to calculate the distance to a fault by some other means, the Gray equation has consistently given better results. The Davis-Hawkins equation usually calculates distances that are conservative.

Problem 16-7:

A well was drilled in a remote area, and was acidized to remove skin damage due to drilling. A production test was conducted, and it was noticed that the pressure dropped at a rate somewhat faster than expected. Therefore, a pressure buildup test was conducted to determine the following:

1. the magnitude of the skin effect,

2. the approximate permeability of the formation,

3. the static reservoir pressure, and

4. the probable reason for the apparent rapid decrease in flowing pressure.

Basic Data:

Pay thickness	115 ft
Porosity	0.25
Wellbore radius	0.226 ft
Connate water saturation	0.25
Oil compressibility	$20 \times 10^{-6} \ psi^{-1}$
Formation compressibility	$4 \times 10^{-6} \ psi^{-1}$
Water compressibility	$3.5 \times 10^{-6} \ psi^{-1}$
Bubblepoint pressure of oil	2532 psia
Oil viscosity	1.2 cp
Oil formation volume factor	1.45 BBL/STB
Producing rate prior to shut-in	165 STBO/D
Cumulative Production (prior to shut-in)	1250 STB

Shut-in Pressure Data:

Time Hr	Pressure, Psia	Time, Hr	Pressure Psia
0.00	3022	8.0	3219
0.167	3035	10.0	3225
0.333	3053	12.0	3229
0.500	3074	15.0	3235
0.667	3094	20.0	3243
0.833	3163	24.0	3249
1.00	3182	30.0	3255
1.50	3190	36.0	3261
2.0	3195	42.0	3265
4.0	3207	48.0	3269
5.0	3211	72.0	3280
6.0	3214	120.0	3292

Problem solution:

Producing time: t = (1250)(24) / 165 = 181.8 hours

Calculation of the Horner Shut-in Time Ratio: $(t + \Delta t) / \Delta t$

Shut-in Time Hrs	Time Ratio	Pressure Psia	Pressure Rise, Psi
0	—	3022	0
0.167	1089.	3035	13
0.333	546.9	3053	31
0.500	364.6	3074	52
0.667	273.6	3097	75
0.833	219.2	3163	141
1.00	182.8	3182	160
1.50	122.2	3190	168
2.00	91.9	3195	173
4.00	46.5	3207	185
5.00	37.4	3211	189
6.00	31.3	3214	192
8.00	23.7	3219	197
10.0	19.2	3225	203
12.0	15.2	3229	207
15.0	12.1	3235	213
20.0	10.1	3243	221
24.0	8.6	3249	227
30.0	7.1	3255	233
36.0	6.1	3261	239
42.0	5.3	3265	243
48.0	4.8	3269	247
72.0	3.5	3280	258
120.0	2.5	3292	270

Figure 16-36 is a Horner plot of these data.

Calculation of Flow Capacity:

$$kh = 162.6 \frac{q_o \, \mu_o \, B_o}{m}$$

$$= \frac{(162.6)(165)(1.2)(1.45)}{40.5} = 1152.65 \ md\text{–}ft$$

$k = kh / h = 1152.65 / 115 = 10.02 \ md$

Calculation of Effective Compressibility, c_e:

$$c_e = c_o S_o + c_w S_w + c_f$$

$$= [\,(20.0)(0.75) + (3.5)(0.25) + 4.0\,] \times 10^{-6}$$

$$c_e = 19.875 \times 10^{-6} \ psi^{-1}$$

Calculation of Skin Effect:

$$s = 1.151 \left[\frac{\Delta p_1}{m} - Log \left(\frac{k}{\phi \ \mu \ c_e \ r_w} \right) + 3.23 \right]$$

$$= 1.151 \left[\left(\frac{160}{40.5} \right) - Log \left(\frac{10.02}{(.25)(1.2)(19.875 \times 10^{-6})(.226)^2} \right) + 3.23 \right]$$

$$s = -0.386$$

$$\Delta p_{skin} = 0.87 \ m \ s = (0.87)(40.50)(-0.386) = -13.6 \ psi$$

By considering the Horner plot of the pressure buildup data, it is apparent that two straight line segments are present. Furthermore, the ratio of the two slopes is close to 2.0 (actually 82/40.5 = 2.025). This suggests the possibility of a permeability barrier near this well. Further considerations of geologic and seismic data suggest that faults are possible in the area. Therefore, it appears that the productivity reduction is probably due to a fault near the well.

Distance to the fault:

Gray Equation

$$D = 0.0328 \ \sqrt{\frac{k \ \Delta t_t}{\phi \ \mu \ c_e}}$$

$$= 0.0328 \ \sqrt{\frac{(10.02)(8)}{(0.25)(1.2)(19.875)(10^{-6})}}$$

$$= 120 \ ft$$

Davis-Hawkins Equation:

$$D = 0.0122 \ \sqrt{\frac{(10.02)(10.8)}{(0.25)(1.2)(19.875)(10^{-6})}}$$

$$= 52 \ ft$$

Assuming the Gray equation is the more accurate, it is likely that the fault is between 100 and 150 feet from the well. This is the probable reason for the rapid reduction in flowing pressure. The false pressure, p*, is obtained by extrapolating the second straight line to:

$$(t + \Delta t) / \Delta t = 1.0$$

Since this is a new well, p* = p_i. This pressure is 3325 psia.

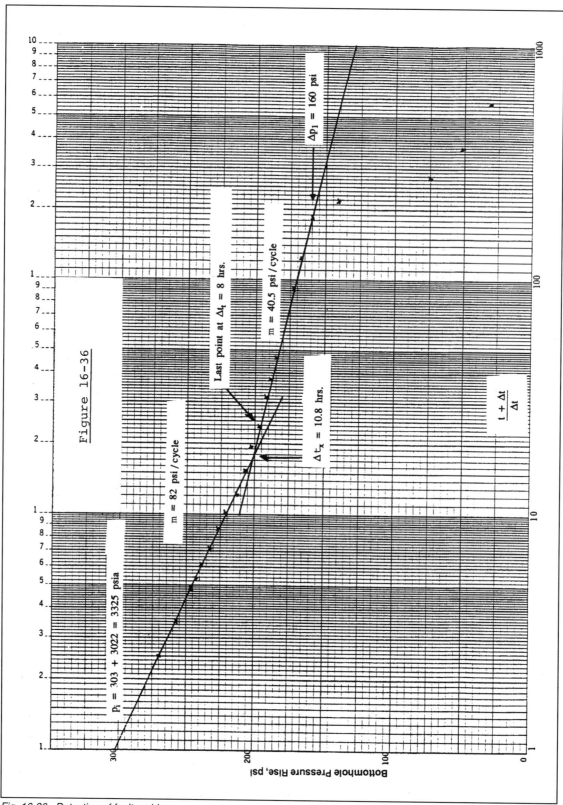

Fig. 16-36. Detection of fault problem.

DETECTING MULTIPLE FAULTS

If a test well is located near two faults which intersect at some angle, then the performance of this well can be modeled mathematically by using a number of image wells. To illustrate, consider a test well with two faults intersecting at 90 degrees. In this case, three image wells are needed. For other angles, a more complicated arrangement of image wells would be required.

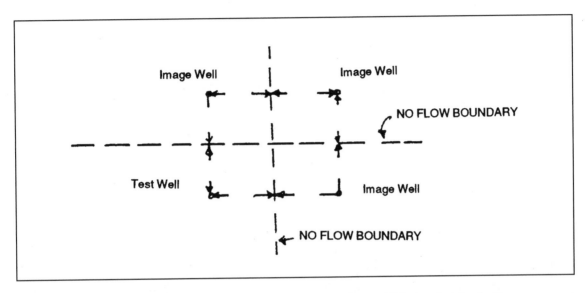

Why are three image wells needed for the 90 degree angle case? Notice that the fault arrangement will only leave about one-fourth of the reservoir area for the test well to drain. Therefore, three image wells are needed (one for each of the other quarters). Typical semilog drawdown and build-up (Horner) plots are shown for a well in the vicinity of multiple barriers.

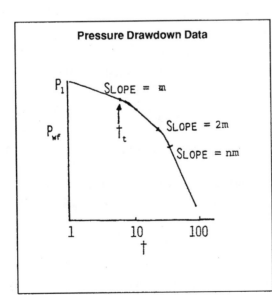

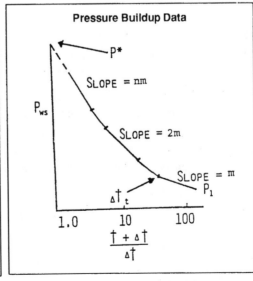

As with the single-barrier case, the first straight line is the pressure response of the test well before the barriers are felt. Notice that in the previous diagrams, this is the line with slope, m. It is this first straight line that is used to calculate flow capacity and skin effect:

$$k\,h = \frac{162.6\,q\,\mu\,B}{m}$$

Buildup:

$$s = 1.151\left[\left(\frac{p_{1hr} - p_{wf}}{m}\right) - Log\left(\frac{k}{\phi\,\mu\,c_e\,r_w^2}\right) + 3.23\right]$$

Drawdown:

$$s = 1.151\left[\left(\frac{p_i - p_{1hr}}{m}\right) - Log\left(\frac{k}{\phi\,\mu\,c_e\,r_w^2}\right) + 3.23\right]$$

If the first fault is closer than the second, then the second semilog straight line that is seen will have a slope approximately twice that of the initial straight line. This is the pressure response indicating that the first barrier has been felt. Then, when the second fault or barrier is encountered, another straight line with an even steeper slope will result. The angle between the two barriers (or faults) is related to the ratio of the straight line slopes. Consider the previous diagram. Then:

Angle between faults = 360 / n   (16-120)

where:

   n = ratio of the slopes (last straight line to first one).

If the two barriers are approximately the same distance from the test well, then the second straight line (described previously with a doubling in slope) will not be seen. After the initial straight line, when the two faults are felt, the slope will increase to more than two times the initial slope. In this case, the second straight line is indicating the presence of more than one fault, and the ratio of the two slopes is related to the angle between the two faults as given in Equation 16-120.

Theoretically, the presence of three barriers in the vicinity of the test well can be analyzed on the basis of the semilog pressure response. However, practically, this is not likely. Whether there is one fault or more, it is the last semilog straight line that should be extrapolated to obtain p*. And normally, this is taken to be the well's approximate static pressure. The distance to the nearest fault can still be calculated by using techniques described earlier. The Gray equation is recommended:

$$D = 0.0328\sqrt{\frac{k\,\Delta t_t}{\phi\,\mu\,c_e}}$$

Unfortunately, there is no theory currently available to allow the calculation of the distance to faults other than the nearest one.

Problem 16-8:

A well is drilled in the proximity of at least one fault. It is possible that a second fault is also near this well. A pressure buildup test was conducted to answer certain questions:

(1) Does there appear to be at least one fault near this well?

(2) If so, about how far away is this fault?

(3) Is there more than one fault near the well?

(4) If there is a second fault, approximately what is the angle between the faults?

Basic Data:

Porosity	21%
Wellbore radius	0.25 ft
Effective compressibility	20×10^{-6} psi^{-1}
Oil Viscosity	1 cp
Pay thickness	10 ft
Producing rate prior to shut-in	80 STB/D
Oil formation volume factor	1.23 bbl/STB
Cumulative production prior to shut-in	240 STB.

Pressure Data:

Time Hrs	Pressure Psia	Time Hrs	Pressure Psia
0	3050	16	3156
0.25	3112	24	3162
0.5	3118	32	3169
0.75	3122	40	3179
1.0	3124	50	3186
2.0	3130	60	3193
4.0	3137	70	3198
7.0	3144	80	3202
10.0	3150	90	3205
12.0	3152		

Calculation of Producing Time, t:

$$t = \frac{Cum\ Prod \times 24}{q} = \frac{(240)(24)}{80} = 72\ hours$$

The Horner plot is shown in Figure 16-37

Calculation of Flow Capacity and Permeability:

$$k\,h = 162.6\ \frac{q\ \mu\ B}{m} = \frac{(162.6)(80)(1)(1.23)}{20} = 800.0\ md-ft$$

Calculation of Skin Effect:

$$s = 1.151 \left[\left(\frac{p_{1hr} - p_{wf}}{m} \right) - Log \left(\frac{k}{\phi \, \mu \, c_e \, r_w^2} \right) + 3.23 \right]$$

$$= 1.151 \left[\left(\frac{3124 - 3050}{20} \right) - Log \left(\frac{80}{(0.21)(1)(20)(10^{-6})(0.25)^2} \right) + 3.23 \right]$$

$$s = 1.151 \left[3.70 - 8.48 + 3.23 \right] = -1.8$$

$$\Delta p_{skin} = 0.87 \; m \, s = (0.87)(20)(-1.8) = -31 \; psi$$

From the shape of the buildup Horner plot, two faults appear to be near this well.

Distance to Nearest Fault:

Gray Equation

$$D = 0.0328 \sqrt{\frac{k \, \Delta t_t}{\phi \, \mu \, c_e}} .$$

$$= 0.0328 \sqrt{\frac{(80)(2)}{(0.21)(1)(20)(10^{-6})}}$$

$$= 202 \; ft$$

Angle Between Faults:

$$\alpha = 360 \, (m_1 / m_3) = \frac{(360)(20)}{(139)} \cong 52 \; deg$$

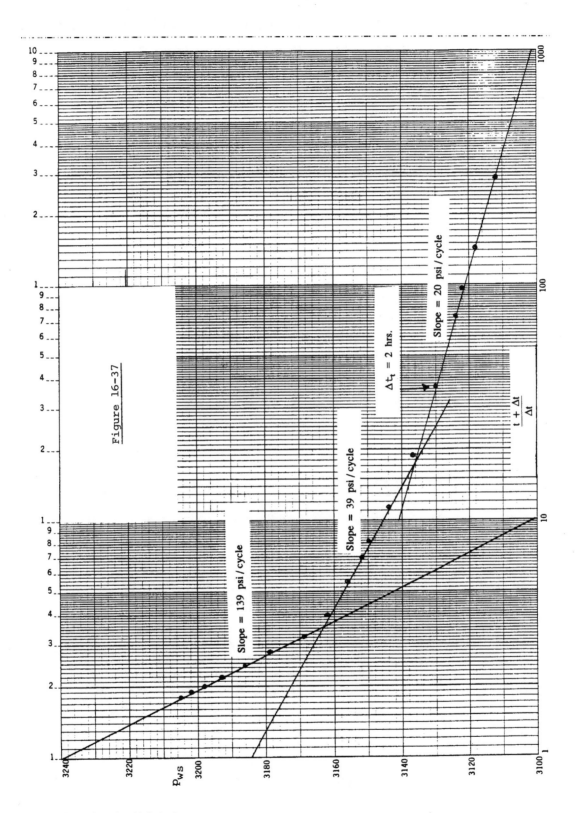

Fig. 16-37. Detection of multiple faults.

PRESSURE BUILDUP DATA FROM A WELL PRODUCING FROM A LONG, NARROW RESERVOIR, SUCH AS A CHANNEL SAND

The pressure transient data collected from a well producing from a long, narrow reservoir have characteristics that are related to the pressure transients of wells with a man-made fracture. In both cases, a combination of radial flow and linear flow exists.

In the case of a long narrow reservoir, the radial flow occurs close to the well, while the linear flow regime is situated at distance from the well. (Note the diagram that illustrates the well position, reservoir boundary, and the isopotential lines at various producing times.) Therefore, the pressure buildup reflects radial flow earlier than linear flow.

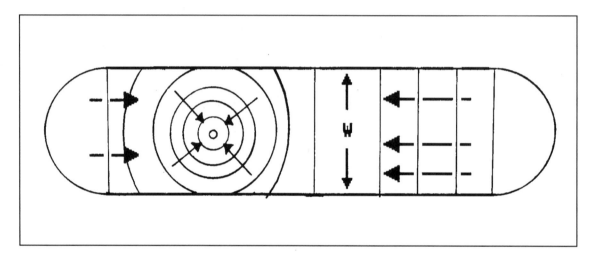

Since the pressure buildup characteristics are related to the state of the isopotential lines at the time of shut-in, inspection of the geometry of such lines will yield insight to the nature of the pressure buildup data. Notice that the isopotential lines near the well are circular in shape. Thus, radial flow is occurring in the area close to the well. At distance from the well, the isopotential lines are parallel straight line segments (linear flow).

The pressure buildup data from such a well would be similar to the following:

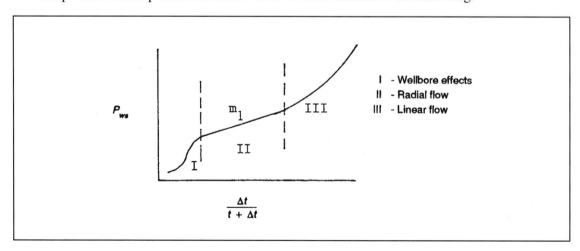

P_{ws} vs $\frac{\Delta t}{t + \Delta t}$

I - Wellbore effects
II - Radial flow
III - Linear flow

Note that the region II (radial flow) straight line slope is denoted as m_1. In region III, where either a straight line or a flattening of the slope is normally expected, an ever increasing slope is seen. If this response is obtained and a long narrow reservoir is suspected, then it is appropriate to make the following linear flow plot.

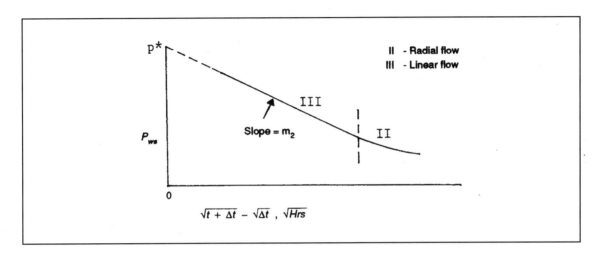

If the data in region III become a straight line on this plot, then this indicates that linear flow is occurring at distance from the well. The slope of this straight line is denoted as m_2 and has units of psi/$\sqrt{hr}$.

For the analysis of such a well, the flow capacity and skin effect are determined on the basis of the semilog plot straight line data (region II). For instance:

$$k\,h \;=\; 162.6\,\frac{q\,\mu\,B}{m_1} \qquad\qquad (16\text{-}121)$$

The pressure, p* is determined from the linear plot by extrapolating the region III data back to:

$$\sqrt{t + \Delta t} \;-\; \sqrt{\Delta t} \;=\; 0$$

which corresponds to infinite shut-in time.

The width of the linear flow channel (in this case, width of the reservoir) can be calculated by using theory that involves a combination of the two slopes m_1 and m_2. The oil well equation is:

$$w \;=\; \frac{0.638}{m_2}\,\sqrt{\frac{q_o\,B_o\,m_1}{h\,\phi\,c_e}} \qquad\qquad (16\text{-}122)$$

where:

w = width of the linear flow channel, ft,

m_1 = slope of the straight line portion of the radial plot; i.e., the plot of p_{ws} vs. log [(t + Δt)/Δt] or p_{ws} vs. log Δt, psi/cycle,

m_2 = slope of the straight line portion of the linear flow plot; i.e., the plot of p_{ws} vs. $\sqrt{t + \Delta t} - \sqrt{\Delta t}$ or p_{ws} vs $\sqrt{\Delta t}$, psi / $\sqrt{hr}$

q_o = oil flow rate, STB/D,

B_o = oil-formation volume factor, bbl/STB,

h = net pay thickness, ft,

ϕ = porosity, fraction, and

c_e = effective or total compressibility, psi^{-1}

For a well in a long narrow gas reservoir, the reservoir width is:

$$w = \frac{2.02}{m_2} \sqrt{\frac{q\,T\,z\,m_1\,p_{avg}}{h\,\phi\,(1 - S_w)}}$$

(16-123)

where:

w = the width of the linear flow region, ft,

m_1 = slope of the straight line portion of data on the radial flow plot:

$$p_{ws}^2 \;\; vs. \;\; \log\left[\frac{t + \Delta t}{\Delta t}\right], \;\; or$$

$$p_{ws}^2 \;\; vs. \;\; \log(\Delta t), \; psi^2 \,/\, cycle$$

m_2 = the slope of the straight line portion of data on the linear flow plot:

$$p_{ws}^2 \;\; vs. \;\; (\sqrt{t + \Delta t} - \sqrt{\Delta t}), \;\; or$$

$$p_{ws}^2 \;\; vs. \;\; \sqrt{\Delta t}, \; psi^2 \,/\, \sqrt{hr}$$

q = gas-producing rate prior to shut-in, MCF/D,

T = reservoir temperature, °R,

p_{avg} = average pressure in the neighborhood of the well, psia,

z = gas deviation factor associated with p_{avg},

ϕ = porosity, fraction, and

S_w = connate water saturation, fraction.

Problem 16-9:

To illustrate the use of Equation 16-123, consider the following data for a gas well that is producing from a sand stringer.

$$q = 3.6 \text{ MMCF/D}$$
$$T = 170 \text{ °F} = 630 \text{ °R}$$
$$z = 0.85$$
$$p_{avg} = 4520 \text{ psia}$$
$$h = 50 \text{ ft}$$
$$\phi = 0.15$$
$$S_w = 0.25$$

From the pressure buildup data plots, the slope values are:

$$m_1 = 3.43 \times 10^6 \text{ psi}^2 / \text{cycle}$$
$$m_2 = 1.57 \times 10^5 \text{ psi}^2 / \sqrt{hr}$$

Then:

$$w = \frac{2.02}{m_2} \sqrt{\frac{q\, T\, z\, p_{avg}\, m_1}{h\, \phi\, (1 - S_w)}} = \frac{2.02}{(1.57)(10^5)} \sqrt{\frac{(3600)(630)(0.85)(4520)(3.43)(10^6)}{(50)(0.15)(1 - 0.25)}}$$

$$w = 938 \text{ ft.}$$

PRESSURE BUILDUP DATA FROM A VERTICALLY FRACTURED WELL

The following figure shows the general characteristics of a pressure buildup curve collected from a vertically fractured well.

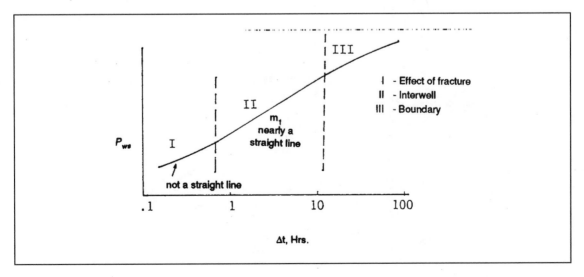

If the early time data are replotted in the following manner, a straight line results:

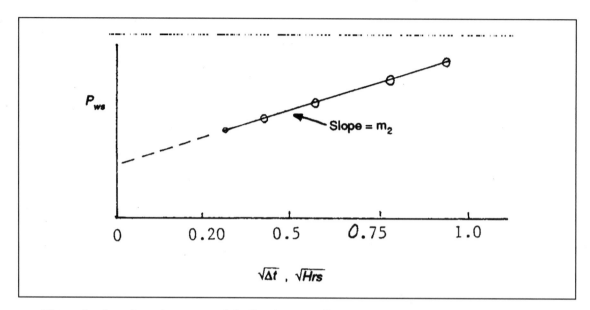

The early time data plot as a straight line because flow close to the well is basically linear flow into the fracture (pressure in the fracture will be close to that in the wellbore).

Whenever a vertical fracture extends outward from the well, the classical straight line (intermediate time data): p_{ws} vs. $\log(\Delta t)$ or p_{ws} vs. $\log[(t + \Delta t) / \Delta t]$ is flatter (slope m_1 is smaller) than it should be to reflect the true reservoir flow capacity. This is due to the presence of the fracture. Therefore, a correction must be made to the slope, m_1, in order to compute the correct value of the flow capacity of the actual reservoir rock. Russell and Truitt[42] have developed a curve to aid in correcting the slope of the radial semilog plot straight line for a vertically fractured well. The following is a representation of this curve. (Also, see Figure 16-38.)

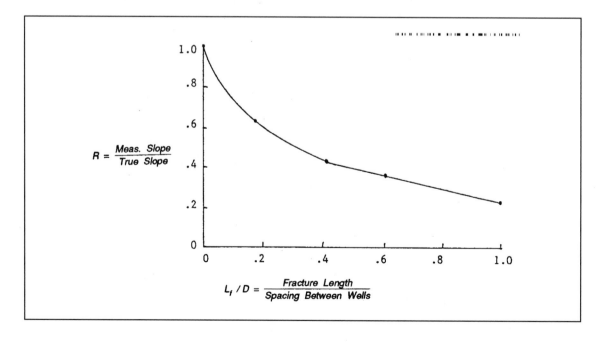

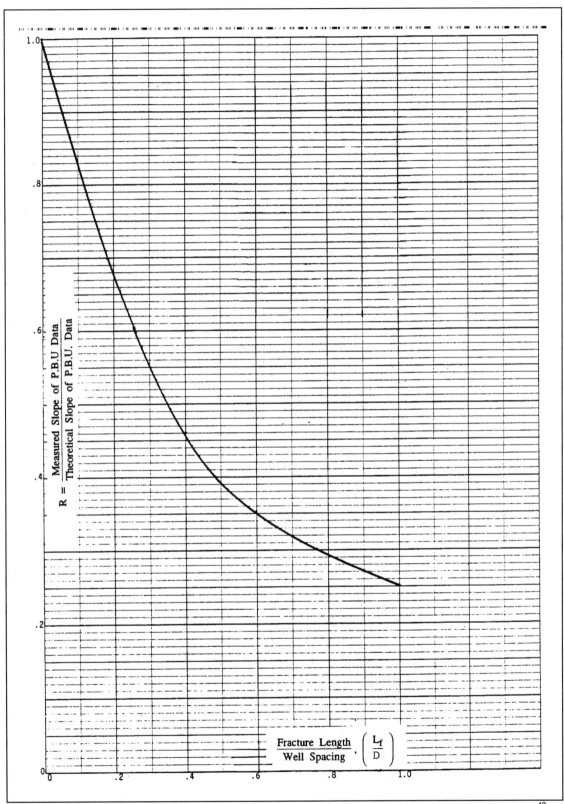

*Fig. 16-38.Correction factor for slope of pressure buildup data because of vertical fracture. (From Russell & Truitt[42])..
Permission to publish by JPT.*

For an oil well, the equation relating fracture length with reservoir and test parameters is:

$$L_f = \frac{0.638}{m_2} \sqrt{\frac{q_o B_o m_1}{h \, \phi \, c_e \, R}} \qquad (16\text{-}124)$$

where:

D = spacing between adjacent wells, ft,

L_f = fracture length (tip to tip), ft,

m_1 = slope of the p_{ws} vs. log Δt plot, psi/cycle, or slope of the

p_{ws} vs. log $\left[\dfrac{t + \Delta t}{\Delta t} \right]$ plot, psi/cycle,

m_2 = slope of the p_{ws} vs. $\sqrt{\Delta t}$ plot, psi / $\sqrt{hr}$,

ϕ = porosity, fraction

c_e = effective compressibility, psi^{-1}

$q_o B_o$ = reservoir withdrawal rate, bbl/D, and

R = correction factor relating the ratio of the measured slope (m_1) to the "true" slope (the "true" slope is the slope that would have been measured if no fracture had been present), dimensionless.

By inspection of Equation 16-124 and the correction curve, R vs. L_f / D, it should be apparent that there is no direct way to solve for both R and L_f. Thus the solution technique normally used is a trial-and-error procedure. To begin, some short fracture length, such as 50 to 100 ft, is assumed. This is divided by the distance between wells, and from the correction curve an R is determined. Then a fracture length is calculated with the equation. This is compared with the assumed value. The calculated value is more accurate and therefore is used to begin the second pass through the procedure. Convergence should occur within two to four iterations. Then, after convergence:

$$k_o h = 162.6 \, \frac{q_o \, B_o \, \mu_o \, R}{m_1} \qquad (16\text{-}125)$$

The Russell and Truitt correction factor, R, is accurately depicted in Figure 16-38.

Problem 16-10:

A well has been hydraulically stimulated causing a vertical fracture. After a long producing time, the well was shut-in for a buildup test.

Basic reservoir data:

Porosity	15 %
Pay Thickness	50 ft
Oil Compressibility	20 x 10^{-6} psi^{-1}
Oil Saturation	0.75
Water Compressibility	3.0 x 10^{-6} psi^{-1}
Water Saturation	0.25
Formation Compressibility	4.25 x 10^{-6} psi^{-1}
Oil Viscosity	0.8 cp
Oil Formation Volume Factor	1.261 bbl/STB
Oil Rate Prior to Shut-in	460 STB/D
Distance Between Adjacent Wells	1320 ft

Pressure - time data:

Time, hrs	Pressure, psia	Δp, psi	$\sqrt{\Delta t}$, $\sqrt{hrs}$
0	2500	0	0.
0.0333	2549	49	0.183
0.0667	2569	69	0.258
0.1000	2585	85	0.316
0.1333	2597	97	0.365
0.1667	2609	109	0.408
0.2500	2634	134	0.500
0.3333	2654	154	0.577
0.4167	2672	172	0.645
0.500	2690	190	0.707
0.750	2733	233	0.867
1.00	2767	267	1.000
2.0	2850	350	1.414
3.	2901	401	
5.	2970	470	
10.	3067	567	
15.	3129	629	
20.	3172	672	
25.	3210	710	
30.	3238	738	
35.	3260	760	
40.	3282	782	
45.	3300	800	
48.	3313	813	

An MDH plot of these data is shown in Figure 16-39.

Figure 16-40 contains the early time data linear plot.

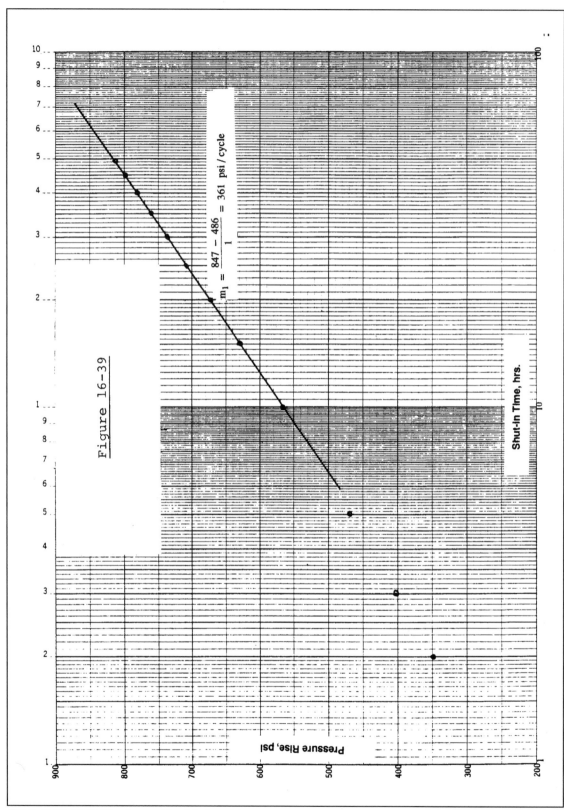

Figure 16-39. Vertical fracture problem late time data.

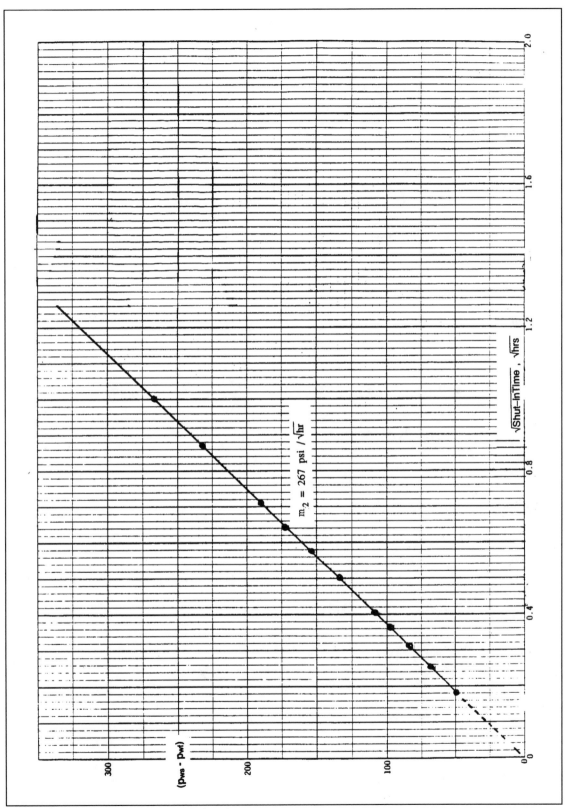

Fig. 16-40. Vertical fracture problem early time data.

Calculations of Effective Compressibility:

$$c_e = c_o\, S_o + c_w\, S_w + c_f$$
$$= [\,(20)(0.75) + (3)(0.25) + 4.25\,] \times 10^{-6}$$
$$= (15 + 0.75 + 4.25) \times 10^{-6} = 20 \times 10^{-6}\ \text{psi}^{-1}$$

The early time plot of the pressure buildup data is shown as p_{ws} vs. $\sqrt{\Delta t}$. The later data are plotted as an MDH plot: p_{ws} vs. log Δt. From these plots the respective slopes of the straight line portions of data were determined to be:

$$m_1 = 361\ \text{psi / cycle} \qquad \text{(Radial flow plot)}$$
$$m_2 = 267\ \text{psi / }\sqrt{\text{hour}} \qquad \text{(Linear flow plot)}$$

Trial and error calculations can be made to determine the probable fracture length as well as the correction factor, R.

Trial and Error Calculations:

Trial 1:

Assume L_f = 50 ft

$$\frac{L_f}{D} = \frac{50}{1320} = 0.038 \rightarrow R = 0.932 \qquad \textit{(from Figure 16-38)}$$

$$L_f = \frac{0.638}{m_2}\sqrt{\frac{q_o\, B_o\, m_1}{h\, \phi\, c_e\, R}} = \frac{0.638}{267}\sqrt{\frac{(460)\,(1.261)\,(361)}{(50)\,(0.15)\,(20)\,(10^{-6})\, R}}$$

$$= \frac{89.3}{\sqrt{R}} = \frac{89.3}{\sqrt{0.932}} = 92.5\ ft$$

Trial 2:

Assume L_f = 92.5 ft.

$$\frac{L_f}{D} = \frac{92.5}{1320} = 0.070 \rightarrow R = 0.878$$

$$L_f = \frac{89.3}{\sqrt{0.878}} = 95.3\ ft$$

Trial 3:

Assume L_f = 95.3 ft.

$$\frac{L_f}{D} = \frac{95.3}{1320} = 0.072 \rightarrow R = 0.870$$

$$L_f = \frac{89.3}{\sqrt{0.870}} = 95.7 \, ft$$

Using the results of trial 3 as representative,

$$R = 0.87 \quad \text{and} \quad L_f = 96 \, ft$$

Calculations of Flow Capacity and Permeability:

$$kh = \frac{162.6 \, q \, \mu \, B \, R}{m_1}$$

$$= \frac{(162.6) \, (460) \, (0.8) \, (1.261) \, (0.87)}{361}$$

$$= 181.8 \, md\text{--}ft.$$

$$k = kh / h = 181.8 \, / \, 50 = 3.64 \, md.$$

PRESSURE TEST DATA FROM WELLS COMPLETED IN MULTIPLE POROSITY PAY ZONES

NATURALLY FRACTURED RESERVOIRS[43,44]

Pressure test data measured on wells producing from naturally fractured reservoirs can be divided into four different regions:

(1) Wellbore effects,

(2) Pressure change in the fissure system connected to the wellbore,

(3) Transition period (time interval where the flow of fluids from the matrix rock to the fractures changes from negligible to significant),

(4) Flow of fluids between the rock matrix and the fissures is occurring to the extent that equalization of pressure throughout the drainage area eventually is attained.

These time regions are illustrated in the Horner plot on the next page for a typical well in a naturally fractured reservoir.

Naturally fractured reservoirs usually have most of their conductivity or flow capacity in the fracture system. However, it is likely that some of the rock matrix material will be in direct contact with the wellbore. Of course, there will usually be some fissures that intersect the wellbore. So, the region II data, shown in the following diagram are labeled "combination" to indicate the combined flow capacity of the fractures and matrix blocks. Practically speaking, because the conductivity of the fissures is much higher than that of the matrix blocks, analysis of the region II data yields primarily the flow capacity of the fissure system:

$$kh = 162.6 \, \frac{q\mu B}{m}$$

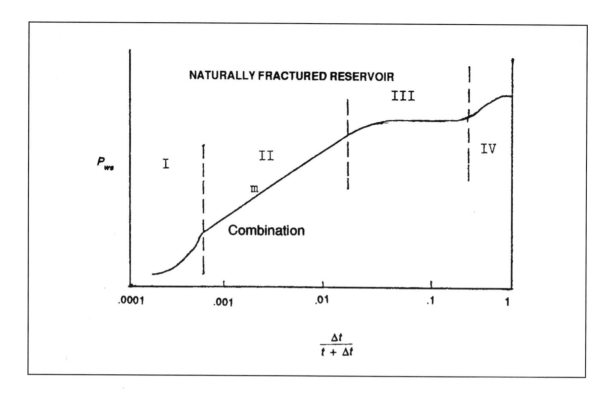

For the situation where there is minimal, if any, pore volume in the matrix portion of the formation, then region III will be quite small or nonexistent. In this case, the reservoir will appear as a single-porosity, highly permeable system.

Normally, most of the pore volume in a naturally fractured reservoir is contained within the matrix blocks. In these systems, the pressure buildup test may be terminated after reaching region III because the measured pressures become nearly constant. This is often interpreted as approaching the static pressure. Of course, this false plateau is below the reservoir's true static pressure. Such an interpretation is likely, particularly in the early history of a well's operation. This is common especially if the double porosity nature of the reservoir is not known. The difference between this false plateau and the true static pressure is greatest early in the reservoir life and decreases with time.

MULTIPLE LAYERED RESERVOIRS[45]

The shape of the pressure buildup data from a well in a reservoir with multiple layers without crossflow is similar to data collected from a well in a naturally fractured reservoir. Once again, four different pressure response regions are seen:

    (1) Wellbore effects,

    (2) Pressure buildup in all of the zones connected with the wellbore,

    (3) Circulation via the wellbore of fluids from the high pressure zones to the zones of lower pressures,

    (4) Pressure equalization between the zones, i.e., final buildup to a common average pressure for all zones.

The total or combined flow capacity of all the zones is calculated using the slope of the straight line data in region II:

$$k\,h \;=\; 162.6\;\frac{q\,\mu\,B}{m}$$

This equation is not strictly correct if there is crossflow between the layers; however, it is usually a reasonable approximation.

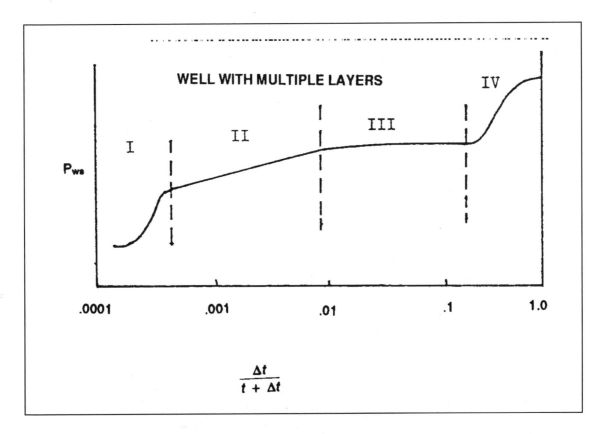

With no crossflow between zones, region III represents the low-pressure zones being recharged by the high-pressure (lower permeability) zones through the wellbore only. Therefore, this pressure rise is usually quite slow. Of course, to the well analyst not thinking about the effect of layers, this slow pressure rise appears to be a complete flattening, and is often interpreted to be the static pressure. Especially in the early life of the well, this pressure is too low. As time goes on, the difference between the true static pressure and the region III pressure diminishes.

DRILLSTEM TESTS

A drillstem test is a well test consisting of multiple producing and shut-in periods that is performed prior to the completion of the well. The primary objective usually is to determine the type of fluids that a zone will produce if a completion is made in the interval of the test. Bottomhole pressures are measured during the test which can allow the determination of:

(1) Initial reservoir pressure,

(2) Approximate value for formation flow capacity, kh,

(3) Magnitude of the skin effect,

(4) Distance out from the well that was "seen" (radius of investigation) during the test.

(5) Evidence of depletion.

The modern drillstem test usually consists of two flowing (or open) periods and two shut-in periods. It is important that sufficient time be allowed for each period.

The recommended time range for the initial flow is 3 to 20 minutes. Except in very high permeability formations, at least 10 minutes should be allowed to reduce the residual effects of "supercharging." This occurs due to leak-off of drilling fluid filtrate because of the higher pressure of the mud column. The loss of mud filtrate can cause the pressure near the well to be higher than the initial pressure. If this excess pressure has not dissipated, the indicated initial reservoir pressure will be higher than the true value.

Following the initial flow period is the initial shut-in (buildup) period. The rule of thumb for the length of the initial shut-in is five to six times as long as the initial flow period. For example, if the initial "open" period is 10 minutes, then the first shut-in should last at least 50 minutes; 60 minutes would be better. The reason for this is practical: the buildup pressure response for the initial shut-in on a Horner plot is quite steep. A long extrapolation of a steep curve can result in a considerable error. If the shut-in period is 6 times as long as the flowing period, the extrapolation required to obtain p_i will be short.

The second and final flow period usually lasts for one hour or more. One of the purposes is to flow the well at a nearly constant rate to get an idea of well productivity. During this flow period samples of the produced fluids are collected. The final shut-in period should be at least as long as the final flow period. A longer shut-in usually allows better analysis.

A schematic drawing of a typical oilwell DST pressure - time trace is shown as:

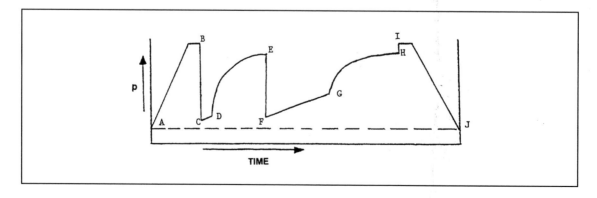

The most important features are:

> A to B: DST tool (including pressure gauge) going into the well. The increasing pressure during this phase indicates the weight of the mud column. After reaching the test location, the tool packer is set, which isolates the formation and tool flow system from the mud system.
>
> B to C: Initial opening of the DST tool.
>
> C to D: Initial open (flow) period.
>
> D to E: Initial shut-in period.
>
> E to F: Second opening of the DST tool.
>
> F to G: Final flow period.
>
> G to H: Final shut-in period.
>
> H to I: Unseating the tool packer; pressure gauge now exposed to the mud column pressure.
>
> I to J: DST tool being withdrawn from the well.

For analysis of either of the shut-in periods (pressure buildup), superposition is required because the producing time (t) is small: normally less than or equal to the shut-in time (Δt). A Horner plot is the usual means of analysis. Each buildup needs the corresponding producing time which is calculated as follows:

Initial shut-in (buildup):

> t = initial open period (typically 3 to 20 min.)

Final shut-in (buildup):

> t = sum of both flow periods (For example, first open period was 10 minutes; and second flow period was 120 minutes. Thus, t, for the final buildup, = 130 min.)

CALCULATION OF PRODUCING RATE

To calculate the well's flow capacity, a producing rate is needed. This rate is normally determined in one of three ways, listed in order of preference:

(1) Measure the rate at the surface. This is the best method; but only applicable to those tests which achieve flow to the surface.

(2) Compute the reservoir flow rate based on the slope of the pressure-time relationship during the final flow period (segment F to G in the schematic drawing discussed earlier). The applicable equation is:

$$q_o B_o = (V/G)(1440)(S)$$

(16-126)

where:

$q_o B_o$ = reservoir flow rate, bbl/D,

V = volume of the drillpipe or tubing in which the test fluids are collected, bbl/ft,

G = average gradient of the fluids collected in the drillpipe or tubing, psi/ft, and

S = average slope of the pressure - time data in the final flow period, psi/min.

(3) Compute the reservoir rate by dividing the total volume of fluids collected in the drillpipe by the total open flow time (sum of both flow periods). The computation is:

$$q_o B_o = \frac{(Total \ Recovery, \ Bbls)(1440)}{(Producing \ Time, \ Min.)} \qquad (16\text{-}127)$$

Both the initial shut-in data as well as the final shut-in data are usually graphed on the same set of Horner plot axes. As stated earlier, the initial shut-in data uses a different producing time (to calculate the Horner time ratio) than is used for the final shut-in data. These two Horner plots will resemble:

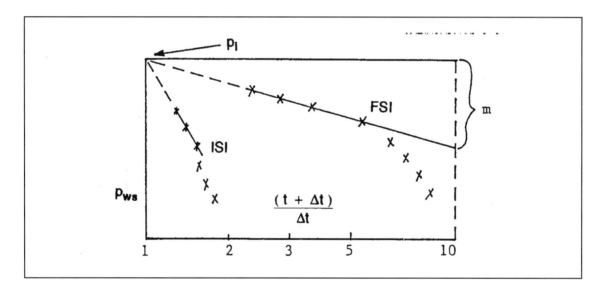

Extrapolating the two sets of data to infinite shut-in time; i.e.:

$$(t + \Delta t) / \Delta t = 1.0$$

should result in the same p_i. If the final shut-in data extrapolate to a p_i that is significantly less than that obtained with the initial shut-in data, then this may be an indication of depletion during the test. If such behavior is observed, then it is best to review all reservoir and test information that may have influenced this result. For instance, does the geological picture for this formation indicate that the reservoir could be quite small. The pressure gauge should also be tested for reliability. Except for extremely small traps, depletion cannot be seen on a DST.

Because of a number of factors (such as the steepness of the plot and because the flow rate during the initial flow period is typically not constant); the initial shut-in data normally are not analyzed for flow capacity.

Reservoir properties are usually calculated based on the Horner straight line of the final shut-in data. The pertinent oil well equations are:

$$k \, h = 162.6 \, \frac{(q_o \, B_o) \, \mu}{m} \tag{16-128}$$

$$\Delta p_{skin} = (p_i - p_{wf}) - m \left\{ Log \left[\frac{k \, t}{\phi \, \mu \, c_e \, r_w^2} \right] - 3.23 \right\} \tag{16-129}$$

$$r_{inv} = 0.0328 \, \sqrt{\frac{k \, t_{test}}{\phi \, \mu \, c_e}} \tag{16-130}$$

where:

t = total producing time (sum of all flow periods), Hr,
t_{test} = total test time (sum of all producing and shut-in times), Hr.
r_{inv} = radius of investigation during the test, Ft,
p_{wf} = Flowing pressure at end of final flow, psia

The corresponding gas well equations are:

p^2 Analysis

$$k \, h = \frac{1637 \, q_g \, \mu_g \, T \, z}{m} \tag{16-131}$$

$$s = 1.151 \left\{ \frac{p_i^2 - p_{wf}^2}{m} - Log \left[\frac{k \, t}{\phi \, \mu \, c_e \, r_w^2} \right] + 3.23 \right\} \tag{16-132}$$

$$\Delta p_{skin} = \sqrt{p_{wf}^2 + 0.87 \, m \, s} \; - \; p_{wf} \tag{16-133}$$

$$r_{inv} = 0.0328 \, \sqrt{\frac{k \, p_i \, t_{test}}{\phi \, \mu \, (1 - S_w)}} \tag{16-134}$$

where:

q_g = MCF/D,
T = reservoir temperature, °R.

Real Gas Potential Analysis

$$k\,h = \frac{1637\,q_q\,T}{m} \tag{16-135}$$

$$s = 1.151\left\{\frac{\psi_i - \psi_{wf}}{m} - Log\left[\frac{k\,t}{\phi\,\mu\,c_e\,r_w^2}\right] + 3.23\right\} \tag{16-136}$$

$$\Delta\psi_{skin} = 0.87\,m\,s \tag{16-137}$$

$$(\psi_{wf})_{ns} = \psi_{wf} + 0.87\,m\,s \tag{16-138}$$

$$r_{inv} = 0.0328\,\sqrt{\frac{k\,p_i\,t_{test}}{\phi\,\mu\,(1 - S_w)}} \tag{16-134}$$

where:

q_g = MCF/D,

$(\psi_{wf})_{ns}$ = theoretical flowing real gas potential at the well (if there were no damage or stimulation),

T = reservoir temperature, °R.

OILWELL DRILLSTEM CHARTS: EVALUATION OF PRODUCING RATE

FLOWING DRILLSTEM TEST

The next chart illustrates the features of a drillstem test on an oil well that achieves flow to the surface during the second flow period. Notice that after the start of the final flow period, the pressure increases through point A to a maximum at point B, and then declines to a relatively flat level to point C. During the portion of the flow period illustrating the pressure increase, the fluid level is rising within the drillpipe. Point B represents flow just reaching the surface. After point B, the pressure declines because the density of the reservoir fluids is less than that of the fluids initially in the wellbore (includes some drilling mud).

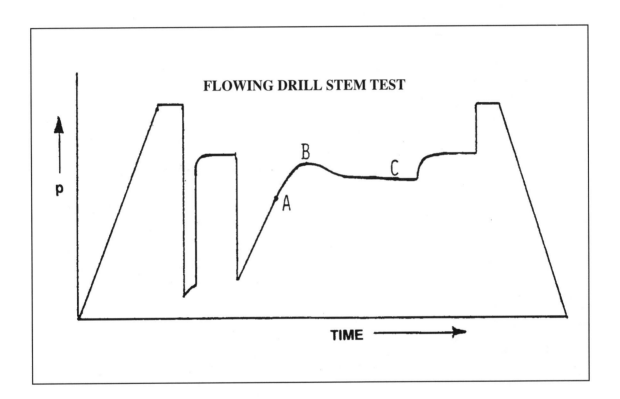

LARGE VOLUME OF FLUIDS ENTER THE DRILLPIPE

The following DST chart shows a characteristic pressure-time curve from a DST where a significant amount of fluids enter the drillpipe during the second flow period. However, there was not enough fluid entry to establish flow to the surface. The reservoir flow rate (needed to calculate reservoir parameters) can still be computed by one of two methods:

(1) An average producing rate can be calculated by drawing a straight line segment between the beginning and end pressures of the final flow period. This line is shown as a dashed line. Determine the slope of this line segment ($\Delta p/\Delta t$) in psi/minute. Then the reservoir flow rate is:

$$q_o B_o = (1440) \left[\frac{V}{G} \right] \left[\frac{\Delta p}{\Delta t} \right]$$

where:

V = volume of the drillpipe or tubing being filled during the flow period, bbl/ft,

G = average gradient of the fluids collected during the flow period, psi/ft, and

$\Delta p/\Delta t$ = the slope of the average pressure-time line, PSI/MIN.

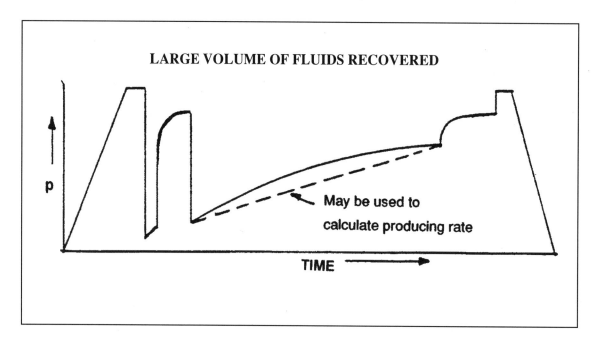

LARGE VOLUME OF FLUIDS RECOVERED

p

May be used to
calculate producing rate

TIME

(2) The second method is more approximate and should be used only when the method above is not workable. The total recovery (total barrels of fluid reversed out minus whatever water cushion) is divided by the total producing time (both flow periods).

$$q_o B_o = \frac{(1440)(\text{Total Recovery, bbls})}{\text{Prod. Time, Min.}}$$

FILL-UP DRILLSTEM TEST

The following illustration shows a typical pressure-time profile for a well that "kills itself" during the final flow period. During the final flow period, as the fluid level is rising within the drillpipe or tubing, the bottomhole pressure opposite the formation is increasing. On a "fill-up" DST, a pressure is reached such that the well can no longer flow against this back pressure, and pressure stabilizes.

In this type of DST, the rate of production is declining during the entire flow period. Because of this and because the shut-in period shows no additional increase in pressure, analysis of such data must be done by superposition or type curve analysis.

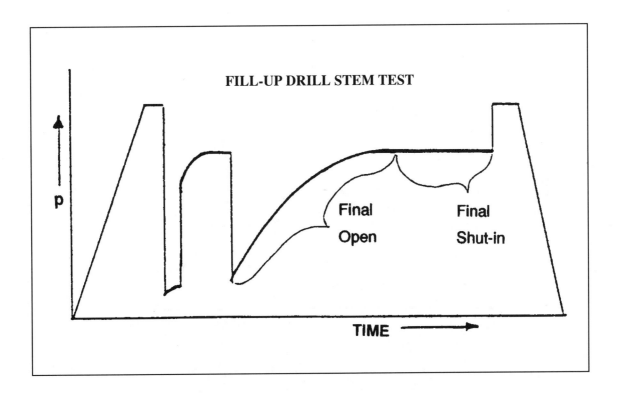

EFFECT OF DIFFERENT SIZE DRILLPIPE

Drillstrings usually include both drillpipe and drill collars. A drill collar often has a smaller internal diameter than does a section of drillpipe. Therefore, the position of the drill collars can affect the rate of pressure rise during the second flow period. Fluid will rise faster when the internal diameter is smaller, so the location of the drill collars in the drillstring should be considered when calculating reservoir rate on the basis of the slope of the pressure-time response during the final flow period.

Notice in the next diagram that from A to B, flow is in the drill collars. From B to C, it is the drillpipe that is being filled. Generally, the reservoir rate would be calculated using the second slope because the well has had more time to stabilize.

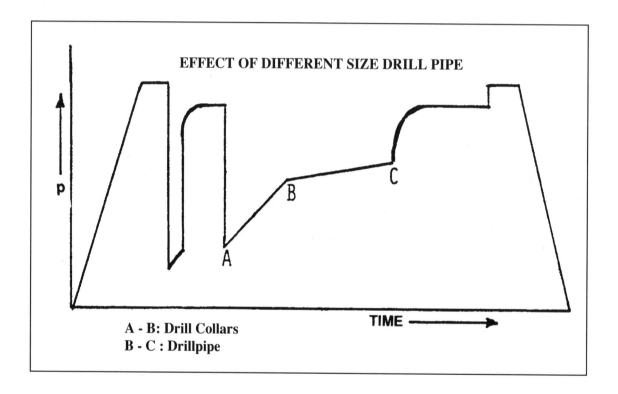

EFFECT OF DIFFERENT SIZE DRILL PIPE

A - B: Drill Collars
B - C : Drillpipe

DRILLSTEM CHARTS: TYPICAL TEST ANOMALIES

I. Good Permeability — Severe Damage

Severe damage is indicated when 90+ % of the pressure increase seen during the shut-in periods occurs right at the beginning. This corresponds to a near vertical pressure-time profile at the beginning of the shut-in. Further, there is little apparent pressure recovery during the flow periods. This also suggests that the major pressure drop is quite close to the well — severe damage.

Notice that the last part of the buildup periods are at first rounded and then fairly flat. Further the flat part of the buildup covers about 90 to 95 % of the total shut-in time. This flat slope is characteristic of good permeability. Another indication of good permeability is the measurement of about the same pressure level at the end of each shut-in period.

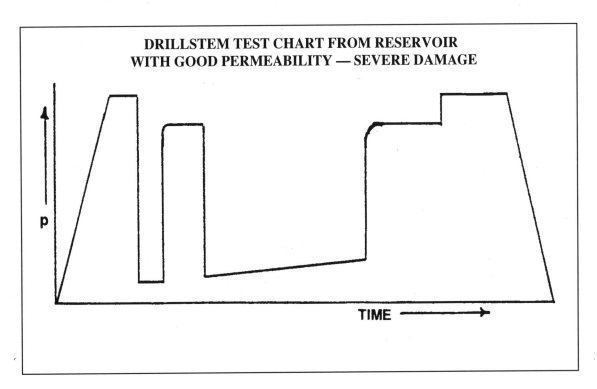

DRILLSTEM TEST CHART FROM RESERVOIR WITH GOOD PERMEABILITY — SEVERE DAMAGE

II. Low Permeability — Negligible Damage

The next chart is one from a well with low permeability and insignificant damage. The negligible damage is indicated by the lack of a near vertical pressure increase at the beginning of the shut-in periods. The entire buildup trace is rounded and fairly steep (but not vertical) with no sections of the response that are at all flat. These characteristics are indicative of low permeability. Also, notice that the final buildup pressures of the two shut-in periods are somewhat different. Generally speaking, in a low permeability reservoir, the longer the shut-in time, then the higher will be the final shut-in pressure.

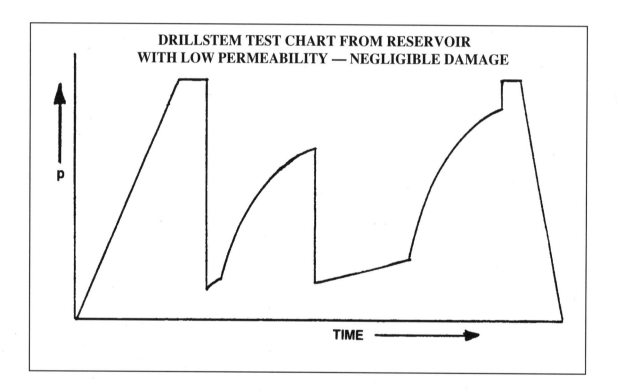

DRILLSTEM TEST CHART FROM RESERVOIR WITH LOW PERMEABILITY — NEGLIGIBLE DAMAGE

III. Depleting Reservoir

A depleting reservoir can be detected by the following characteristics:

(1) The final pressure at the end of the second shut-in period is substantially less than the pressure at the end of the first shut-in period.

(2) The pressure-time profile during the second flow period increases quite slowly and becomes flatter with time. Little pressure recovery is seen during the final buildup.

Depletion seen on a DST is evidence of very small reservoir size. If this depletion is real, then the reservoir is too small to be profitable in the zone tested. Given a DST where depletion is indicated, it is advisable to check the reliability of the tool pressure gauges.

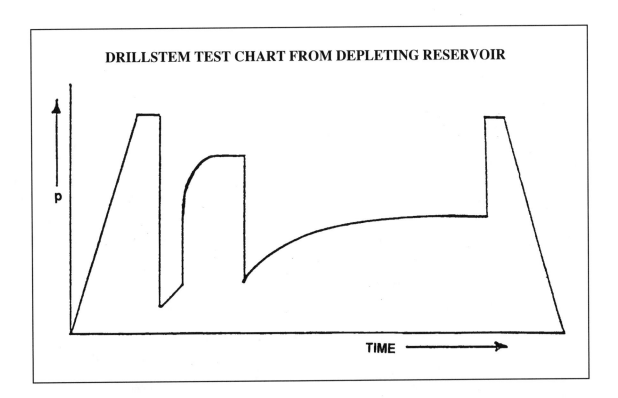

DRILLSTEM TEST CHART FROM DEPLETING RESERVOIR

REFERENCES

1. Muskat, M.: *Flow of Homogeneous Fluids Through Porous Media*, McGraw-Hill Book Co., Inc., New York City (1937) Ch. 10.

2. Matthews, C. S. and Russell, D. G.: *Pressure Buildup and Flow Tests in Wells*, SPE Monograph No. 1 (1967) Ch. 2.

3. Earlougher, R. C., Jr.: *Advances in Well Test Analysis*, SPE Monograph No. 5 (1977).

4. van Everdingen, A. F. and Hurst, W.: "The Application of the LaPlace Transformation to Flow Problems in Reservoirs", *Trans.*, AIME (1949) **186**, 305-324.

5. Reference 2, 14-16, 19-20.

6. van Everdingen, A. F.: "The Skin Effect and its Influence on the Productive Capacity into a Wellbore," *Pet. Eng. Intl.* (Oct. 1953) **25**, B-6.

7. Reference 3, Ch. 2-5.

8. Brons, F. and Marting, V. E.: "The Effect of Restricted Fluid Entry on Well Productivity," *JPT* (Feb. 1961) 172-74; *Trans.*, AIME, **222**.

9. Reynolds, A. C., Chen, J. C. and Raghavan, R.: "Psuedoskin Factor Caused by Partial Penetration," *JPT*, (Dec. 1984) 2197-2210.

10. Reference 2, 18.

11. Reference 3, 42-50.

12. Horner, D. R.: "Pressure Buildup in Wells," *Proc.*, Third World Pet. Cong., The Hague (1951) **II**, 503-523.

13. Miller, C. C., Dyes, A. B. and Hutchinson, C. A. Jr.: "The Estimation of Permeability and Reservoir Pressure from Bottomhole Pressure Buildup Characteristics," *Trans.*, AIME (1950) **189**, 91-104.

14. Reference 2, 19-21.

15. Al-Hussainy, R., Ramey, H. J. and Crawford, P. B.: "The Flow of Real Gas Through Porous Media," *JPT*, **237**, 624-636.

16. Tracy, G. W.: "Why Gas Wells Have Low Productivity," *Oil and Gas J.* (Aug. 6, 1956) 84.

17. *Theory and Practice of Testing Gas Wells*, Oil and Gas Conservation Board of Alberta, Canada (1975), 3rd Edition, 2nd Printing, (1978).

18. Horner, D. R.: "Pressure Buildup in Wells," *Proc.*, Third World Pet. Cong., The Hague, (1951) **II**, 503.

19. Matthews, C. S. Brons, F. and Hazebroek, P.: "A Method of Determination of Average Pressure in a Bounded Reservoir," *Trans.*, AIME (1954) **201**, 182-191.

20. Reference 3, 59-66.

21. Dietz, D. M.: "Determination of Average Reservoir Pressure From Buildup Surveys," *JPT* (Aug. 1965) 955-959; *Trans.*, AIME, **234**.

22. Reference 13, 100-101.

23. Muskat, M.: "Use of Data on the Buildup of Bottomhole Pressures," *Trans.*, AIME (1937) **123**, 44-48.

24. Hurst, L. L. and Guerrero, E. T.: "Calculating Static Bottomhole Pressures in a Finite Reservoir," *Oil and Gas J.* (Sept. 9, 1958) **56**, No. 39.

25. Gladfelter, R. E., Tracy, G. W. and Wilsey, L. E.: "Selecting Wells which will Respond to Production-Stimulation Treatment," *Drill. and Prod. Prac.*, API (1955) 117-129.

26. Agarwal, Ram G., Al-Hussainy, R. and Ramey, H. J., Jr.: "An Investigation of Wellbore Storage and Skin Effect in Unsteady Liquid Flow: I. Analytical Treatment," *SPEJ* (Sept. 1970) 279-290; *Trans.*, AIME, **249**.

27. Bourdet, D. and Gringarten, A. C.: "Determination of Fissure Volume and Block Size in Fractured Reservoirs by Type-Curve Analysis," paper SPE 9293 presented at the 1980 Annual Technical Conference and Exhibition, Dallas (Sept. 21-24).

28. McKinley, R. M.: "Wellbore Transmissibility from Afterflow-Dominated Pressure Buildup Data," *JPT* (July 1971), 863-872; *Trans.*, AIME, **251**.

29. Odeh, A. S.: "Unsteady-State Behavior of Naturally Fractured Reservoirs," *SPEJ* (Sept. 1965).

30. Warren, J. E. and Root, P. J.: "Behavior of Naturally Fractured Reservoirs", *SPEJ* (Sept. 1963) 245-255.

31. Mavor, M. J. and Cinco, H.: "Transient Pressure Behavior of Naturally Fractured Reservoirs," paper SPE 7977 presented at the 1979 California Regional Meeting of the Society of Petroleum Engineers of AIME, Ventura, CA. (April).

32. Kazemi, H.: "Pressure Transient Analysis of Naturally Fractured Reservoirs with Uniform Fracture Distribution," *SPEJ* (Dec. 1969) 451.

33. de Swaan, A. O.: "Analytic Solutions for Determining Naturally Fractured Reservoir Properties by Well Testing," *SPEJ* (June 1976) 117.

34. Agarwal, Ram G.: "A New Method to Account for Producing Time Effects when Drawdown Type Curves Are Used to Analyze Pressure Buildup and Other Test Data," paper SPE 9289 presented at the 1980 Annual Technical Conference and Exhibition, Dallas (September).

35. Reference 2, 48-57.

36. Jones, P.: "Reservoir Limit Test," *Oil and Gas J.* (June 18, 1957) 184-196.

37. Jones, L. G.: "Reservoir Reserve Tests," *JPT* (March 1963) 333-337; *Trans.*, AIME, **228**.

38. Tracy, G. W.: Private Communication, 1985.

39. Gray, K. E.: "Approximating Well-To-Fault Distance from Pressure Buildup Tests," *JPT* (July 1965) 761-767.

40. Davis, E. Grady Jr. and Hawkins, M. F. Jr.: "Linear Fluid - Barrier Detection by Well Pressure Measurements," *JPT* (Oct. 1963) 1077-1079.

41. van Everdingen, A. F. and Hurst, W.: "The Application of the Laplace Transformation to Flow Problems in Reservoirs," *Trans.*, AIME (1949) **186**, 305-324.

42. Russell, D. G. and Truitt, W. E.: "Transient Pressure Behavior in Vertically Fractured Reservoirs," *JPT* (Oct. 1964) 1159-1170.

43. Warren, J. E. and Root, P. S.: "The Behavior of Naturally Fractured Reservoirs," *SPEJ* (Sept. 1963) 245-255.

44. Odeh, A. S.: "Unsteady-State Behavior of Naturally Fractured Reservoirs," *SPEJ* (March 1965) 60 ff.

45. Lefkovits, H. C. *et al.*: "A Study of the Behavior of Bounded Reservoirs Composed of Stratified Layers," *SPEJ* (March 1961) 43-58.

46. "Review of Basic Formation Evaluation," published by Johnston Testers in 1965.

47. Agarwal, R. G.: "Real Gas PseudoTime — A New Function for Pressure Buildup Analysis of MHF Gas Well," paper SPE 8279 presented at the 1979 Annual Technical Conference and Exhibition (Sept. 23-26).

48. Meunier, D. F., Kabir, C. S. and Wittman, M. J.: "Gas Well Test Analysis: Use of Normalized Pseudovariables," SPE Formation Evaluation (Dec. 1987) 629-636.

49. Reynolds, A. C., Bratvold, R. B. and Ding, W.: "Semilog Analysis of Gas Well Drawdown and Buildup Data," SPE Formation Evaluation (Dec. 1987) 657-670.

50. Crawford, G. E., Pierce, A. E. and McKinley, R. M.: "Type Curves for McKinley Analysis of Drillstem Test Data," paper SPE 6754 presented at the 1977 Annual Technical Conference and Exhibition, Denver (Oct. 9-12).

51. Bourdet, D., Whittle, T. M., Douglas, A. A. and Pirard, Y. M. "A New Set of Type Curves Simplifies Well Test Analysis," *World Oil*, (May 1983) 95-106.

52. Bourdet, D., Ayoub, T. M., Whittle, T. M., Pirard, Y. M. and Kniazeff, V.: "Interpreting Well Tests in Fractured Reservoirs", *World Oil* (1983) 77-87.

53. Bourdet, D., Alagoa, A., Ayoub, J. A. and Pirard, Y. M.: "New Type Curves Aid Analysis of Fissured Zone Well Tests", *World Oil*, (April 1984) 111-124.

54. Bourdet. F., Ayoub, J. A. and Pirard, Y. M.: "Use of Pressure Derivative in Well Test Interpretation", paper SPE 12777 presented at the 1984 Annual Technical Conference and Exhibition, Long Beach (April 11-13).

55. Serra, K., Reynolds, A. C. and Raghavan, R.: "New Pressure Transient Analysis Methods for Naturally Fractured Reservoirs," *JPT* (Dec. 1983) 2271 - 2283.

56. Petak, K. R., Soliman, M. Y. and Anderson, M.F.: "Type Curves for Analyzing Naturally Fractured Reservoirs," paper SPE 15638 presented at the 1986 Annual Technical Conference and Exhibition, New Orleans (Oct. 5-8).

57. Clark, D. G. and Van Golf-Racht, T. D.: "Pressure-Derivative Approach to Transient Test Analysis: A High-Permeability North Sea Reservoir Example," *JPT* (Nov. 1985) 2023-2039.

58. Lee, John: "Well Testing," SPE Textbook Series, SPE of AIME, Dallas (1982) **1**, 68-71.

17

INTRODUCTION TO
COMPUTER RESERVOIR
SIMULATION

INTRODUCTION

The purpose of this chapter is to review the more common computer techniques used to simulate reservoir performance today. Concerning a particular reservoir, the function of the reservoir engineer is to predict ultimate recovery and future performance considering different recovery mechanisms and development methods. Many of the classical tools of reservoir engineering such as tank models (material balance treating the reservoir as a large tank with uniformly average properties) break down when attempting to model reservoirs, hydrocarbon systems, and/or recovery schemes of great complexity. For instance, the material balance - aquifer equation techniques given in the "Water Drive" chapter will yield the cumulative water encroachment into the reservoir at a given time, but the areal distribution of this water is unknown. Hence the optimum location of wells has not been addressed. The "hand-calculation" coning methods discussed in the "Water Coning and Fingering" chapter all neglected the effects of capillary pressure. Therefore, these methods would not be appropriate to use to predict performance of a well with a long transition zone.

By more rigorously formulating the physical system to be modeled, including variable rock properties, fluid properties vs. pressure, material balance, porous media flow equations, and capillary pressure, mathematical equations can be derived which allow complicated reservoir phenomena to be studied. Some of these mathematical relationships are nonlinear, partial differential equations which can only be solved approximately (but usually fairly accurately) with a computer.

This process allows the reservoir to be divided into blocks (Figure 17-1) with each block having its own properties, if needed. With the new generation of "super computers" now available, literally thousands of blocks can be used to model (in a reasonable amount of time) large reservoirs with many wells and considerable lithologic variation. The coordinate system used is normally chosen to fit the problem under study. Some of the common coordinate types are: linear, rectangular, radial, and radial-cylindrical. For a full-scale reservoir study, a 2-D or 3-D rectangular type grid (Figure 17-2A) would normally be used; whereas, for a single well coning prediction, a radial-cylindrical model [Figure 17-2B), (ii) or (iii)] would be the logical choice.

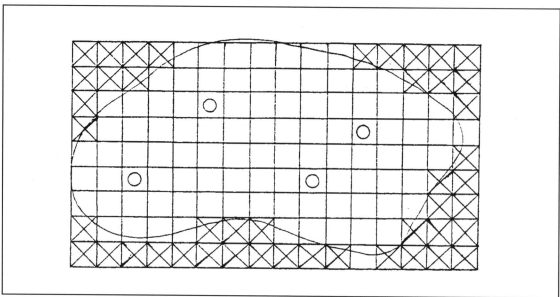

Fig. 17-1. *Example reservoir with four wells and 2-D reservoir simulation grid.*

It is the computer program, designed to simulate reservoir performance, that is termed the "reservoir simulator." According to Aziz and Settari,[1] there are actually three steps involved in developing a computer model. Step one is to derive the mathematical model; i.e., to formulate the physical system to be modeled in terms of appropriate mathematical equations. All physical phenomena that are deemed to be important, such as, capillary pressure, interphase mass transfer, or flow in a multiple porosity system, must be rigorously mathematically described. Step two is to translate the mathematical model into a numerical model. This step is needed because the partial differential equations of the math model are usually too complex to have an analytical solution. The procedure here is to approximate the original equations with finite difference analogs which are amenable to computer solution. Step three is to convert the numerical model into a computer model or program. This step involves all of the computer routines needed to solve the numerical model generated in step two.

DIFFERENT TYPES OF MODELS

There are many different kinds of computer reservoir-simulation programs. "Single phase" models only allow one phase to move within the reservoir. If another phase such as water is present, it is considered to be immobile. A "dry gas" reservoir simulator is one of the more common types of single-phase models. The so-called "black oil" model allows either two (water/ oil or gas/oil) or three (oil/gas/water) flowing phases. Black oil models have the inherent assumption that the compositions of the oil and gas are not changing significantly with depletion.

Compositional models allow interphase mass transfer; i.e., pure components (such as propane) are allowed to go back and forth between the oil and gas phases as conditions change in the reservoir. So, the oil and gas compositions are changing with time. This type of modeling often is needed for volatile-oil and retrograde-condensate reservoirs as well as certain types of gas injection and/or enhanced oil recovery (EOR) processes.

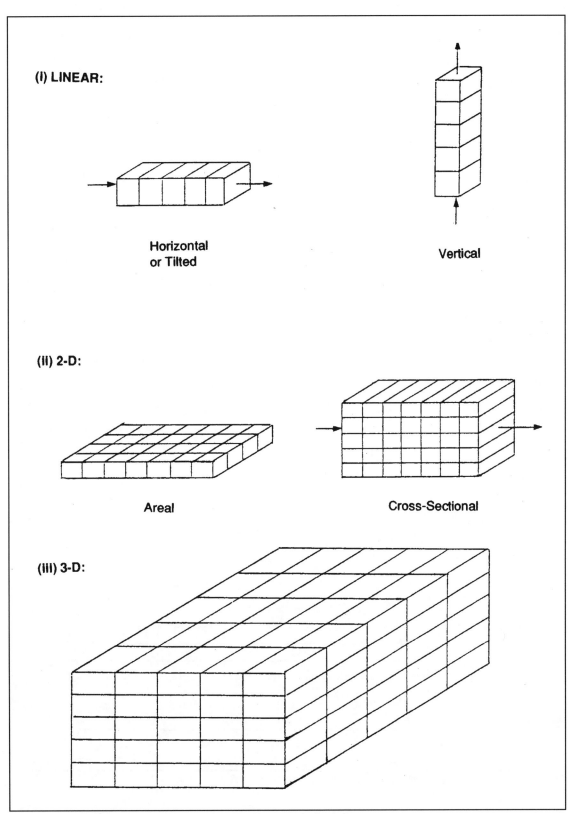

Fig. 17-2A. Cartesian coordinate systems often used in reservoir simulation.

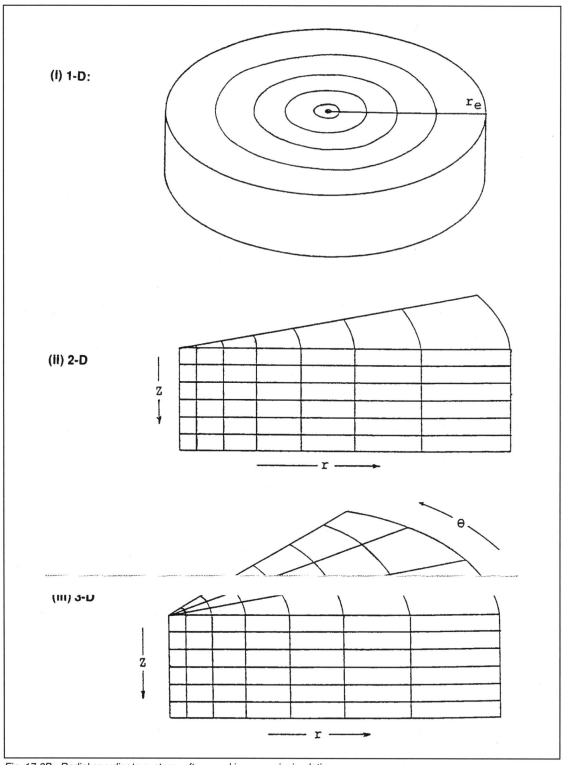

Fig. 17-2B. Radial coordinate systems often used in reservoir simulation.

Dual-porosity models are needed to accurately model performance in naturally fractured reservoirs as well as in some carbonate systems. The flow - pressure behavior of these types of reservoirs can be considerably more complex than a single-porosity (or even a layered) system.

To simulate many of the EOR processes (e.g., fire flooding, steam flooding, etc.), a thermal model is needed which includes in the equation set, an energy balance. There are a number of different chemical flood simulators used to model the chemical EOR processes.

All of these models above can have either one, two, or three dimensions. There are also so-called zero-dimensional models which are really tank models that do not consider flow effects within the reservoir and treat the reservoir with uniform "average" properties. Even though compositional effects are sometimes treated, the O-D models usually are considered to be more in the realm of material balance than reservoir simulation.

Most simulation studies use either two or three dimensions. However, there are some simple problems that are amenable to 1-D analysis. Displacement of oil by water in a system that is essentially linear can be studied with a 1-D model. Some other examples of frequent 1-D studies are gas percolation (free gas/oil segregation) and radial flow into a single well. Whenever possible, simulations are performed in 2-D instead of 3-D due to the considerable savings in cost and time. Examples of situations adequately handled with two dimensions are: (1) areal flow in reservoirs with minimal stratification and little saturation change in the vertical direction; (2) vertical - radial flow considered in a single well coning model; and (3) areal displacement of oil by water. Nevertheless, there are many reservoirs that only can be studied accurately with 3-D modeling.

Why Reservoir Simulation?

The potential benefits of reservoir simulation can be summarized.

(1) Ultimate primary recovery and performance can be studied under different modes of operation such as natural depletion, water injection, and/or gas injection.

(2) When should secondary recovery be initiated to maximize recovery? What type of pattern should be used?

(3) What type of EOR is appropriate for this resource? And what will be the EOR performance and ultimate recovery?

(4) The effect of the locations and spacing of the wells can be investigated.

(5) What is the effect of producing rate on recovery?

(6) What types of data have the most effect on recovery and, therefore, should be studied carefully with physical laboratory experiments?

Simulator-Computed Results and Errors

Because the original mathematical model equations were replaced with finite difference equations, the reservoir simulator steps forward in time using discrete time steps. At each new point in time, results from a black oil reservoir simulator include average pressures and saturations for each block in the model. In a compositional model, concentrations of each modeled component will also be available. In a thermal model, block temperatures or enthalpies probably will be calculated.

Unfortunately, there are several places where errors can creep into the computed results. The original mathematical model involved assumptions which were likely untrue to some extent. Translating the mathematical equations into finite difference analogs introduces a numerical error referred to as the "truncation" error. The truncation error generally decreases with a greater number of blocks. Computer solution, due to the limited number of significant digits that the machine is capable of carrying, causes "round-off" error to enter the results. These three sources of error should be recognized and reasonable effort should be expended to minimize these effects. Perhaps the most serious simulation errors result from inadequate reservoir description. For a black oil simulator, an outline of the input data requirements is listed in Table 17-1. Notice that every block in the model has to have input rock properties and cell dimensions. Tables of fluid properties vs. pressure must be available. And the relative permeability and capillary pressure data required can have a tremendous effect on model performance. So, if a concerted effort has not been made to acquire this data as accurately as possible, then results can be disappointing.

Table 17-1
Summary of Required Simulator Input Data

I. Cell Data

For each block in the simulator grid system, there must be input a value of:

1. Absolute permeability (in each direction)
2. Porosity
3. Thickness
4. Elevation
5. Block or grid dimensions
6. Rock compressibility
7. Initial reservoir pressure
8. Initial phase saturations*

* Note: The initial phase saturations often can be calculated by the simulator if capillary pressure data and the oil/water and gas/oil contacts are input, and capillary equilibrium is assumed.

II. Pressure-Dependent Data

Each of the following fluid properties varies with pressure. Therefore, each of these variables must be input (or calculated by the simulator) as a table versus pressure.

1. Oil, water, and gas formation volume factors
2. Oil, water, and gas viscosities
3. Oil and water solution gas/oil (/water) ratios
4. Oil, gas, and water densities

III. Saturation-Dependent Data

The following variables are functions of saturation and normally are input in tabular form.

1. Relative permeability for each phase
2. Oil/water capillary pressure
3. Gas/oil capillary pressure

IV. Well Data

1. Areal location and producing interval
2. Observed oil, gas, and water rates versus time
3. Observed bottomhole pressures versus time
4. Future production rates versus time, or
5. Future bottomhole pressures.

The Inverse Problem: History Matching

Reservoir simulators are perhaps best known for their ability to predict future reservoir behavior. But how is confidence attained that the prediction will have any conformance to what the actual reservoir will do? The procedure is normally to calibrate or "history match" the model parameters (input data) using reservoir performance history. If the model accurately predicts what has happened in the past, then there is the likelihood that the predictions are somewhat meaningful. Traditionally, the history matching phase of a model study has been a trial-and-error procedure. Using pressure buildups, DST's, logs, seismic data, conjectures, and whatever else; the simulation analyst prepares an initial guess of the input data. Then, the model is run and results compared with observed history. It is usually necessary to modify some of the input data until calculated and observed pressures and production data compare favorably.

Throughout the 30 year history of reservoir simulation, applied mathematicians and research engineers have attempted mathematically to find a direct solution to the inverse problem. That is, without the analyst's trial-and-error manual manipulation of the input data, use the historical reservoir pressures and production data to calculate directly the model input data, such as, block permeabilities, thicknesses, etc. Unfortunately, such solutions have been plagued with non-uniqueness problems: more than one set of input data capable of yielding the same model response. However, a new technique using "geostatistics"[8] seems to have some promise in dealing with the inverse modeling problem. In the procedure, a least squares objective function (the sum of the squares of the differences between observed and calculated pressures) is minimized iteratively. The new part of this method is the use of geostatistics to describe the distribution of a parameter, such as transmissibility (kh / μ), both initially and during the modifications required during the minimization procedure.

After the history match has been made, the predictive runs are performed. Here, the effects of different exploitation schemes, numbers of wells, rates, etc., may be evaluated. A general rule of thumb is not to put too much faith in predictions beyond about twice the amount of time used to history-match the model. However, if the recovery mechanism changes, then the amount of "safe" predictive time can be even less.

Critical Data Identification

One important use of reservoir simulation is to identify which portions of the input data have the greatest effect on recovery. There are reservoirs (e.g., segregation drive) where the relative permeability relationships are extremely important, and capillary pressure has little effect. In other reservoirs, the opposite could be true. By performing a sensitivity analysis (with 10 to 30% variation) on the input data, the effects of the different rock and fluid properties on ultimate recovery can be determined. With this knowledge in hand, the analyst can decide whether further laboratory (or field) studies of the critical relationships are needed.

Black Oil Mathematical Development

In this section, a sketch will be given for the development of the equations used in black oil reservoir simulation. The intent here is to be descriptive, not rigorous. For the details, the references of Breitenbach *et al.*,[2] Aziz & Settari,[1] and Thomas[3] are suggested.

Consider the flow of a single liquid in the linear system shown in Figure 17-3. By performing a mass balance over the representative reservoir volume element, we can write the law of mass conservation or the continuity equation as:

(Mass rate into the element) - (Mass rate out of the element) =
(Mass rate of accumulation within the element)

(17-1)

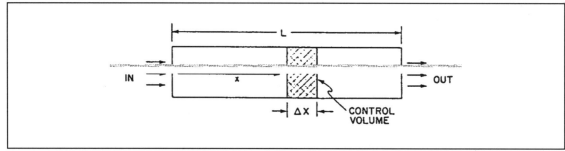

Fig. 17-3. *Porous-medium linear flow considering volume element of length Δx. (From Aziz and Settari[1]).*

Now, if Equation 17-1 is expressed in mathematical terms and the limit is taken as the volume element approaches zero, then the 1-D, linear continuity equation results:

$$- \frac{\partial (\rho u)}{\partial x} = \frac{\partial (\rho \phi)}{\partial t} + \bar{q}$$

(17-2)

where:

     u = flow velocity in the x-direction,
     ρ = fluid density,
     ϕ = fractional porosity,
     $\bar{q}$ = source or sink term allowing for the injection or withdrawal of mass from the control volume, mass per unit volume per unit time.

Equation 17-2 has allowed for the possibility of production from or injection into the reservoir volume element.

Next, the fluid velocity will be represented using Darcy's Law:

$$u = - \frac{k}{\mu} \left[\frac{\partial p}{\partial x} - \gamma \frac{\partial D}{\partial x} \right]$$

(17-3)

where:

     u = flow velocity in the x-direction,
     k = absolute permeability in the x-direction,
     μ = fluid viscosity,
     p = system pressure,
     D = elevation of the center of the flow element,
     $\gamma \partial D / \partial x$ = gravitational gradient due to non-horizontal flow.

For liquid flow, the equation of state for a slightly compressible fluid is often utilized:

$$\rho = \left[\rho_{ref} \right] \left[e^{c_1 (p - p_{ref})} \right] \qquad (17\text{-}4)$$

where:

p_{ref} = reference pressure (constant),
p = system pressure
c_1 = fluid (liquid) compressibility,
ρ_{ref} = fluid density at the reference pressure,
ρ = fluid density at the system pressure.

Fluid density can also be described using the concept of formation volume factor as:

$$\rho = \frac{(\rho_{ref})(B_{ref})}{B} \qquad (17\text{-}5)$$

where:

B = formation volume factor at system pressure, res. vol. / surf. vol.,
B_{ref} = formation volume factor at the reference pressure.

Porosity may be represented as a function of pressure as:

$$\phi = \phi_{ref} \left[1 + c_f (p - p_{ref}) \right] \qquad (17\text{-}6)$$

where:

ϕ = porosity at system pressure,
ϕ_{ref} = porosity at reference pressure,
c_f = pore volume compressibility.

Substituting Equations 17-3 through 17-6 into Equation 17-2 and simplifying, then Equation 17-7 results:

$$\frac{\partial}{\partial x} \left\{ \frac{k}{\mu B} \left[\frac{\partial p}{\partial x} - \gamma \frac{\partial D}{\partial x} \right] \right\} = \frac{\partial}{\partial t} \left[\frac{\phi}{B} \right] + q \qquad (17\text{-}7)$$

where:

q = source or sink (well) rate expressed as: stock-tank fluid volume per unit time per unit reservoir volume.

Of course, Equation 17-7 depended on the form of the velocity equation and the representation of the fluid equation of state. For gas reservoirs, often a more complicated velocity equation than Equation 17-3 is used to consider turbulent flow effects. A different equation of state is needed (so that gas compressibility is not treated as constant) which is often based on the real gas law. Thus, the conservation equation for a single phase gas reservoir model will be somewhat different than Equation 17-7. A detailed treatment of the development and use of the equations for gas systems has been published.[9]

The 1-D linear, slightly compressible fluid, conservation Equation 17-7 may be generalized to three Cartesian dimensions by treating absolute permeability as a diagonal tensor:

$$K = \begin{bmatrix} k_x & & \\ & k_y & \\ & & k_z \end{bmatrix} \qquad (17\text{-}8)$$

where:

k_x, k_y, and k_z are absolute formation permeabilities in the x-, y-, and z- directions, respectively.

To keep the notation simple, the "del" (∇) operator, which is a three dimensional vector differential operator, will be used:

$$\nabla \equiv i\frac{\partial}{\partial x} + j\frac{\partial}{\partial y} + k\frac{\partial}{\partial z} \qquad (17\text{-}9)$$

Then, using the mathematical notions represented by Equations 17-8 and 17-9, Equation 17-7 may be written in three dimensions as:

$$\nabla \cdot \left\{ \frac{1}{\mu B}\, K\, [\, \nabla p - \gamma \nabla D\,] \right\} = \frac{d}{dp}\left[\frac{\phi}{B}\right]\frac{\partial p}{\partial t} + q \qquad (17\text{-}10)$$

So, Equation 17-7 represents single phase liquid flow in one linear dimension, while one liquid flowing in three dimensions is described by Equation 17-10. Strictly speaking, Equation 17-10 holds for any valid coordinate system. Expanding Equation 17-10 into three dimensional Cartesian coordinates:

$$\frac{\partial}{\partial x}\left[\frac{k_x}{\mu B}\left(\frac{\partial p}{\partial x} - \gamma\frac{\partial D}{\partial x}\right)\right] + \frac{\partial}{\partial y}\left[\frac{k_y}{\mu B}\left(\frac{\partial p}{\partial y} - \gamma\frac{\partial D}{\partial y}\right)\right]$$

$$+ \frac{\partial}{\partial z}\left[\frac{k_z}{\mu B}\left(\frac{\partial p}{\partial z} - \gamma\frac{\partial D}{\partial z}\right)\right] = \frac{d}{dp}\left[\frac{\phi}{B}\right]\frac{\partial p}{\partial t} + q \qquad (17\text{-}11)$$

To mathematically account for more than one flowing phase, each component must have its own conservation equation. This entails introducing the concepts of saturation and relative permeability. Since the pore volume is no longer filled with just one phase, a saturation equation is needed. Assuming three phases (oil, water, and gas) at each point in the reservoir, or for each block in the reservoir grid system, we must have:

$$S_o + S_w + S_g = 1.0 \qquad (17\text{-}12)$$

Thus, considering more than one phase, Equation 17-10 written for the oil phase becomes:

$$\nabla \cdot \left\{ \frac{k_{ro}}{\mu_o B_o} K \left[\nabla p_o - \gamma_o \nabla D \right] \right\} = \frac{\partial}{\partial t} \left[\frac{\phi S_o}{B_o} \right] + q_o \qquad (17\text{-}13)$$

where:

k_{ro} = relative permeability to the oil phase, fraction.

The water equation would be similar except the "o" subscripts would be replaced with "w". The gas equation is slightly more complicated since some of the gas is in solution, and some exists as a free-gas phase.

Also needed for the solution of a black oil reservoir model are the capillary pressure relationships, which are normally input to the simulator in the form of tables. The oil/water capillary pressure relationship relates the saturation conditions to the difference between the oil phase pressure and the water phase pressure. The gas/oil capillary pressure function, again through saturation, gives the difference between the oil pressure and the free-gas pressure.

For three-phase flow, there are six equations to be solved: three conservation equations, the saturation equation, and the two capillary pressure functional relationships. Fortunately, this is just enough to yield the six dependent variables involved: p_o, p_w, p_g, S_o, S_w, and S_g.

Black Oil Numerical Model

Unfortunately, the phase conservation equations, such as 17-13 for the oil, are nonlinear partial differential equations with no known analytical solutions. There are, however, a number of techniques, such as finite difference and variational methods (finite element, Galerkin, collocation, etc.), that exist to obtain approximate solutions to Equation 17-13. Finite difference techniques seem to be the most practical for use in large reservoir problems.

For a finite difference approach, the coordinate system (space and time) must be discretized; i.e., no longer regarded as continuous. The space and time dimensions must be divided into a finite number of discrete points or blocks. Some of the different discretized spatial coordinate systems were illustrated in Figure 17-2. The time domain must also be considered at distinct points or in steps. For instance, if a 30-day time step were being used, then new pressures and saturations would be solved for every 30 days; i.e., at 30 days, 60 days, 90 days, etc.

As an example, consider Equation 17-7, the single phase flow conservation equation in one dimension. This equation can be "linearized" by using finite differences to represent the derivatives such as $\partial p / \partial x$. Using the grid system shown in Figure 17-4, the first derivative of pressure at point x_i can be approximated as:

$$\frac{\partial p_i}{\partial x} \approx \frac{p_{i+1} - p_{i-1}}{2\Delta x} \qquad (17\text{-}14)$$

where:

p_{i+1} = the pressure at point x_{i+1}, and
p_{i-1} = the pressure at point x_{i-1}.

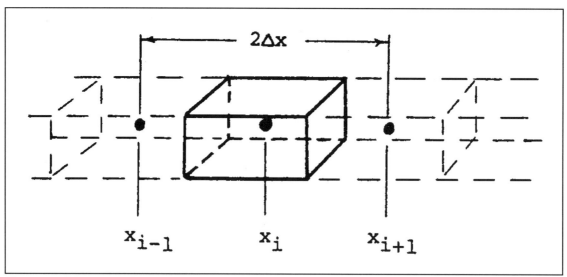

Fig. 17-4. 1-D linear finite difference grid.

Equation 17-14 is known as a "central difference" approximation for the derivative. There are also other types of difference approximations such as "forward" and "backward" equations. Finite difference approximations can be shown through Taylor series expansions to involve an error (known as the truncation error) that is dependent on the block size or Δx. The smaller the grid, or the greater the number of blocks, the better the finite difference approximation becomes. Second order derivatives, such as $\partial p^2 / \partial x^2$, can also be approximated using finite differences. Notice that one way to do this would be to use a formula similar to Equation 17-14:

$$\frac{\partial^2 p_i}{\partial x^2} \approx \frac{\dfrac{\partial p_{i+1/2}}{\partial x} - \dfrac{\partial p_{i-1/2}}{\partial x}}{\Delta x} \approx \frac{p_{i+1} - 2p_i + p_{i-1}}{(\Delta x)^2} \tag{17-15}$$

Finite differences are also used in multidimensional simulations, although the procedure is somewhat more complicated. It is the finite difference representation of $\partial p / \partial t$ that creates the need for the time domain to be discretized with the result that time steps must be used to propagate model behavior forward in time.

Using finite difference approximations, the phase (conservation) pressure equations can be translated into algebraic equations. Then at each new time level, for each block in the reservoir grid, a set of different equations must be solved simultaneously. For the three phase simulation, there are six equations: three pressure equations, a saturation equation, and two capillary pressure relationships. For a two phase model, there are four equations that must be satisfied: two pressure equations, a saturation equation, and a capillary pressure relationship.

As these equations must be solved simultaneously for each block in the reservoir, the problem can be viewed as a system of simultaneous equations to be solved with each new time step. Since the total number of equations to be solved is large, they are normally analyzed mathematically as a matrix equation or as a collection of matrix equations. Such a matrix problem for a reservoir with

1000 blocks would take years to solve by hand for a single time step. However, the computer can do it in seconds. Nonetheless, it significantly reduces the solution time if the matrix size (total number of equations in the system) can be reduced.

Historically, one of the more popular ways that has been used to reduce the number of equations to be solved is to use the IMPES (implicit pressure-explicit saturation) method. By assuming that capillary pressure and mobilities over the time step can be considered as constant and controlled by the saturations at the beginning of the time step, a single pressure equation can be obtained. Further, after the solution of the pressure equation, the saturations are evaluated explicitly. One large disadvantage of the IMPES method is that it has problems (solutions become unstable) when large throughputs occur in a grid block during a single time step. Thus, for certain types of problems (e.g., coning simulations), the use of IMPES requires ridiculously small time steps.

The sequential solution method (SEQ[6]) is much like IMPES in that the procedure is initiated by solving a single pressure equation. However, the second step of the SEQ procedure involves implicit calculation of the saturations. In the IMPES saturations solution, relative permeabilities are evaluated explicitly (by using the saturations of the last time step); while the SEQ saturations step uses implicit relative permeabilities. This involves an additional matrix problem, but yields larger permissible time steps. While IMPES preserves material balance, SEQ can yield material balance errors for some problems.

There are some simulators that are formulated to be fully implicit; i.e., nothing is evaluated based on the conditions of the last time step. For the same size problem, these simulators involve much larger matrix dimensions than those of the IMPES or SEQ procedures. Hence, storage requirements are greater, and much more computer time and money are spent per time step using a fully implicit simulator. There are times, particularly with difficult coning problems, that the fully implicit simulator is needed for the most reliable results. Material balance is preserved.

The adaptive implicit method[10] shows promise for accurate solutions with modest computer requirements. Although difficult to implement, with this technique, part of the reservoir is treated implicitly (such as near wells), while most of the grid is handled explicitly. The matrix sizes are greatly reduced compared to those of the fully implicit method. The two main difficulties are: (1) the switching criteria (when to switch blocks from implicit to explicit and vice-versa), and (2) the matrix solving method. The structure of the matrices changes as different blocks change their implicit or explicit status. Thus, a general matrix "solver" is needed that is capable of rapidly obtaining solutions to the changing matrix problems.

Three Dimensional Simulation Using Two Dimensions

Generally speaking, the more dimensions that are used to model a reservoir, the larger the matrices become that must be solved for each time step. There are many reservoirs that are sufficiently complex (e.g., many wells, significant vertical and areal lithology variation) that simulation must be performed with three dimensions to be able to model adequately all of the reservoir mechanisms involved. If needed, then three dimensional simulation is warranted; but it is expensive. Hence, one should attempt to perform a reservoir study in two dimensions, if possible. For some reservoirs this can be achieved with the use of pseudofunctions. That is, using altered relative permeability and capillary pressure relationships, it is sometimes possible to preserve 3-D detail (e.g., vertical equilibrium) while performing the simulation in two dimensions. Dake,[4] Aziz & Settari,[1] and Starley[7] have discussed the construction and application of these pseudofunctions.

Solving the Matrix Problems

As mentioned earlier, for each time step, there is a large system of simultaneous equations to be solved. This results in at least one, and usually more, matrix problem(s) per time step. For 2- and 3-D problems with many blocks in the grid system, the matrices are large. The optimum solution method for such matrix problems depends on the structure of the matrix. And, the structure depends on the number of dimensions, the number of phases being modeled, and the numerical formulation used (e.g., IMPES, SEQ, etc.).

There are "direct" solution methods, which use a form of Gaussian elimination, and different types of iterative procedures. A general rule is that for large problems, iterative methods will typically require less work; whereas, for coarse grids, a direct method will probably be better. Of course, most of the questions still remain. Which iterative method? Where is the cross-over point between small and large problems? The answers are different for different types of simulations: areal, cross-sectional, full scale, single well, etc. This is one of the areas where an experienced reservoir simulation engineer/analyst can certainly demonstrate his or her worth.

Single Well Performance Simulation

To model the performance of a single well, as with an entire reservoir, one, two or three dimensions can be used. However, with a single well, normally either radial (r) or radial-cylindrical (r-z or r-z-θ) coordinates are chosen. These systems are illustrated in Figure 17-2B.

Two important applications of single well models are coning prediction and well testing. The industry-standard Horner analysis of buildup and drawdown tests involves many simplifying assumptions which, for many wells, are not valid. The single-well reservoir simulator may be used to analyze buildups and drawdowns in such wells, or it may be used to analyze more complicated well tests involving sequences of different rates and shut-ins. The successful outcome of such a simulator analysis, based on a well test, is the prediction of basic reservoir data, which may be used to predict the well's productivity and future performance.

Single-well models in r-z or r-z-θ coordinates are used to model movement of water and/or free gas towards producing wells. Some techniques were given in the "Coning" chapter for calculating a well's "critical" rate and time to breakthrough. A method also was given to predict after-breakthrough performance for wells in bottomwater-drive reservoirs. However, none of these "hand-calculation" techniques allowed the consideration of the effects of capillary pressure, heterogeneity (in particular, layering), or actual well completion scheme. Although coning simulations are expensive to perform, sometimes the answers that can be provided are worth it.

Recall that Equation 17-13 is the conservation equation for the oil phase in a multidimensional, multiphase model. Expanding this equation in 3-D radial-cylindrical coordinates:

$$\frac{1}{r} \frac{\partial}{\partial r} \left\{ \frac{r \, k_{ro}}{\mu_o B_o} \, k_r \left[\frac{\partial p_o}{\partial r} - \gamma_o \, \frac{\partial D}{\partial r} \right] \right\} + \frac{1}{r^2} \frac{\partial}{\partial \theta} \left\{ \frac{k_{ro}}{\mu_o B_o} \, k_\theta \left[\frac{\partial p_o}{\partial \theta} - \gamma_o \, \frac{\partial D}{\partial \theta} \right] \right\}$$

$$+ \frac{\partial}{\partial z} \left\{ \frac{k_{ro}}{\mu_o B_o} \, k_z \left[\frac{\partial p_o}{\partial z} - \gamma_o \, \frac{\partial D}{\partial z} \right] \right\} = \frac{\partial}{\partial t} \left[\frac{\phi \, S_o}{B_o} \right] + q_o \qquad (17\text{-}16)$$

It should be noted that most coning simulations are 2-D (r-z), 2-phase (oil/water or oil/gas) studies with reservoir properties averaged in the angular direction. Each phase being modeled has a conservation equation similar to Equation 17-16.

As in the full reservoir (Cartesian coordinates) black-oil models, a saturation equation and capillary-pressure relationship(s) also are needed to be able to solve for individual phase pressures and saturations.

Notice in Figure 17-2 that the radial-cylindrical coordinate system blocks are usually unequally spaced in the radial dimension. The blocks away from the well are large; while the block next to the well is small. The significance here is that the saturation state in the blocks close to the well can change significantly over a single time step. Hence, the IMPES formulation, when attempted with a coning model, usually results in maximum permissible time steps that are uneconomically small. Coning simulators often need a fully implicit formulation; however, some coning situations are modeled quite successfully with the SEQ method of coupling multiphase flow equations.

Misuse of Reservoir Simulation Models

A reservoir simulation model is an extremely powerful tool. An actual reservoir can be exploited only once, but the simulation model can be run many different times considering different methods of development. In the hands of an experienced simulation engineer, such a model can provide valuable answers to reservoir management.

Unfortunately, reservoir simulation models also have the power to provide very expensive results that are worthless. More than once, using a reservoir simulator, people have proved preconceived notions about a reservoir. Objectivity is absolutely essential for a simulation analyst who must alter much of the input data during the history-matching phase of the study. Experience and engineering judgement also are two necessary components to properly control a simulator and interpret its results. Which model should be chosen? How many dimensions should be used? And within a particular model, which options should be used? An analyst with experience in simulation and reservoir engineering can most effectively and economically answer these questions.

Several years ago when reservoir simulators first became widely available, there was a tendency to take any reservoir study that was needed and go straight to a 3-D, 3-phase model. Some engineers did not even want to first study the reservoir with a material balance model. Fortunately, this trend of jumping first to the most complex model available has largely been reversed. It would be nice to think that engineers have gotten smarter, however, it is probably more of a sign of today's oil industry economic conditions.

The general rule has always been to begin a reservoir study with the simple tools of classical reservoir engineering: Darcy's law, material balance, Buckley-Leverett, etc. If the complexity of the system is such that the reservoir cannot be adequately investigated with the simple tools, then move up to a more complicated tool: perhaps, a 1-D simulator. After the 1-D simulation, a 2-D study may be required. But, even the simple reservoir engineering techniques will usually lend some insight into the mechanisms important in a particular reservoir. This insight gained as the study progresses to more complex methods will aid in the interpretation of each new level of results. So, keep it as simple as possible, and, as needed, work upwards in complexity in a step-wise fashion.

REFERENCES

1. Aziz, K. and Settari, A.: *Petroleum Reservoir Simulation*, Applied Science Publishers Ltd., London (1979).

2. Breitenbach, E. A., Thurnau, D. H. and van Poolen, H. K.: "The Fluid Flow Simulation Equations," paper SPE 2020 presented at the 1968 SPE Annual Meeting, Dallas (April 22-23).

3. Thomas, G. W.: *Principals of Hydrocarbon Reservoir Simulation*, IHRDC Publishers, Boston (1982).

4. Dake, L. P.: *Fundamentals of Reservoir Engineering*, Elsevier Scientific Publishing Company, Amsterdam (1978).

5. Coats, K. H.: "Elements of Reservoir Simulation," Lecture Notes, Univ. of Texas, reprint by Intercomp Resources Devel. and Engr. Inc., Houston (1968).

6. Spillette, A. G., Hillestad, J. G. and Stone, H. L.: "A High-Stability Sequential Solution Approach to Reservoir Simulation," paper SPE 4542 presented at the 1973 SPE Annual Technical Conference and Exhibition, Las Vegas.

7. Starley, G. P.: "A Material Balance Method for Deriving Interblock Water/ Oil Pseudofunctions for Coarse Grid Reservoir Simulation," paper SPE 15621 presented at the 1986 SPE Annual Technical Conference and Exhibition, New Orleans (October 5-8).

8. Fasanino, G., Molinard, J. E., de Marsily, G. and Pelce, V.: "Inverse Modeling in Gas Reservoirs," paper SPE 15592 presented at the 1986 Annual Technical Conference and Exhibition, New Orleans.

9. Energy Resources Conservation Board: *Theory and Practice of the Testing of Gas Wells*, Third Edition, ERCB, Calgary, Alberta, Canada (1975).

10. Thomas, G. W., Thurnau, D. H.: "Reservoir Simulation Using an Adaptive Implicit Method," *SPEJ* (1983) **23**, 759-768.

NOMENCLATURE

A = area

A = dimensionless time conversion constant (van Everdingen & Hurst aquifer model)

B = Fetkovich aquifer model constant

B_g = gas formation volume factor (reservoir volume divided by volume at standard conditions)

B_o = oil formation volume factor (volume at reservoir conditions divided by stock tank oil volume)

B_w = water formation volume factor

c = isothermal compressibility

c_e = effective (total system) compressibility

c_f = formation (pore volume) compressibility

c_g = gas compressibility

c_o = oil compressibility

c_{oe} = effective oil compressibility (applied to oil pore volume, undersaturated oil)

c_{pr} = pseudoreduced compressibility

c_w = water compressibility

c_{we} = effective (total) aquifer compressibility

C = coefficient of gas well back-pressure equation

C = wellbore storage coefficient

$\overline{C}$ = dimensionless wellbore storage coefficient

C_A = shape factor (well drainage geometry)

C_d = dimensionless wellbore storage coefficient

C_s = Schilthuis aquifer constant

C_v = van Everdingen & Hurst aquifer coefficient

d = diameter

D = non-Darcy flow coefficient

D = completion interval thickness

D = distance to closest well

D = distance to no-flow barrier

e = 2.7182 . . .

exp = e

$E_{f,w}$ = term relating expansion of connate water and rock in the material balance equation

E_g = term in the material balance equation relating to the expansion of the gas in the gas cap

E_o= term in the material balance equation relating to the expansion of the original oil including liberated solution gas

f= friction factor

f_g= fraction of the total flowing stream composed of gas

f_o= fraction of the total flowing stream composed of oil

f_w= fraction of the total flowing stream composed of water

F= cumulative production term in the material balance equation

F= wellbore storage factor of the McKinley type curve method

g= acceleration of gravity

g_c= conversion factor in Newton's Second Motion Law

G= original gas in place (OGIP)

G_a= apparent gas in place

G_p= cumulative gas production

G_{pc}= cumulative gas production from the gas cap

G_{poz}= cumulative gas production from the oil zone

h= formation thickness

I= injectivity index

I_s= specific injectivity index

J= J-function (capillary pressure) of Leverett

J= productivity index

J_s= specific productivity index

k= absolute permeability

k_f= fracture permeability

k_g= effective permeability to gas

k_m= matrix permeability

k_o= effective permeability to oil

k_{rg}= relative permeability to gas

k_{ro}= relative permeability to oil

k_{rw}= relative permeability to water

k_w= effective permeability to water

k_v= vertical permeability

L= length

L_f= fracture length (tip to tip)

Log= logarithm, base 10

Ln= logarithm, base e

m= mass

m= ratio of the original free gas reservoir volume to the original oil reservoir volume

m= slope

M= mobility ratio

M=molecular weight

n= number of samples

n= reciprocal of the slope of the gas well deliverability plot

n= total number of moles

N= stock tank oil originally in place

N_a= apparent oil in place

N_p= cumulative stock tank oil production

p= pressure

p_a= abandonment pressure

p_a= average aquifer pressure

p_b= bubble point pressure

p_c= capillary pressure

p_c= critical pressure

p_d= dew point pressure

p_e= external boundary pressure

p_g= pressure in the gas phase

p_i= initial pressure

p_o= pressure in the oil phase

p_{pc}= pseudocritical pressure

p_{pr}= pseudoreduced pressure

p_r= reduced pressure

p_{sc}= pressure at standard conditions

p_{sep}= separator pressure

p_w= bottom hole well pressure

p_w= pressure in the water phase

p_{wf}= flowing bottom hole pressure

p_{wp}= producing bottomhole pressure

p_{ws}= shut-in bottomhole pressure

p_{1hr}= pressure on straight line portion of semilog plot corresponding to a test time of one hour

$\bar{p}$= average (static) reservoir pressure

p^*= "false" pressure obtained by extrapolating on the Horner plot to a Horner time ratio of one

Δp= pressure change, drawdown, or pressure drop

ΔP_d= dimensionless pressure

q = flow rate or production rate

q= fluid flow rate per unit of cross-sectional area

q_g= gas production rate

q_{gt}= net gas cap expansion rate

q_o= oil production rate

q_w= water production rate

q_{wt}= net water influx rate

Q= gas flow rate

Q= van Everdingen & Hurst cumulative influx function

Q_i= pore volumes of water injected

Q_{gin}= cumulative gas injection

r= radial distance

r_e= external boundary radius

r_i= radius of investigation

r_w= well radius

R= producing gas-oil ratio

R= universal gas constant

R_d= dimensionless radius (r / r_w)

R_p= cumulative produced gas-oil ratio

R_{poz}= cumulative produced GOR from the oil zone

R_s= solution gas-oil ratio

R_{sb}= solution gas-oil ratio at the bubble point

R_{si}= solution gas-oil ratio at original conditions

R_{sw}= solution gas-water ratio

s= skin effect or skin factor

S= saturation

S_D= displacing phase saturation

S_{Df}= displacing phase saturation at the front

$\bar{S}_D$= average displacing phase saturation behind the front

S_g= gas saturation

S_{gc}= critical gas saturation

S_o= oil saturation

S_{or}= residual oil saturation

S_w= water saturation

S_{wc}= connate water saturation

S_{wi}= irreducible water saturation

t= time

t= effective producing time

t_d= dimensionless time

t_s= producing time to stabilization

Δt= shut-in time

T= temperature

T= transmissibility ($k h / \mu$)

T_c= critical temperature

T_{pc}= pseudocritical temperature

T_{pr}= pseudoreduced temperature

T_{sc}= temperature at standard conditions

T_{sep}= separator temperature

u= linear flow direction

v= velocity

V= moles of vapor phase

V= volume

V_b= bulk volume

V_p= pore volume

V_w= wellbore volume

w= width

W_e= cumulative water influx

W_{in}= cumulative water injected

W_p= cumulative water produced

x= horizontal dimension

x= mole fraction of a component in the liquid phase

y= mole fraction of a component in the vapor phase

z= gas deviation factor

z= mole fraction of a component in a mixture

z= vertical dimension

α= angle of formation dip (from horizontal)

α

= angle of flow direction (from horizontal)

γ= specific gravity

γ_g= gas specific gravity

γ_o= oil specific gravity

γ_w= water specific gravity

Δ= difference

λ= mobility (k / μ)

λ= inter-porosity flow coefficient (double porosity reservoirs)

λ_g= gas mobility

λ_o= oil mobility

λ_w= water mobility

μ= viscosity

μ_g= gas viscosity

μ_o= oil viscosity

μ_w= water viscosity

ρ= density

ρ_g= gas density

ρ_o= oil density

ρ_w= water density

σ= surface tension

σ= interfacial tension

ψ= real gas potential (pseudopressure)

τ= pseudo-time

ϕ= porosity

ω= fracture to total system storativity ratio (double porosity reservoirs)

INDEX